PERGAMON INTERNATIONAL LIBRARY
of Science, Technology, Engineering and Social Studies
The 1000-volume original paperback library in aid of education, industrial training and the enjoyment of leisure
Publisher Robert Maxwell

Single-case Experimental Designs: Strategies for Studying Behavior Change

PERGAMON GENERAL PSYCHOLOGY SERIES

Other Titles in the Series

COHEN
Psych City

LIEBERT
The Early Window

WEINER
Personality–The Human Potential

CATTELL
A New Morality from Science–Beyondism

FREDERIKSEN
Prediction of Organizational Behavior

BURCK et al.
Counseling and Accountability

ASHEM & POSER
Adaptive Learning

FELDMAN
College and Student

O'LEARY & O'LEARY
Classroom Management

ROSSI
Dreams and the Growth of Personality

NEWBOLD
The Psychiatric Programming of People

Single-case Experimental Designs: Strategies for Studying Behavior Change

by

MICHEL HERSEN, Ph.D.
Department of Psychiatry
Western Psychiatric Institute and Clinic
University of Pittsburgh School of Medicine

and

DAVID H. BARLOW, Ph.D.
Section of Psychiatry and Human Behavior
Brown University
and Butler Hospital

with an invited chapter on statistical analyses
by ALAN E. KAZDIN, Ph.D.
The Pennsylvania State University

PERGAMON PRESS
NEW YORK · TORONTO · OXFORD · SYDNEY
PARIS · FRANKFURT

U.K.	Pergamon Press Ltd., Headington Hill Hall, Oxford OX3 0BW, England
U.S.A.	Pergamon Press Inc., Maxwell House, Fairview Park, Elmsford, New York 10523, U.S.A.
CANADA	Pergamon of Canada Ltd., 75 The East Mall, Toronto, Ontario, Canada
AUSTRALIA	Pergamon Press (Aust.) Pty. Ltd., 19a Boundary Street, Rushcutters Bay, N.S.W. 2011, Australia
FRANCE	Pergamon Press SARL, 24 rue des Ecoles, 75240 Paris, Cedex 05, France
WEST GERMANY	Pergamon Press GmbH, 6242 Kronberg-Taunus, Pferdstrasse 1, West Germany

First edition 1976

Reprinted 1977

Library of Congress Cataloging in Publication Data

Hersen, Michel.
Single case experimental designs.

(Pergamon general psychology series; 56)
1. Psychological research. 2. Experimental design
I. Barlow, David H., joint author. II. Title.
[DNLM: 1. Behavior. 2. Psychology, Experimental.
3 Research design. BF181 H572s]
BF76.5.H47 1975 150'.7'24 75-4791
ISBN 0-08-019512-1 (Hardcover)
ISBN 0-08-019511-3 (Flexicover)

Printed in Great Britain by Biddles Ltd., Guildford, Surrey

Contents

3. General Procedures in Single-case Research

4. Repeated Measurement Techniques

5. Basic A-B-A Designs

6. Extensions of the A-B-A Design, Uses in Drug Evaluation, and Interaction Design Strategies

The Authors

Michel Hersen (Ph.D., State University of New York at Buffalo) is Professor of Clinical Psychiatry and Director of the Resocialization Treatment Center in the Department of Psychiatry, Western Psychiatric Institute and Clinic, University of Pittsburgh School of Medicine. In addition, Dr. Hersen is Director of the Clinical Psychology Internship Training Program in the Department of Psychiatry, University of Pittsburgh School of Medicine. He also holds an appointment as Professor in the Psychology Department at the University of Pittsburgh. Dr. Hersen is Editor-in-Chief of *Progress in Behavior Modification*, Editor-in-Chief of *Behavior Modification*, Associate Editor of *Addictive Behaviors*, and is on the editorial boards of *Behavior Therapy* and *Behavior Therapy and Experimental Psychiatry*. He is consultant to the Veterans Administration Medical Research Service Merit Review Board in Behavioral Sciences and is editorial consultant for numerous journals.

Dr. Hersen has published over 90 papers, chapters, and books in the fields of clinical psychology and psychiatry. His professional interests are in the area of behavior modification and social skills training.

David H. Barlow (Ph.D., University of Vermont) is Professor in the section of Psychiatry and Human Behavior, Brown University, and Director, Division of Education and Training and Director of Psychology at Butler Hospital, Providence, Rhode Island. He also directs the Clinical Psychology Internship Program in the Brown Medical School. He is on the editorial board of several journals and since 1973 has been Associate Editor of the *Journal of Applied Behavior Analysis*. Dr. Barlow holds a Diplomate in clinical psychology from the American Board of Professional Psychology and his professional interests are in the area of training and behavior therapy. In addition, he is a consultant to the National Institute of Mental Health and the National Heart and Lung Institute. He has published widely on the experimental analysis of therapeutic techniques as applied to sexual deviation and neurotic disorders.

Preface

This book is a response to the growing popularity of single-case experimental designs in applied settings. Discussions with colleagues and students in recent years convinced us that a source book of single-case experimental design strategies would be useful. The original intent was to present the design alternatives along with some examples of their applications in applied settings—in other words, an applied research "cookbook." As we began writing, we realized that it would be extremely difficult to present design alternatives without discussing issues relevant to applied research in general. Similarly, it was difficult to discuss issues such as variability or generality of findings without putting single-case experimental designs in perspective against the entire field of experimental design.

As a result, chapters were added on the history of single-case designs in basic and applied research and on the relation of basic issues in experimental design to the single-case approach. This historical perspective, in turn, persuaded us that the issue of generality of findings from single-case experiments has been a much misunderstood problem. Consequently, we devote a chapter to a discussion of replication techniques.

Finally, the growing controversies over statistical approaches to analyzing data from single-case experiments convinced us that a chapter in this area would be useful. We were fortunate in having Alan Kazdin write this chapter, which we think is a substantial contribution to the book. In the process, we hope not to have lost sight of the original goal of the book—to provide a source book of single-case designs with guidelines for their use in applied settings.

We do not expect this book to be the final statement on single-case designs. We learned at least as much as we already knew in analyzing the variety of innovative and creative applications of these designs to varying applied problems. The unquestionable appropriateness of these designs in applied settings should ensure additional design innovations in the future. We do hope, however, that this book will assist those who are struggling in the area of applied research by providing a statement on the basic goals and procedures necessary for successful execution of single-case designs.

Many people contributed to this book in one way or another. Our debt to Murray Sidman's pioneering work will be obvious. Much of what we brought to this book was the result of years of interaction and discussion with Harold Leitenberg and Stewart Agras, who were teachers (DHB's) and later our colleagues. In our own setting, discussions with many colleagues and students at the University of Mississippi Medical Center influenced our ideas. We would especially like to thank Tom Sajwaj and Len Epstein, who read earlier drafts of several chapters, and Ed Blanchard, who helped us solve some historical puzzles.

Many people have secretaries and research assistants, but very few have the dedication to search out seemingly endless articles and type draft after draft of chapters containing figures and descriptions of experimental designs. Fortunately, we had such people. To Margie Leiberton, Sallie Morgan, and Sue Noblin, who did most of the typing, and to Harriet Alford, Susan Pinkston, Ann Stein, Carol Thames, and John Watts, who made significant contributions as research assistants, we extend our sincere gratitude.

Although each of us assumed responsibility for individual chapters, our close collaboration ensured that all chapters represented an integration of our thinking. As such, our individual contributions to the book were equal and order of authorship was determined by a coin-toss (Hersen's coin). Finally, the dedication of this book signifies our hope that students in the clinical and behavioral sciences will benefit from this volume and will use these tools to begin to answer the difficult questions in the area of human behavior disorders. It is to these students that we dedicate our book.

MICHEL HERSEN
Pittsburgh, Pennsylvania

DAVID H. BARLOW
Providence, Rhode Island

Epigram
Conversation between Tolman and Allport

TOLMAN: "I know I should be more idiographic in my research, but I just don't know how to be."

ALLPORT: "Let's learn!"

CHAPTER 1

The Single-case in Basic and Applied Research: An Historical Perspective

1.1. INTRODUCTION

The individual is of paramount importance in the clinical science of human behavior change. Until recently, however, this science lacked an adequate methodology for studying behavior change in individuals. This gap in our methodology has retarded the development and evaluation of new procedures in clinical psychology and psychiatry as well as educational fields.

Historically, the intensive study of the individual held a preeminent place in the fields of psychology and psychiatry. In spite of this background, an adequate experimental methodology for studying the single case was very slow to develop in applied research.[1] Equally puzzling, from an historical point of view, are comments by noted experts in the field of applied research who still say of the experimental study of the individual, "... idiographic study has little place in the confirmatory aspects of scientific activity which looks for laws applying to individuals generally" (Kiesler, 1971, p. 66). To uncover the bases for these notions, it is useful to gain some perspective on the historical development of methodology in the broad area of psychological research.

The purpose of this chapter is to provide such a perspective, beginning with the origins of methodology in the basic sciences of physiology and experimental psychology in the middle of the last century. Since most of this early work was performed on individual organisms, reasons for the development of between-group comparison methodology in basic research (which did not occur until the turn of the century) are outlined. The rapid development of inferential statistics and sampling theory during the early 20th century enabled greater sophistication in the research methodolgy of experimental psychology. The manner in which this affected research methods in applied areas during the middle of the century is discussed.

[1] In this book applied research refers to experimentation in the area of human behavior change relevant to the disciplines of clinical psychology, psychiatry, social work, and education.

In the meantime, applied research was off to a shaky start in the offices of early psychiatrists with a technique known as the case study method. The separate development of applied research is traced from those early beginnings through the grand collaborative group comparison studies proposed in the 1950s. The subsequent disenchantment with this approach in applied research forced a search for alternatives. The rise and fall of the major alternatives—process research and naturalistic studies—is outlined near the end of the chapter. This disenchantment also set the stage for a renewal of interest in the scientific study of the individual. The multiple origins of single case experimental designs from the laboratories of experimental psychology and the offices of clinicians complete the chapter. Descriptions of single case designs and guidelines for their use as they are evolving in applied research comprise the remainder of this book.

1.2. BEGINNINGS IN EXPERIMENTAL PHYSIOLOGY AND PSYCHOLOGY

The scientific study of individual human behavior has roots deep in the history of psychology and physiology. When psychology and physiology became sciences, the initial experiments were performed on individual organisms and the results of these pioneering endeavors remain relevant to the scientific world today. The science of physiology began in the 1830s, with Johannes Müller and Claude Bernard, but an important landmark for applied research was the work of Paul Broca in 1861. At this time, Broca was caring for a man who was hospitalized due to an inability to speak intelligibly. Before the man died, Broca examined him carefully; subsequent to death, he performed an autopsy. The finding of a lesion in the third frontal convolution of the cerebral cortex convinced Broca, and eventually the rest of the scientific world, that this was the speech center of the brain. Broca's method was the clinical extension of the newly developed experimental methodology called extirpation of parts, introduced to physiology by Marshall Hall and Pierre Flouren in the 1850s. In this method brain function was mapped out by systematically destroying parts of the brain in animals and noting the effects on behavior.

The importance of this research in the context of the present discussion lies in the demonstration that important findings with wide generality were gleaned from single organisms. This methodology was to have a major impact on the beginnings of experimental psychology.

Boring (1950) fixes the beginnings of experimental psychology in 1860 with the publication of Fechner's *Elemente der Psychophysik*. Fechner is most famous for developing measures of sensation through several psychophysical methods. With these methods, Fechner was able to determine sensory thresholds and just noticeable differences (JNDs) in various sense modalities. What is common to these methods is the repeated measurement of a response at different intensities or different locations of a given stimulus in an individual subject. For example, when stimulating skin with two points in a certain region to determine the minimal separation which the subject reliably recognizes as two stimulations, one may use the method of constant stimuli. In this method the two points repeatedly stimulate two areas of skin at five to seven fixed separations, in random order, ranging from a few millimeters apart to the relatively large separation of 10 millimeters. During each stimulation, the subject reports whether he senses one point or two. After repeated trials, the point at which the subject "notices" two separate points can be determined. It is interesting to note that Fechner was one of the first to apply statistical methods to psychological problems. Fechner noticed that judgments of just noticeable differences in the sensory modalities varied somewhat from trial to trial. To quantify this variation, or "error" in judgment, he borrowed the normal law of error and demonstrated that these "errors" were normally distributed around a mean, which then became the "true" sensory threshold. This use of descriptive statistics anticipated the application of these procedures to groups of individuals at the turn of the century when traits of capabilities were also found to be normally distributed around a mean. It should be noted, however, that Fechner was concerned with variability *within* the subject and he continued his remarkable work on series of individuals.

These traditions in methodology were continued by Wilhelm Wundt. Wundt's contributions and those of his students and followers, most notably Titchener, had an important impact on the field of psychology, but it is the scientific methodology he and his students employed that most interests us.

To Wundt, the subject matter of psychology was immediate experience, such as how a subject experiences light and sound. Since these "experiences" were private events and could not be directly observed, Wundt created a new method called introspection. Mention of the procedure may strike a responsive chord in some modern-day clinicians, but in fact this methodology is quite different from the "introspection" technique of free association, etc., often used in clinical settings to uncover repressed or unconscious material. Nor did introspection bear any relation to armchair dreams or reflections that are so frequent a part of experience. Introspection, as Wundt employed it, was a highly specific and rigorous procedure that was used with individual subjects who were highly trained. This training involved learning to describe experiences in an objective manner, free from emotional or language restraints. For example, the experience of seeing a brightly colored object would be described in terms of shapes and hues without recourse to aesthetic appeal. To illustrate the objectivity of this system, introspection of emotional experiences where scientific calm and objectivity might be disrupted was not allowed. Introspection of this experience was to be done at a later date when the scientific attitude returned. This method, then, became retrospection, and the weaknesses of this approach were accepted by Wundt to preserve objectivity. Like Fechner's psychophysics, which is essentially an introspectionist methodology, the emphasis hinges on the study of a highly trained individual with the clear assumption, after some replication on other individuals, that findings would have generality to the population of individuals. Wundt and his followers comprised a school of psychology known as the Structuralist School, and many topics important to psychology were first studied with this rather primitive but individually oriented form of scientific analysis. The major subject matter, however, continued to be sensation and perception; borrowing Fechner's psychophysical methods, the groundwork for the study of sensation and perception was laid. Perhaps because of these beginnings, a strong tradition of studying individual organisms has ensued in the fields of sensation and perception and physiological psychology. This tradition has not extended to other areas of experimental psychology, such as learning, or to the more clinical areas of investigation that are broadly based on learning principles or theories. This course of events is surprising since the efforts to study principles of learning comprise one of the more famous examples of the scientific study of the single case. This effort was made by Hermann Ebbinghaus, one of the towering figures in the development of psychology. With a belief in the scientific approach to psychology, and heavily influenced by Fechner's methods (Boring, 1950), Ebbinghaus established principles of human learning that remain basic to work in this area.

Basic to Ebbinghaus's experiments was the invention of a new instrument to measure learning and forgetting—the nonsense syllable. With a long list of nonsense syllables and himself as the subject, he investigated the effects of different variables (such as the amount of material to be remembered) on the efficiency of memory. Perhaps his best-known discovery was the retention curve, which illustrated the process of forgetting over time. Chaplin and Kraweic (1960) note that he "worked so carefully that the results of his experiments have never been seriously questioned" (p. 180). But what is most relevant and remarkable about his work was his emphasis on repeated measures of performance in one individual over time (see Chapter 4). As Boring (1950) points out, Ebbinghaus made repetition the basis for the experimental measurement of memory. It would be some 70 years before a new approach, called the experimental analysis of behavior, was to employ repeated measurement in individuals to study complex animal and human behaviors.

One of the best-known scientists in the fields of physiology and psychology during these early years was Pavlov (Pavlov, 1928). Although Pavlov considered himself a physiologist, his work on principles of association and learning were his greatest contribution and, along with his basic methodology, are so well known that summaries are not required. What is often overlooked, however, is that Pavlov's basic findings were gleaned from single organisms and strengthened by replication on other organisms. In terms of scientific yield, the study of the individual organism reached an early peak with Pavlov, and Skinner would later cite this approach as an important link and a strong bond between himself and Pavlov (Skinner, 1966b).

1.3. ORIGINS OF THE GROUP COMPARISON APPROACH

Important research in experimental psychology and physiology using single cases did not stop with these efforts, but the turn of the century witnessed a new development which would have a marked effect on basic and, at a later date, applied research. This development was the discovery and measurement of individual differences. The study of individual differences can be traced to Adolphe Quetelet, a Belgian astronomer, who discovered that human traits (e.g., height) followed the normal curve (Stilson, 1966). Quetelet interpreted these findings to mean that nature strove to produce the "average" man but, due to various reasons, failed, resulting in errors or variances in traits that grouped around the average. As one moved further from this average, fewer examples of the trait were evident, following the well-known normal distribution. This approach, in turn, had its origins in Darwin's observations on individual variation within a species. Quetelet viewed these variations or errors as unfortunate since he viewed the average man, which he termed "l'homme moyen," as a cherished goal rather than a descriptive fact of central tendency. If nature were "striving" to produce the average man, but failed due to various accidents, then the average, in this view, was obviously the ideal. Where nature failed, however, man could pick up the pieces, account for the errors, and estimate the average man through statistical techniques. The influence of this finding on psychological research was enormous as it paved the way for the application of sophisticated statistical procedures to psychological problems. Quetelet would probably be distressed to learn, however, that his concept of the average individual would come under attack during the 20th century by those who observed that there is no average individual (e.g., Dunlap, 1932; Sidman, 1960).

This viewpoint notwithstanding, the study of individual differences and the statistical approach to psychology became prominent during the first half of the 20th century and changed the face of psychological research. With a push from the American functional school of psychology and a developing interest in the measurement and testing of intelligence, the foundation for comparing groups of individuals was laid.

Galton and Pearson expanded the study of individual differences at the turn of the century and developed many of the descriptive statistics still in use today, most notably the notion of correlation, which led to factor analysis, and significant advances in construction of intelligence tests first introduced by Binet in 1905. At about this time, Pearson, along with Galton and Weldon, founded the journal *Biometricka* with the purpose of advancing quantitative research in biology and psychology. Many of the newly devised statistical tests were first published there. Pearson was highly enthusiastic about the statistical approach and seemed to believe, at times, that inaccurate data could be made to yield accurate conclusions if the proper statistics were applied (Boring, 1950). While this view was rejected by more conservative colleagues, it points up a confidence in the power of statistical procedures that reappears from time to time in the execution of psychological research.

One of the best-known psychologists to adopt this approach was James McKeen Cattell. Cattell, along with Farrand, devised a number of simple mental tests that were administered to freshmen at Columbia University to determine the range of individual differences. Cattell also devised the order of merit method, whereby a number of judges would rank items or people on a given quality and the average response of the judges constituted the rank of that item *vis-à-vis* other items. In this way, Cattell had ten scientists rate a number of eminent colleagues. The scientist with the highest score (on the average) achieved the top rank.

It may seem ironic at first glance that a concern with individual differences led to an emphasis on groups and averages, but differences among individuals, or inter-subject variability, and the distribution of these differences necessitate a comparison among individuals and a concern for a description of a group or population as a whole. In this context observations from a single organism are irrelevant. Darwin, after all, was concerned with survival of a species and not the survival of individual organisms.

The invention of many of the descriptive statistics and some crude statistical tests of comparison made it easier to compare performance in large groups of subjects. From 1900 to 1930, much of the research in experimental psychology, particularly learning, took advantage of these statistics to compare groups of subjects (usually rats) on various performance tests (e.g., see Birney and Teevan, 1961). Crude statistics that could attribute differences between groups to something other than chance began to appear, such as the critical ratio test (Walker and Lev, 1953). The idea that the variability or error among organisms could be accounted for or averaged out in large groups was a common-sense notion emanating from the new emphasis on variability among organisms. The fact that this research resulted in an average finding from the hypothetical average rat drew some isolated criticism. For instance, in 1932, while reviewing research in experimental psychology, Dunlap pointed out that there was no average rat, and Lewin (1933) noted that "... the only situations which should be grouped for statistical treatment are those which have for the individual rats or for the individual children the same psychological structure and only for such period of time as this structure exists" (p. 328). The new emphasis on variability and averages, however, would have pleased Quetelet, whose slogan could have been "Average is Beautiful."

The influence of inferential statistics

During the 1930s, the work of R. A. Fisher, which subsequently excited considerable influence on psychological research, first appeared. Most of the sophisticated statistical procedures in use today for comparing groups were invented by Fisher. It would be difficult to pick up psychological or psychiatric journals concerned with behavior change and not find research data analyzed by the ubiquitous analysis of variance. It is interesting, however, to consider the origin of these tests. Early in his career, Fisher, who was a mathematician interested in genetics, made an important decision. Faced with pursuing a career at a biometrics laboratory, he chose instead a relatively obscure agricultural station on the grounds that this position would offer him more opportunity for independent research. This personal decision at the very least changed the language of experimental design in psychological research, introducing agricultural terms to describe relevant designs and variables (e.g., split plot analysis of variance). While Fisher's statistical innovations were one of the more important developments of the century for psychology, the philosophy underlying the use of these procedures is clearly in line with Quetelet's notion of the importance of the average. As a good agronomist, Fisher was concerned with the yield from a given area of land under various soil treatments, plant varieties, or other agricultural variables. Much as in the study of individual differences, the fate of the individual plant is irrelevant in the context of the yield from the group of plants in that area. Agricultural variables are important to the farm and society if the yield is better *on the average* than a similar plot treated differently. The implications of this philosophy for applied research will be discussed in Chapter 2.

The work of Fisher was not limited to the invention of sophisticated statistical tests. An equally important contribution was the consideration of the problem of induction or inference. Essentially, this issue concerns generality of findings. If some data are obtained from a group or a plot of land, this information is not very valuable if it is relevant only to that particular group or plot of land since similar data must be collected from each new plot. Fisher (1925) worked out the properties of statistical tests, which made it possible to estimate the relevance of data from one small group with certain characteristcs to the universe of individuals with those characteristics. In other words, inference is made from the sample to the population. This work and the subsequent developments in the field of sampling theory made it possible to talk in terms of psychological principles with broad generality and applicability—a primary goal in any science. This type of estimation, however, was based on appropriate statistics, averages, and inter-subject variability in the sample, which further reinforced the group comparison approach in basic research.

As the science of psychology grew out of its infancy, its methodology was largely determined by the lure of broad generality of findings made possible through the brilliant work of Fisher and his followers. Because of the emphasis on averages and inter-subject variability required by this design in order to make general statements, the intensive study of the single organism, so popular in the early history of psychology, fell out of favor. By the 1950s, when investigators began to consider the possibility of doing serious research in applied settings, the group comparison approach was so entrenched that anyone studying single organisms was considered something of a revolutionary by no less of an authority than Underwood (1957). The zeitgeist in psychological research was group comparison and statistical estimation. While an occasional paper was published during the 1950s defending the study of the single case (Beck, 1953; Rosenzweig, 1951) or at least pointing out its place in psychological research (duMas, 1955), very little basic research was carried out on single cases. A notable exception was the work of B. F. Skinner and his students and colleagues, who were busy developing an approach known as the experimental analysis of behavior, or operant conditioning. This work, however, did not have a large impact on methodology in other areas of psychology during the 1950s, and applied research was just beginning. Against this background, it is not surprising that applied researchers in the 1950s employed the group comparison approach, despite the fact that the origins of the study of clinically relevant phenomena were quite different from the origin of more basic research described above.

1.4. THE DEVELOPMENT OF APPLIED RESEARCH: THE CASE STUDY METHOD

As the sciences of physiology and psychology were developing during the late 19th and 20th centuries, people were suffering from emotional and behavioral problems and were receiving treatment. Occasionally, patients recovered and therapists would carefully document their procedures and communicate them to colleagues. Hypotheses attributing success or failure to various assumed causes emanated from these cases and these hypotheses gradually grew into theories of psychotherapy. Theories proliferated and procedures based on observations of cases and inferences from these theories grew in number. As Paul (1969) notes, those theories or procedures that could be communicated clearly or that presented new and exciting principles tended to attract followers to the organization, and schools of psychotherapy were formed. At the heart of this process is the "case study" method of investigation (Bolger, 1965). This method (and its extensions) was, with few exceptions, the sole methodology of clinical investigation through the first half of the 20th century.

The case study method, of course, is the clinical base for the experimental study of single cases and, as such, it retains an important function in present-day applied research (see Section 1.7). Unfortunately, during this period clinicians were unaware, for the most part, of the basic principles of applied research, such as definition of variables and manipulation of independent variables. Thus, it is noteworthy from an historical point of view that several case studies reported during this period came tantalizingly close to providing the basic scientific ingredients of experimental single case research. The most famous of these, of course, is the Watson and Rayner (1920) study of an analogue of clinical phobia in a young boy where a prototype of a "withdrawal" design was attempted (see Chapter 5). These investigators unfortunately suffered the fate of many modern-day clinical researchers in that the subject moved away before the "reversal" was complete.

Any time that a treatment produced demonstrable effects on an observable behavior disorder, the potential for scientific investigation was there. An excellent example, among many, was Breuer's classic description of the treatment of hysterical symptoms in Anna O. through psychoanalysis in 1895 (Breuer and Freud, 1957). In a series of treatment sessions, Breuer dealt with one symptom at a time through hypnosis and subsequent "talking through", where each symptom was traced back to its hypothetical causation in circumstances surrounding the death of her father. One at a time, these behaviors disappeared, but only when treatment was administered to each respective behavior. This process of treating one behavior at a time fulfills the basic requirement for a multiple baseline experimental design described in Chapter 7, and the clearly observable success indicated that Breuer's treatment was effective. Of course, Breuer did not define his independent variables in that there were several components to his treatment (e.g., hypnosis, interpretation, etc.); but, in the manner of a good scientist as well as a good clinician, Breuer admitted that he did not know which component or components of his treatment were responsible for success. He noted at least two possibilities, the suggestion inherent in the hypnosis or the interpretation. He then described events discovered through his talking therapy as possibly having etiological significance and wondered about the reliability of the girl's report as he hypothesized various etiologies for the symptoms. However, he did not, at the time, firmly link successful treatment with the necessity of discovering the etiology of the behavior disorder. One wonders if the early development of clinical techniques, including psychoanalysis, would have been different if careful observers like Breuer had been cognizant of the experimental implications of their clinical work. Of course, this small leap from uncontrolled case study to scientific investigation of the single case did not occur because of a lack of awareness of basic scientific principles in early clinicians. The result was an accumulation of successful individuals case studies with clinicians from varying schools claiming that their techniques were indispensable to success. In many cases their claims were grossly exaggerated. Brill noted in 1909 on psychoanalysis that "The results obtained by the treatment are unquestionably very gratifying. They surpass those obtained by simpler methods in two chief respects; namely, in permanence and in the prophylactic value they have for the future" (Brill, 1909). Much later, in 1935, Kessel and Hyman observed, "this patient was saved from an inferno and we are convinced that this could have been achieved by no other method" (Kessell and Hyman, 1933). From an early behavioral standpoint, Max (1935) noted that electrical aversion therapy produced "95 percent relief" from the compulsion of homosexuality.

These kinds of statements did little to endear the case study method to serious applied researchers when they began to appear in the 1940s and 1950s. In fact, the case study method, if anything, deteriorated somewhat over the years in terms of the amount and nature of publicly observable data available in these reports. Frank (1961) noted the difficulty in even collecting data from a therapeutic hour in the 1930s due to lack of necessary equipment, reluctance to take detailed notes, and concern about confidentiality. The advent of the phonograph record at this time made it possible at least to collect raw data from those clinicians who would cooperate, but this method did not lead to any fruitful new ideas on research. With the advent of serious applied research in the 1950s, investigators tended to reject reports from uncontrolled case studies due to an inability to evaluate the effects of treatment. Given the extraordinary claims by clinicians after successful case studies, this attitude is understandable. However, from the viewpoint of single case experimental designs, this rejection of the careful observation of behavior change in a case report had the effect of throwing out the baby with the bathwater.

Percentage of success in treated groups

A further development in applied research was the reporting of collections of case studies in terms of percentage of success. Many of these reports have been cited by Eysenck (1952). However, reporting of results in this manner probably did more harm than good to the evaluation of clinical treatment. As Paul (1969) notes, independent and dependent variables were no better defined than in most case reports and techniques tended to be fixed and "school" oriented. Since all procedures achieved some success, practitioners within these schools concentrated on the positive results, explained away the failures, and decided that the overall results confirmed that their procedures, as applied, were responsible for the success. Due to the strong and overriding theories central to each school, the successes obtained were attributed to theoretical constructs underlying the procedure. This precluded a careful analysis of elements in the procedure or the therapeutic intervention that may have been responsible for certain changes in a given case and had the effect of reinforcing the application of a global, ill-defined treatment from whatever theoretical orientation, to global definitions of behavior disorders, such as "neurosis." This, in turn, led to statements such as "psychotherapy works with neurotics." Although applied researchers later rejected these efforts as unscientific, one carryover from this approach was the notion of the average response to treatment; that is, if a global treatment is successful on the average with a group of "neurotics," then this treatment will probably be successful with any individual "neurotic" who requests treatment.

Intuitively, of course, descriptions of results from 50 cases provide a more convincing demonstration of the effectiveness of a given technique than separate descriptions of 50 individual cases. The major difficulty with this approach, however, is that the category in which these clients are classified most always becomes unmanageably heterogeneous. The "neurotics" described in Eysenck's (1952) paper may have less in common than any group of people one would choose randomly. When cases are described individually, however, a clinician stands a better chance of gleaning some important information, since specific problems and specific procedures are usually described in more detail. When one lumps cases together in broadly defined categories, individual case descriptions are lost and the ensuing report of percentage success becomes meaningless. This unavoidable heterogeneity in any group of patients is an important consideration that will be discussed in more detail below and in Chapter 2.

The group comparison approach in applied research

By the late 1940s, clinical psychology and, to a lesser extent, psychiatry began to produce the type of clinician who was also aware of basic research strategies. These scientists were quick to point out the drawbacks of both the case study and reports of percentages of success in groups in evaluating the effects of psychotherapy. They noted that any adequate test of psychotherapy would have to include a more precise definition of terms, particularly outcome criteria or dependent variables (e.g., Knight, 1941). Most of these applied researchers were trained as psychologists, wherein a new emphasis was placed on the "scientist practitioner" model. Thus, the source of research methodology in the newly developing areas of applied research came from experimental psychology. By this time, the predominant methodology in experimental psychology was the between-subjects group design.

The group design also was a logical extension of the earlier clinical reports of percentage success in a large group of patients, since the most obvious criticism of this endeavor is the absence of a control group of untreated patients. The appearance of Eysenck's (1952) notorious article comparing percentage success of psychotherapy in large groups to rates of "spontaneous" remission gleaned from discharge rates at state hospitals and insurance company records had two effects. *First*, it reinforced the growing conviction that the effects of psychotherapy could not be evaluated from case reports or "percentage success groups" and sparked a new flurry of interest in evaluating psychotherapy through the scientific method. *Second*, the emphasis on comparison between groups and quasi-control groups in Eysenck's review strengthened the notion that the logical way to evaluate psychotherapy was through the prevailing methodology in experimental psychology—the between-groups comparison designs.

This approach to applied research did not suddenly begin in the 1950s, although interst certainly increased at this time. Scattered examples of research with clinically relevant problems can be found in earlier decades. One interesting example is a study reported by Kantorovich (1928), who applied aversion therapy to one group of twenty alcoholics in Russia and compared results to a control group receiving hypnosis or medication. The success of this treatment (and the direct derivation from Pavlov's work) most likely ensured a prominent place for aversion therapy in Russian treatment programs for alcoholics. Some of the larger group comparison studies typical of the 1950s also began before Eysenck's celebrated paper. One of the best known is the Cambridge–Somerville youth study, which was reported in 1951 (Powers and Witmer, 1951) but was actually begun in 1937. Although this was an early study, it is quite representative of the later group comparison studies in that many of the difficulties in execution and analysis of results were repeated again and again as these studies accumulated.

The major difficulty, of course, was that these studies did not prove that psychotherapy worked. In the Cambridge–Somerville study, despite the advantages of a well-designed experiment, the discouraging finding was that "counseling" for delinquents or potential delinquents had no significant effect when compared to a well-matched control group.

When this finding was repeated in subsequent studies (e.g., Barron and Leary, 1955), the controversy over Eysenck's assertion on the ineffectiveness of psychotherapy became heated. Most clinicians rejected the findings outright since they were convinced that psychotherapy was useful, while scientists such as Eysenck hardened their convictions that psychotherapy was at best ineffective and at worst some kind of great hoax perpetrated on unsuspecting clients. This controversy, in turn, left serious applied researchers groping for answers to difficult methodological questions on how to even approach the issue of evaluating effectiveness in psychotherapy. As a result, major conferences on research in psychotherapy were called to discuss these questions (e.g., Rubenstein and Parloff, 1959). It was not until Bergin reexamined these studies in a very important article (Bergin, 1966) that some of the discrepancies between clinical evidence from uncontrolled case studies and experimental evidence from between-subject group comparison designs were clarified. Bergin noted that some clients *were* improving in these studies, but others were getting worse. When subjected to statistical averaging of results, these effects canceled each other out, yielding an overall result of no effect when compared to the control group. Furthermore, Bergin pointed out that these therapeutic effects had been described in the original articles, but only as afterthoughts to the major statistical findings of no effect. Reviewers such as Eysenck, approaching the results from a methodological point of view, concentrated on the statistical findings. These studies did not, however, prove that psychotherapy was ineffective for a given individual. What these results demonstrated is that people, particularly clients with emotional or behavioral disorders, are quite different from each other. Thus, attempts to apply an ill-defined and global treatment such as psychotherapy to a heterogeneous group of clients classified under a vague diagnostic category such as neurosis are incapable of answering the more basic question on the effectiveness of a specific treatment for a specific individual.

The conclusion that psychotherapy was ineffective was premature, based on this reanalysis, but the overriding conclusion from Bergin's review was that "Is psychotherapy effective?" was the wrong question to ask in the first place, even when appropriate between-group experimental designs are employed. During the 1960s, scientists (e.g., Paul, 1967) began to realize that any test of a global treatment such as psychotherapy would not be fruitful and that clinical researchers must start defining the independent variables more precisely and must ask the question: "What specific treatment is effective with a specific type of client under what circumstances?"

1.5. LIMITATIONS OF THE GROUP COMPARISON APPROACH

The clearer definition of variables and the call for experimental questions that were precise enough to be answered was a major advance in applied research.

The extensive review of psychotherapy research by Bergin and Strupp (1972), however, demonstrated that even under these more favorable conditions, the application of the group comparison design to applied problems posed many difficulties. These difficulties, or objections, which tend to limit the usefulness of a group comparison approach in applied research, can be classified under five headings: (1) ethical objectives, (2) practical problems in collecting large numbers of patients, (3) averaging of results over the group, (4) generality of findings, and (5) inter-subject variability.

Ethical objectives

An oft-cited issue, usually voiced by clinicians, is the ethical problem in withholding treatment from a no-treatment control group. This notion, of course, is based on the assumption that the therapeutic intervention, in fact, works, in which case there would be little need to test it at all. Despite the seeming illogic of this ethical objection, in practice many clinicians and other professional personnel react with distaste at withholding some treatment, however inadequate, from a group of clients who are undergoing significant human suffering. This attitude is reinforced by scattered examples of experiments where control groups did endure substantial harm during the course of the research, particularly in some pharmacological experiments.

Practical problems

On a more practical level, the collection of large numbers of clients homogeneous for a particular behavior disorder is often a very difficult task. In basic research in experimental psychology most subjects are animals (or college sophomores) where matching of relevant behaviors or background variables such as personality characteristics is feasible. When dealing with severe behavior disorders, however, the possibility of obtaining sufficient clients suitably matched to constitute the required groups in the study is often impossible. As Isaac Marks, who is well known for his applied research with large groups, notes:

> Having selected the technique to be studied, another difficulty arises in assembling a homogeneous sample of patients. In uncommon disorders this is only possible in centers to which large numbers of patients are regularly referred, from these a tiny number are suitable for inclusion in the homogeneous sample one wishes to study. Selection of the sample can be so time consuming that it severely limits research possibilities. Consider the clinician who wishes to assemble a series of obsessive-compulsive patients to be assigned at random into one of two treatment conditions. He will need at least 20 such cases for a start, but obsessive-compulsive neuroses (not personality) make up only 0.5-3 percent of the psychiatric outpatients in Britain and the USA. This means the clinician will need a starting population of about 2000 cases

> to sift from before he can find his sample, and even then this assumes that all his colleagues are referring every suitable patient to him. In practice, at a large center such as the Maudsley Hospital, it would take up to two years to accumulate a series of obsessive compulsives for study. (Bergin and Strupp, 1972, p. 130)

To Marks' credit, he has successfully undertaken this arduous venture on several occasions (Marks, 1972), but the practical difficulties in executing this type of research in settings other than the enormous clinical facility at the Maudsley are apparent.

Even if this approach is possible in some large clinical settings, or in state hospital settings where one might study various aspects of schizophrenia, the related economic considerations are also inhibiting. Activities such as gathering and analyzing data, following patients, paying experimental therapists, etc., require large commitments of research funds, which are often unavailable.

Recognizing the practical limitations on conducting group comparison studies in one setting, an initial goal of the Bergin and Strupp review of the state of psychotherapy research was to explore the feasibility of large collaborative studies among various research centers. One advantage, at least, was the potential to pool adequate numbers of patients to provide the necessary matching of groups. Their reluctant conclusion was that this type of large collaborative study was not feasible due to differing individual styles among researchers and the extraordinary problems involved in administering such an endeavor (Bergin and Strupp, 1972).

Averaging of results

A third difficulty noted by many applied researchers is the obscuring of individual clinical outcome in group averages. This issue was cogently raised by Sidman (1960) and Chassan (1967) and repeatedly finds its way into the informal discussions with leading researchers conducted by Bergin and Strupp and published in their book, *Changing Frontiers in the Science of Psychotherapy* (1972). Bergin's (1966) review of large outcome studies where some clients improved and others worsened highlighted this problem. As noted above, a move away from tests of global treatments on ill-defined variables with the implicit question "Is psychotherapy effective?" was a step in the right direction. But even when specific questions on effects of therapy in homogeneous groups are approached from the group comparison point of view, the problem of obscuring important findings remains because of the enormous complexities of any individual patient included in a given treatment group. The fact that patients are seldom truly "homogeneous" has been described by Kiesler (1966) in his discussion of the patient uniformity myth. To take Marks' example, ten patients, homogeneous for obsessive-compulsive neurosis, may bring entirely different histories, personality variables, and environmental situations to the treatment setting and will respond in varying ways to treatment. That is, some patients will improve and others will not. The average response, however, will not represent the performance of any individual in the group. In relation to this problem, Bergin (Bergin and Strupp, 1972) notes that he consulted a prominent statistician about a therapy research project who dissuaded him from employing the usual inferential statistics applied to the group as a whole and suggested instead that individual curves or descriptive analyses of small groups of highly homogeneous patients might be more fruitful.

Generality of findings

Averaging and the complexity of individual patients also bring up some related problems. Since results from group studies do not reflect changes in individual patients, these findings are not readily translatable or generalizable to the practicing clinician since Chassan (1967) points out that the clinician cannot determine which particular patient characteristics are correlated with improvement. In ignorance of the responses of individual patients to treatment, the clinician does not know to what extent a given patient under his care is similar to patients who improved or perhaps deteriorated within the context of an overall group improvement. Furthermore, as groups become more homogeneous, which most researchers agree is a necessary condition to answer specific questions about effects of therapy, one loses the ability to make inferential statements to the population of patients with a particular disorder since the individual complexities in the population will not have been adequately sampled. Thus, it becomes difficult to generalize findings at all beyond the specific group of patients in the experiment. These issues of averaging and generality of findings will be discussed in greater detail in Chapter 2.

Inter-subject variability

A final issue bothersome to clinicians and applied researchers is variability. Between-subject group comparison designs consider only variability between subjects as a method of dealing with the enormous differences among individuals in a group. Progress is assessed usually only once (in a post-test). This large inter-subject variability is often responsible for the "weak" effect obtained in these studies where some clients show considerable improvement, others deteriorate, and the average improvement is statistically significant but clinically weak. Ignored in these studies is within-subject variability or the clinical course of a specific patient during treatment, which is of great practical interest to clinicians. This issue will also be discussed more fully in Chapter 2.

1.6. ALTERNATIVES TO THE GROUP COMPARISON APPROACH

Many of these practical and methodological difficulties seemed overwhelming to clinicians and applied researchers. Some investigators wondered if serious, meaningful research on evaluation of psychotherapy was even possible (e.g., Hyman and Berger 1966) and the gap between clinician and scientist widened. One difficulty here was the restriction placed on the type of methodology and experimental design applicable to applied research. For many scientists, a group comparison design was the only methodology capable of yielding important information in psychotherapy studies. In view of the dearth of alternatives available and against the background of case study and "percentage success" efforts, these high standards were understandable and correct. Since there were no clearly acceptable scientific alternatives, however, applied researchers failed to distinguish between those situations where group comparison designs were practical, desirable, and necessary (see Section 2.9) and situations where the development of alternative methodology was required. During the 1950s and 1960s, several alternatives were tested.

Many applied researchers reacted to the difficulties of the group comparison approach with a "flight into processes" where components of the therapeutic process, such as relationship variables, were carefully studied (Hoch and Zubin, 1964). A second approach, favored by many clinicians, was the "naturalistic study," which was very close to actual clinical practice but had dubious scientific underpinnings. As Kiesler (1971) notes, these approaches are quite closely related since both are based on "correlational" methods where dependent variables are correlated with therapist or patient variables either within therapy or at some point after therapy. This is distinguished from the experimental approach where independent variables are systematically manipulated.

Naturalistic studies

The advantage of the naturalistic study for most clinicians was that it did little to disrupt the typical activities engaged in by the clinician in his day-to-day practice. Unlike the experimental group comparison design, he was not restricted by precise definitions of an independent variable (treatment, time limitation, or random assignment of patients to groups). Kiesler (1971) notes that naturalistic studies involve "... live, unaltered, minimally controlled, unmanipulated 'natural' psychotherapy sequences—so-called experiments of nature" (p. 54).

Naturally this approach had great appeal to clinicians since it dealt directly with their activities and in doing so promised to consider the complexities inherent in treatment. Typically, measures of multiple therapist and patient behaviors are taken so that all relevant variables (based on a given clinician's conceptualization of which variables are relevant) may be examined for interrelationships with every other variable.

Perhaps the best-known example of this type of study is the project at the Menninger Foundation (Kernberg, 1973). Begun in 1954, this was truly a mammoth undertaking involving 38 investigators, ten consultants, three different project leaders, and 18 years of planning and data collection. Forty-two patients were studied in this project. This group was broadly defined, although overtly psychotic patients were excluded. Assignment of patient to therapist and to differing modes of psychoanalytic treatment was not random but based on clinical judgments of which therapist or mode of treatment was most suitable for the patient. In other words, the procedures were those normally in effect in a clinical setting. In addition, other treatments, such as pharmacological or organic interventions, were administered to certain patients as needed. Against this background, the investigators measured multiple patient characteristics (such as various components of ego strength) and correlated these variables, measured periodically throughout treatment by referring to detailed records of treatment sessions, with multiple therapeutic activities and modes of treatment. As one would expect, the results are enormously complex and contain many seemingly contradictory findings. At least one observer (Malan, 1973) notes that the most important finding is that purely supportive treatment is ineffective with borderline psychotics, but working through of the transference relationship under hospitalization with this group is effective. Notwithstanding the global definition of treatment and broad diagnostic categories (borderline psychotic) also present in early group comparison studies, this report was generally hailed as an extremely important breakthrough in psychotherapy research. Methodologists, however, were not so sure. While admitting the benefits of a clearer definition of psychoanalytic terms emanating from the project, May (1973) wondered about the power and significance of the conclusions. Most of this criticism concerns the purported strength of the naturalistic study—that is, the lack of control over factors in the naturalistic setting. If subjects are assigned to treatments based on certain characteristics, were these characteristics responsible for improvement rather than treatment? What is the contribution of additional treatments received by certain patients? Did nurses and other therapists possibly react differently to patients in one group or another? What was the contribution of "spontaneous remission?"

In its pure state, the naturalistic study does not advance much beyond the uncontrolled case study in the power to isolate the effectiveness of a given treatment, as severe critics of the procedure point out (e.g., Bergin and Strupp, 1972), but this process is an improvement over case studies or reports of "percentage success" in groups since measures of relevant variables are constructed and administered, sometimes repeatedly. However, to increase confidence in any correlational findings from naturalistic studies, it would seem necessary to undermine the stated strengths of the study–that is, the "unaltered, minimally controlled, unmanipulated" condition prevailing in the typical naturalistic project–by randomly assigning patients, limiting access to additional confounding modes of treatment, and observing deviation of therapists from prescribed treatment forms. But if this were done, the study would no longer be naturalistic.

A further problem is obvious from the example of the Menninger project. The practical difficulties in executing this type of study seem very little less than those inherent in the large group comparison approach. The one exception is that the naturalistic study, in retaining close ties to the actual functioning of the clinic, requires less structuring or manipulating of large numbers of patients and therapists. The fact that this project took 18 years to complete makes one consider the significant administrative problem inherent in maintaining a research effort of this length of time. This factor is most likely responsible for the admission from one prominent member of the Menninger team, Robert S. Wallerstein, that he would not undertake such a project again (Bergin and Strupp, 1972).

Correlational studies, of course, do not have to be quite so "naturalistic" as the Menninger study. Kiesler (1971) reviews a number of studies without experimental manipulation that contain adequate definitions of variables and experimental attempts to rule out obvious confounding factors. Under such conditions, and if practically feasible, correlational studies may expose heretofore unrecognized relationships among variables in the psychotherapeutic process. But the fact remains that correlational studies by their nature are incapable of determining causal relationships on the effects of treatment. As Kiesler points out, the most common error in these studies is the tendency to conclude that a relationship between two variables indicates that one variable is causing the other. For instance, the conclusion in the Menninger study that working through transference relationships is an effective treatment for borderline psychotics (assuming other confounding factors were controlled or randomized) is open to several different interpretations. One might alternatively conclude that certain behaviors subsumed under the classification "borderline psychotic" caused the therapist to behave in such a way that transference variables changed or that a third variable, such as increased therapeutic attention during this more directive approach, was responsible for changes.

Process research

The second alternative to between-group comparison research was the process approach so often referred to in the APA conferences on psychotherapy research (e.g., Strupp and Luborsky, 1962). Hoch and Zubin's (1964) popular phrase "flight into process" was an accurate description of the reaction of many clinical investigators to the practical and methodological difficulties of the large group studies. Typically, process research has concerned itself with what goes on *during* therapy between an individual patient and therapist instead of the final outcome of any therapeutic effort. In the late 1950s and early 1960s, a large number of studies appeared on such topics as relation of therapist behavior to certain patient behaviors in a given interview situation (e.g., Rogers, Gendlin, Kiesler, and Truax, 1967). As such, process research held much appeal to clinicians and scientists alike. Clinicians were pleased by the focus on the individual and the resulting ability to study actual clinical processes. In some studies repeated measures during therapy gave clinicians an idea of the patient's course during treatment. Scientists were intrigued by the potential of defining variables more precisely within one interview without concerning themselves with the complexities involved before or after the point of study. The increased interest in process research, however, led to an unfortunate distinction between process and outcome studies that persists to this day (see Kiesler, 1966). This distinction was well stated by Luborsky (1959), who noted that process research was concerned with how changes took place in a given interchange between patient and therapist, whereas outcome research was concerned with what change took place as a result of treatment. As Paul (1969) and Kiesler (1966) point out, the dichotomization of process and outcome led to an unnecessary polarity in the manner in which measures of behavior change were taken. Process research collected data on patient changes at one or more points during the course of therapy, usually without regard for outcome, while outcome research was concerned only with pre-post measures outside of the therapeutic situation. Kiesler notes that this was unnecessary because measures of change within treatment could be continued throughout treatment until an "outcome" point is reached. He also quotes Chassan (1962) on the desirability of determining what transpired between the beginning and end of therapy in addition to outcome. Thus, the major concern of the process researchers, perhaps as a result of this imposed distinction, continued to be changes in patient behavior at points within the therapeutic endeavor. The discovery of meaningful clinical changes as a result of these processes was left to the prevailing experimental stategy of the group comparison approach. This reluctance to relate process variables to outcome and the resulting inability of this approach to evaluate the effects of psychotherapy led to a decline of process research. Matarazzo noted that in the 1960s the number of people interested in process studies of psychotherapy had declined and their students were nowhere to be seen (Bergin and Strupp, 1972). Because process and outcome were dichotomized in this manner, the notion eventually evolved that changes during treatment are not relevant or legitimate to the important question of outcome. Largely overlooked at this time was the work of Shapiro (e.g., 1961) at the Maudsley Hospital in London, begun in the 1950s. Shapiro was repeatedly administering measures of change to individual cases during therapy and also continuing these measures to an endpoint, thereby relating "process" changes to "outcome" and closing the artificial gap which Kiesler was to describe so cogently some years later.

1.7 THE SCIENTIST-PRACTITIONER SPLIT

The state of affairs of clinical practice and research in the 1960s satisfied few people. Clinical procedures were largely judged as unproven (Bergin and Strupp, 1972; Eysenck, 1965), and the prevailing naturalistic research was unacceptable to most scientists concerned with precise definition of variables and cause-effect relationships. On the other hand, the elegantly designed and scientifically rigorous group comparison design was seen as impractical and incapable of dealing with the complexities and idiosyncrasies of individuals by most clinicians. Somewhere in between was process research, which dealt mostly with individuals but was correlational rather than experimental. In addition, the method was viewed as incapable of evaluating the clinical effects of treatment since the focus was on changes within treatment rather than on outcome.

These developments were a major contribution to the well-known and oft-cited scientist-practitioner split (e.g., Joint Commission on Mental Illness and Health, 1961). The notion of an applied science of behavior change growing out of the optimism of the 1950s did not meet expectations and many clinician-scientists stated flatly that applied research had no effect on their clinical practice. Prominent among them was Matarazzo, who noted, "Even after 15 years, few of my research findings affect my practice. Psychological science *per se* doesn't guide me one bit. I still read avidly but this is of little direct practical help. My clinical experience is the only thing that has helped me in my practice to date. . . ." (Bergin and Strupp, 1972, p. 340). This opinion was echoed by one of the most productive and best-known researchers of the 1950s, Carl Rogers, who as early as the 1958 APA conference on psychotherapy noted that research had no impact on his clinical practice and by 1969 advocated abandoning formal research in psychotherapy altogether (Bergin and Strupp, 1972). Since this view prevailed among prominent clinicians who were well acquainted with research methodology, it follows that clinicians without research training or expertise were largely unaffected by the promise or substance of scientific evaluation of behavior change procedures.

Although the methodological difficulties outlined above were only one contribution to the scientist-practitioner split (see Leitenberg, 1974, for a detailed analysis), the concern and pessimism voiced by leading researchers in the field during Bergin and Strupp's comprehensive series of interviews led these commentators to reevaluate the state of the field. Voicing dissatisfaction with the large-scale group comparison design, Bergin and Strupp concluded:

> Among researchers as well as statisticians, there is a growing disaffection from traditional experimental designs and statistical procedures which are held inappropriate to the subject matter under study. This judgment applies with particular force to research in the area of therapeutic change, and our emphasis on the value of experimental case studies underscores this point. We strongly agree that most of the standard experimental designs and statistical procedures have exerted and are continuing to exert, a constricting effect on fruitful inquiry, and they serve to perpetuate an unwarranted overemphasis on methodology. More accurately, the exaggerated importance accorded experimental and statistical dicta cannot be blamed on the techniques proper—after all, they are merely tools—but their veneration mirrors a prevailing philosophy among behavioral scientists which subordinates problems to methodology. The insidious effects of this trend are tellingly illustrated by the typical graduate student who is often more interested in the details of a factorial design than in the problem he sets out to study; worse, the selection of a problem is dictated by the experimental design. Needless to say, the student's approach faithfully reflects the convictions and teachings of his mentors. With respect to inquiry in the area of psychotherapy, the kinds of effects we need to demonstrate at this point in time should be significant enough so that they are readily observable by inspection or descriptive statistics. If this cannot be done, no fixation upon statistical and mathematical niceties will generate fruitful insights, which obviously can come only from the researcher's understanding of the subject matter and the descriptive data under scrutiny. (Bergin and Strupp, 1972, p. 440)

1.8. A RETURN TO THE INDIVIDUAL

Bergin and Strupp were harsh in their comments on group comparison design and failed to specify those situations where between-group methodology is practical and desirable (see Chapter 2). However, their conclusions on alternative directions, outlined in a paper appropriately titled "New Directions in Psychotherapy Research" (Bergin and Strupp, 1970), had radical and far-reaching implications for the conduct of applied research. Essentially, Bergin and Strupp advised against investing further effort in process and outcome studies and proposed the experimental single case approach for the purpose of isolating mechanisms of change in the therapeutic process. Isolation of these mechanisms of change would then be followed by construction of new procedures based on a combination of variables whose effectiveness was demonstrated in single case experiments. As the authors note, "As a general paradigm of inquiry, the individual experimental case study and the experimental analogue approaches appear to be the primary strategies which will move us forward in our understanding of the mechanisms of change at this point" (Bergin and Strupp, 1970, p. 19). The hope is also expressed that this approach would tend to bring research and practice closer together.

With the recommendations emerging from Bergin and Strupp's comprehensive analysis, the philosophy underlying applied research methodology had come full circle in a little over 100 years. The disillusionment with large-scale between-group comparisons observed by Bergin and Strupp and their subsequent advocacy of the intensive study of the individual is an historical repetition of a similar position taken in the middle of the last century. At that time, the noted physiologist, Claude Bernard, in *An Introduction to the Study of Experimental Medicine* (1957), attempted to dissuade colleagues who believed that physiological processes were too complex for experimental inquiry within a single organism. In support of this argument, he noted that the site of processes of change is in the individual organism and group averages and variance might be misleading. The intensive scientific study of the individual in physiology then flourished. But methodology in physiology and experimental psychology is not directly applicable to the complexities present in applied research. Although the splendid isolation of Pavlov's laboratories allowed discovery of important psychological processes without recourse to sophisticated experimental design, it is unlikely that the same results would have obtained with a household pet in his natural environment. Yet these are precisely the conditions under which most applied researchers must work.

The plea of applied researchers for appropriate methodology grounded in the scientific method to investigate complex problems in individuals is never more evident than in the writings of Gordon Allport. Allport argues most eloquently that the science of psychology should attend to the uniqueness of the individual (e.g., Allport, 1961, 1962). In terms commonly used in the 1950s, Allport became the champion of the idiographic (individual) approach, which he considered superior to the nomothetic (general or group) approach.

> Why should we not start with individual behavior as a source of hunches (as we have in the past) and then seek our generalization (also as we have in the past) but finally come back to the individual not for the mechanical application of laws (as we do now) but for a fuller and more accurate assessment then we are now able to give? I suspect that the reason our present assessments are now so often feeble and sometimes even ridiculous, is because we do not take this final step. We stop with our wobbly laws of generality and seldom confront them with the concrete person. (Allport, 1962, p. 407)

Lacking a practical, applied methodology with which to study the individual, however, most of Allport's own research was nomothetic. The increase in the intensive study of the individual in applied research led to a search for appropriate methodology and several individuals or groups began developing ideas during the 1950s and 1960s.

The role of the case study

One result of the search for appropriate methodology was a reexamination of the role of the uncontrolled case study so strongly rejected by scientists in the 1950s. Recognizing its inherent limitations as an evaluation tool, Lazarus and Davison (1971) rightly suggest that the case study can make important contributions to an experimental effort. One of the more important functions of the case study is the generation of new hypotheses, which later may be subjected to more rigorous experimental scrutiny. Lazarus and Davison (1971) also agree with Dukes (1965) that the case study can occasionally be used to shed some light on extremely rare phenomena or cast doubt on well-established theoretical assumptions. Nevertheless, as Leitenberg (1973) points out, the case study is not capable of isolating therapeutic mechanisms of change, and the inability of many scientists and clinicians to discriminate the critical difference between the uncontrolled case study and the experimental study of an individual case has most likely retarded the implementation of single case experimental design (see Chapter 5).

The representative case

During this period, other theorists and methodologists were attempting to formulate viable approaches to the experimental study of single cases. Shontz (1965) proposed the study of the representative case as an alternative to traditional approaches in experimental personality research. Essentially, Shontz was concerned with validating previously established personality constructs or measurement instruments on individuals who appear to possess the necessary behavior appropriate for the research problem. Shontz's favorite example is a study of the contribution of psychodynamic factors to epilepsy described by Bowdlear (1955). After reviewing the literature on the presumed psychodynamics in epilepsy, Bowdlear chose a patient who closely approximated the diagnostic and descriptive characteristics of epilepsy presented in the literature (i.e., the representative case). Through a series of questions, Bowdlear then correlated seizures with a certain psychodynamic concept in this patient—acting out dependency. Since this case was "representative," Bowdlear assumed some generalization to other similar cases.

Shontz's contribution was not methodological since the experiments he cites were largely correlational and in the tradition of process research. Shontz also failed to recognize the value of the single case study in isolating effective therapeutic variables or building new procedures as suggested later by Bergin and Strupp (1972). Rather, he proposed the use of a single case in a deductive manner to test previously established hypotheses and measurement instruments in an individual who is known to be stable in certain personality characteristics such that he is "representative" of these characteristics. Conceptually, Shontz moved beyond Allport, however, in noting that this approach was not truly idiographic in that he was not proposing to investigate a subject as a self-contained universe with its own laws. To overcome this objectionable aspect of single case research, he proposed replication on subjects who differed in some significant way from the first subject. If the general hypothesis were repeatedly confirmed, this would begin to establish a generally applicable law of behavior. If the hypothesis were sometimes confirmed and sometimes rejected, he noted that ". . . the investigator will be in a position either to modify his thinking or to state more clearly the conditions under which the hypothesis does and does not provide a useful model of psychological events" (Shontz, 1965, p. 258). With this statement, Shontz anticipated the applied application of the methodology of direct and systematic replication in basic research (see Chapter 9) suggested by Sidman (1960).

Shapiro's methodology in the clinic

One of the most important contributions to the search for a methodology came from the pioneering work of Shapiro in London. As early as 1951, Shapiro was advocating a scientific approach to the study of individual phenomena, an advocacy that continued through the 1960s (e.g., Shapiro, 1961, 1966).

Unlike Allport, however, Shapiro went beyond the point of noting the advantages of applied research with single cases and began the difficult task of constructing an adequate methodology. One important contribution by Shapiro was the utilization of carefully constructed measures of clinically relevant responses administered repeatedly over time in an individual. Typically, Shapiro would examine fluctuations in these measures and hypothesize on the controlling effects of therapeutic or environmental influences. Many of these studies were correlational in nature, or what Shapiro refers to as simple or complex descriptive studies (1966). As such, these efforts bear a striking resemblance to process studies mentioned above in that the effect of a therapeutic variable was correlated with a target response. Shapiro attempted to go beyond this correlational approach, however, by defining and manipulating independent variables within single cases. One good example is the systematic alteration of two therapeutic approaches in a case of paranoid delusions (Shapiro and Ravenette, 1959). In a prototype of what was later to be called the **A-B-A** design, the authors measured paranoid delusions by asking the patient to rate the "intensity" of a number of paranoid ideas on a one-to-five scale. The sum of the score across 18 different delusions then represented his paranoid "score." Treatments consisted of "control" discussion concerning guilt feelings about situations in the patient's life, unrelated to his paranoid ideation, and rational discussion aimed at exposing the falseness of the patient's paranoid beliefs. The experimental sequence consisted of 4 days of "guilt" discussion followed by 8 days of rational discussion and a return to 4 days of "guilt" discussion. The authors observed an overall decline in paranoid scores during this experiment, which the authors rightly note as correlational and thus potentially due to a variety of causes. Close examination of the data revealed, however, that on weekends when no discussions were held the patient worsened during the "guilt" control phase and improved during the rational discussion phase. These fluctuations around the regression line were statistically significant. This effect, of course, is weak and of dubious importance since overall improvement in paranoid scores was not functionally related to treatment. Furthermore, several guidelines for a true experimental analysis of the treatment were violated. Examples of experimental error include the absence of baseline measurement to determine the pre-treatment course of the paranoid beliefs and the simultaneous withdrawal of one treatment and introduction of a second treatment (see Chapter 3). The importance of the case and other early work from Shapiro, however, is not the knowledge gained from any one experiment, but the beginnings of the development of a scientifically based methodology for evaluating effects of treatment within a single case. To the extent that Shapiro's correlational studies were similar to process research, he broke the semantic barrier which held that process criteria were unrelated to outcome. He demonstrated clearly that repeated measures within an individual could be extended to a logical endpoint and that this endpoint *was* the outcome of treatment. His more important contribution from our point of view, however, was the demonstration that independent variables in applied research could be defined and systematically manipulated within a single case, thereby fulfilling the requirements of a "true" experimental approach to the evaluation of therapeutic technique (Underwood, 1957).

Quasi-experimental designs

In the area of research dealing with broad-based educational or social change, Campbell and Stanley (1963) proposed a series of important methodological innovations, which they termed quasi-experimental designs. These designs, many of which are applicable to either groups or individuals, also are directly relevant to applied research. The two designs most appropriate for analysis of change in the individual are termed the time series design and the equivalent term series design. From the perspective of applied clinical research, the time series design is similar to Shapiro's effort to extend process observation throughout the course of a given treatment to a logical endpoint or outcome. This design goes beyond observations within treatment, however, to include observations from repeated measures in a period preceding and following a given intervention. Thus, one can observe changes from a baseline as a result of a given intervention. While the inclusion of a baseline is a distinct methodological improvement, this design is basically correlational in nature and is unable to isolate effects of therapeutic mechanisms or establish cause-effect relationships. Basically, this design is the A-B design described in Chapter 5. The equivalent time series design, however, involves experimental manipulation of independent variables through alteration of treatments, as in the Shapiro and Ravenette study (1959), or introduction and withdrawal of one treatment in an A-B-A fashion. Approaching the study of the individual from a different perspective than Shapiro, Campbell and Stanley arrived at similar conclusions on the possibility of manipulation of independent variables and establishment of cause-effect relationships in the study of a single case.

What is perhaps the more important contribution of these methodologists, however, is the description of various limitations of this design in their ability to rule out alternative plausible hypotheses (internal validity) or the extent to which one can generalize conclusions obtained from this design (external validity) (see Chapter 2).

Chassan and intensive designs

It remained for Chassan (1967) to pull together many of the methodological advances in single case research to this point in a book that made clear distinctions between the advantages and disadvantages of what he terms extensive (group) design and intensive (single case) design. Drawing on long experience in applied research, Chassan outlines the desirability and applicability of single case designs evolving out of applied research in the 1950s and early 1960s. While most of his own experience in single case design concerned the evaluation of pharmacologic agents on behavior disorders, Chassan also illustrated the uses of single case designs in psychotherapy research, particularly psychoanalysis. As a statistician rather than a practicing clinician, he emphasized the various statistical procedures capable of establishing relationships between therapeutic intervention and dependent variables within the single case. He concentrated on the correlation type of design using trend analysis but made occasional use of a prototype of the A-B-A design (e.g., Bellak and Chassan, 1964) which, in this case, extended the work of Shapiro to evaluation of drug effects but, in retrospect, contained some of the same methodological faults. Nevertheless, the sophisticated theorizing in the book on thorny issues in single case research, such as generality of findings from a single case, provided the most comprehensive treatment of these issues to this time. Many of Chassan's ideas on this subject will appear repeatedly in later sections of this book.

1.9. THE EXPERIMENTAL ANALYSIS OF BEHAVIOR

While innovative applied researchers such as Chassan and Shapiro made methodological advances in the experimental study of the single case, their advances did not have a major impact on the conduct of applied research outside of their own settings. As late as 1965, Shapiro noted in an invited address to the Eastern Psychological Association that a large majority of research in prominent clinical psychology journals involved between-group comparisons with little and, in some cases, no reference to the individual approach that he advocated. He hoped that his address might presage the beginning of a new emphasis on this method. In retrospect there are several possible reasons for the lack of impact. First, as Leitenberg (1973) was later to point out, many of the measures used by Shapiro in applied research were indirect and subjective (e.g., questionnaires), precluding observation on direct behavioral effects that gained importance with the rise of behavior therapy (see Chapter 4). Second, Shapiro and Chassan, in studies of psychotherapy, did not produce the strong, clinically relevant changes that would impress clinicians, perhaps due to inadequate or weak independent variables or treatments, such as instructions within interview procedures. Finally, the advent of the work of Shapiro and Chassan was associated with the general disillusionment during this period concerning the possibilities of research in psychotherapy. Nevertheless, Chassan and Shapiro demonstrated that meaningful applied research was possible and even desirable in the area of psychotherapy. These investigators, along with several of Shapiro's students (e.g., Davidson and Costello, 1969; Inglis, 1966; Yates, 1970), had an important influence on the development and acceptance of more sophisticated methodology, which was beginning to appear in the 1960s.

It is significant that it was the rediscovery of the study of the single case in basic research coupled with a new approach to problems in the applied area that marked the beginnings of a new emphasis on the experimental study of the single case in applied research. One indication of the broad influence of this combination of events was the emergence of a journal in 1968 (*Journal of Applied Behavior Analysis*) devoted to single case methodology in applied research and the appearance of this experimental approach in increasing numbers in the major psychological and psychiatric journals. The methodology in basic research was termed the experimental analysis of behavior; the new approach to applied problems became known as behavior modification or behavior therapy.

Some observers have gone so far as to define behavior therapy in terms of single case methodology (Yates, 1970) but, as Leitenberg (1973) points out, this definition is without empirical support since behavior therapy is a clinical approach employing a number of methodological strategies (see Krasner, 1971, for a history of behavior therapy). The relevance of the experimental analysis of behavior to applied research is the development of sophisticated methodology enabling intensive study of individual subjects. In rejecting a between-subject approach as the only useful scientific methodology, Skinner (1938, 1953) reflected the thoughts of the early physiologists such as Claude Bernard and emphasized repeated objective measurement in a single subject over a long period of time under highly controlled conditions. As Skinner noted (1966a), ". . . instead of studying a thousand rats for one hour each, or a hundred rats for ten hours each, the investigator is likely to study one rat for a thousand hours" (p. 21), a procedure that clearly recognizes the individuality of an organism. Thus, Skinner and his colleagues in the animal laboratories developed and refined the single case methodology that became the foundation of a new applied science. Culminating with the definitive methodological treatise by Sidman (1960), entitled *Tactics of Scientific Research*, the assumption and conditions of a true experimental analysis of behavior were outlined. Examples of fine grain analyses of behavior and the use of withdrawal, reversal, and multi-element experimental designs in the experimental laboratories began to appear in more applied journals in the 1960s as researchers adapted these strategies to the investigation of applied problems.

It is unlikely, however, that this approach would have had a significant impact on applied clinical research without the growing popularity of behavior therapy. The fact that Shapiro and Chassan were employing rudimentary prototypes of withdrawal designs (independent of influences from the laboratories of operant conditioning) without marked effect on applied research would seem to support this contention. The growth of the behavior therapy approach to applied problems, however, provided a vehicle for the introduction of the methodology on a scale that attracted attention from investigators in applied areas. Behavior therapy, as the application of the principles of general-experimental and social psychology to the clinic, also emphasized direct measurement of clinically relevant target behaviors and experimental evaluation of independent variables or "treatments." Since many of these "principles of learning" utilized in behavior therapy emanated from operant conditioning, it was a small step for behavior therapists to borrow also the operant methodology to validate the effectiveness of these same principles in applied settings. The initial success of this approach (e.g., Ullmann and Krasner, 1965) led to similar evaluations of additional behavior therapy techniques that did not derive directly from the operant laboratories (e.g., Agras, Leitenberg, Barlow, Curtis, Edwards, and Wright, 1971; Barlow, Leitenberg, and Agras, 1969). During this period, methodology originally intended for the animal laboratory was adapted more fully to the investigation of applied problems and "applied behavior analysis" became an important supplementary and, in some cases, alternative approach to between-subjects experimental designs. A subsequent step in the evolution of this methodology was a call by Leitenberg (1973) for utilization of the methodology beyond the evaluation of operant principles in applied settings and beyond techniques subsumed under the title behavior therapy to the entire area of what Leitenberg terms psychotherapy research. Although a similar plea by Shapiro was largely ignored, we believe that the recent methodological developments and the demonstrated effectiveness of this methodology provide a base for the establishment of a true science of human behavior change with a focus on the paramount importance of the individual. A description of the new methodology is the purpose of this book.

References

Agras, W. S., Leitenberg, H., Barlow, D. H., Curtis, N. A., Edwards, J., and Wright, D. Relaxation in systematic desensitization. *Archives of General Psychiatry*, 1971, **25**, 511-514.

Allport, G. W. *Pattern and growth in personality.* New York: Holt, Rinehart and Winston, 1961.

Allport, G. W. The general and the unique in psychological science. *Journal of Personality*, 1962, **30**, 405-422.

Barlow, D. H., Leitenberg, H., and Agras, W. S. Experimental control of sexual deviation through manipulation of the noxious scene in covert sensitization. *Journal of Abnormal Psychology*, 1969, **74**, 596-601.

Barron, F., and Leary, T. Changes in psychoneurotic patients with and without psychotherapy. *Journal of Consulting Psychology*, 1955, **19**, 239-245.

Beck, S. J. The science of personality: Nomothetic or idiographic? *Psychological Review*, 1953, **60**, 353-359.

Bellak, L., and Chassan, J. B. An approach to the evaluation of drug effect during psycotherapy: A double-blind study of a single case. *Journal of Nervous and Mental Disease*, 1964, **139**, 20-30.

Bergin, A. E. Some implications of psychotherapy research for therapeutic practice. *Journal of Abnormal Psychology*, 1966, **71**, 235-246.

Bergin, A. E., and Strupp, H. H. New directions in psychotherapy research. *Journal of Abnormal Psychology*, 1970, **76**, 13-26.

Bergin, A. E., and Strupp, H. H. *Changing frontiers in the science of psychotherapy.* New York: Aldine-Atherton, 1972.

Bernard, C. *An introduction to the study of experimental medicine.* New York: Dover, 1957.

Birney, R. C., and Teevan, R.C. (Eds.) *Reinforcement.* Princeton: Van Nostrand, 1961.

Bolger, H. The case study method. In B. B. Wolman (Ed.), *Handbook of clinical psychology.* Pp. 28-39. New York: McGraw-Hill, 1965.

Boring, E. G. *A history of experimental psychology.* New York: Appleton-Century-Crofts, 1950.

Bowdlear, C. M. Dynamics of idiopathic epilepsy as studied in one case. Unpublished doctoral dissertation. Case Western Reserve University, 1955.

Breuer, J., and Freud, S. *Studies on hysteria.* New York: Basic Books, 1957.

Brill, A. A. Selected papers on hysteria and other psychoneuroses: Sigmund Freud. *Nervous and Mental Disease Monograph Series*, 1909, No. 4.

Campbell, D. T., and Stanley, J. C. Experimental and quasi-experimental designs for research on teaching. In N. L. Gage (Ed.), *Handbook of research on teaching.* Pp. 171-246. Chicago: Rand-McNally, 1963.

Chaplin. J. P., and Krawiec, T. S. *Systems and theories of psychology.* New York: Holt, Rinehart and Winston, 1960.

Chassan, J. B. Probability processes in psychoanalytic psychiatry. In J. Scher (Ed.), *Theories of the mind.* Pp. 598-618. New York: Free Press of Glencoe, 1962.

Chassan, J. B. *Research design in clinical psychology and psychiatry.* New York: Appleton-Century-Crofts, 1967.

Davidson, P. O., and Costello, C. G. *N=1: Experimental studies of single cases.* New York: Van Nostrand Reinhold, 1969.

Dukes, W. F. N=1. *Psychological Bulletin,* 1965, **64**, 74-79.

duMas, F. M. Science and the single case. *Psychological Reports,* 1955, **1**, 65-75.

Dunlap, K. *Habits: Their making and unmaking.* New York: Liveright, 1932.

Eysenck, H. J. The effects of psychotherapy: An evaluation. *Journal of Consulting Psychology,* 1952, **16**, 319-324.

Eysenck, H. J. The effects of psychotherapy. *International Journal of Psychiatry,* 1965, **1**, 97-178.

Fisher, R. A. On the mathematical foundations of the theory of statistics. In *Theory of statistical estimation* (Proceeding of the Cambridge Philosophical Society), 1925.

Frank, J. D. *Persuasion and healing.* Baltimore: Johns Hopkins Press, 1961.

Hoch, P. H., and Zubin, J. (Eds.) *The evaluation of psychiatric treatment.* New York: Grune and Stratton, 1964.

Hyman, R., and Berger, L. Discussion. In H. J. Eysenck (Ed.), *The effects of psychotherapy.* Pp. 81-86. New York: International Science Press, 1966.

Inglis, J. *The scientific study of abnormal behavior.* Chicago: Aldine, 1966.

Joint Commission on Mental Illness and Health. *Action for mental health.* New York: Science Editions, 1961.

Kantorovich, N. V. An attempt of curing alcoholism by associated reflexes. *Novoye Refleksologii Nervoy i Fiziologii Sisterny,* 1928, **3**, 436.

Kernberg, O. F. Summary and conclusions of psychotherapy and psychoanalysis, final report of the Menninger Foundation's psychotherapy research project. *International Journal of Psychiatry,* 1973, **11**, 62-77.

Kessel, L., and Hyman, H. T. The value of psychoanalysis as a therapeutic procedure. *Journal of American Medical Association,* 1933, **101**, 1612-1615.

Kiesler, D. J. Some myths of psychotherapy research and the search for a paradigm. *Psychological Bulletin,* 1966, **65**, 110-136.

Kiesler, D. J. Experimental designs in psychotherapy research. In A. E. Bergin and S. L. Garfield (Eds.), *Handbook of psychotherapy and behavior change: An empirical analysis.* Pp. 36-74. New York: Wiley, 1971.

Knight, R. P. Evaluation of the results of psychoanalytic therapy. *American Journal of Psychiatry,* 1941, **98**, 434-466.

Krasner, L. Behavior therapy. *Annual Review of Psychology,* 1971, **22**, 483-532.

Lazarus, A. A., and Davison, G. C. Clinical innovation in research and practice. In A. E. Bergin and S. L. Garfield (Eds.), *Handbook of psychotherapy and behavior change: An empirical analysis.* Pp. 196-213. New York: Wiley, 1971.

Leitenberg, H. The use of single-case methodology in psychotherapy research. *Journal of Abnormal Psychology,* 1973, **82**, 87-101.

Leitenberg, H. Training clinical researchers in psychology. *Professional Psychology,* 1974, **5**, 59-69.

Lewin, K. Vectors, cognitive processes and Mr. Tolman's criticism. *Journal of General Psychology,* 1933, **8**, 318-345.

Luborsky, L. Psychotherapy. In P. R. Farnsworth and Q. McNemar (Eds.), *Annual review of psychology.* Pp. 317-344. Palo Alto: Annual Review, 1959.

Malan, D. H. Therapeutic factors in analytically oriented brief psychotherapy. In R. H. Gosling (Ed.), *Support, innovation and autonomy.* Pp. 187-205. London:Tavistock, 1973.

Marks, I. M. Flooding (implosion) and allied treatments. In W. S. Agras (Ed.), *Behavior modification: Principles and clinical applications.* Pp. 151-213. Boston: Little, Brown, 1972.
Max, L. W. Breaking up a homosexual fixation by the conditioned reaction technique: A case study. *Psychological Bulletin*, 1935, **32**, 734 (abstract).
May, P. R. A. Research in psychotherapy and psychoanalysis. *International Journal of Psychiatry*, 1973, **1**, 78-86.
Paul, G. L. Strategy of outcome research in psychotherapy. *Journal of Consulting Psychology*, 1967, **31**, 104-118.
Paul, G. L. Behavior modification research: Design and tactics. In C. M. Franks (Ed.), *Behavior therapy: Appraisal and status.* Pp. 29-62. New York: McGraw-Hill, 1969.
Pavlov, I. P. *Lectures on conditioned reflexes.* Trans. W. H. Gantt. New York: International, 1928.
Powers, E., and Witmer, H. *An experiment in the prevention of delinquency.* New York: Columbia University Press, 1951.
Rogers, C. R., Gendlin, E. T., Kiesler, D., and Truax, C. B. *The therapeutic relationship and its impact: A study of psychotherapy with schizophrenics.* Madison: University of Wisconsin Press, 1967.
Rosenzweig, S. Idiodynamics in personality theory with special reference to projective methods. *Psychological Review*, 1951, **58**, 213-223.
Rubenstein, E. A., and Parloff, M. B. Research problems in psychotherapy. In E. A. Rubenstein and M. B. Parloff (Eds.), *Research in psychotherapy.* Vol. 1. Pp. 276-293. Washington, D. C.: American Psychological Association, 1959.
Shapiro, M. B. The single case in fundamental clinical psychological research. *British Journal of Medical Psychology*, 1961, **34**, 255-263.
Shapiro, M. B. The single case in clinical-psychological research. *Journal of General Psychology*, 1966, **74**, 3-23.
Shapiro, M. B., and Ravenette, A. T. A preliminary experiment of paranoid delusions. *Journal of Mental Science*, 1959, **105**, 295-312.
Shontz, F. C. *Research methods in personality.* New York: Appleton-Century-Crofts, 1965.
Sidman, M. *Tactics of scientific research: Evaluating experimental data in psychology.* New York: Basic Books, 1960.
Skinner, B. F. *The behavior of organisms.* New York: Appleton-Century-Crofts, 1938.
Skinner, B. F. *Science and human behavior.* New York: Macmillan, 1953.
Skinner, B. F. Operant behavior. In W. K. Honig (Ed.), *Operant behavior: Areas of research and application.* Pp. 12-32. New York: Appleton-Century-Crofts, 1966a.
Skinner, B. F. Invited Address–Pavlovian Society of America, Boston, Massachusetts, 1966b.
Stilson, D. W. *Probability and statistics in psychological research and theory.* San Francisco: Holden-Day, 1966.
Strupp, H. H., and Luborsky, L. (Eds.) *Research in psychotherapy.* Vol. 2. Washington, D. C.: American Psychological Association, 1962.
Ullmann, L. P., and Krasner, L. (Eds.), *Case studies in behavior modification.* New York: Holt, Rinehart and Winston, 1965.
Underwood, B. J. *Psychological research.* New York: Appleton-Century-Crofts, 1957.
Walker, H. M., and Lev, J. *Statistical inference.* New York: Holt, Rinehart and Winston, 1953.
Watson, J. B., and Rayner, R. Conditioned emotional reactions. *Journal of Experimental Psychology*, 1920, **3**, 1-14.
Yates, A. J. *Behavior therapy.* New York: Wiley, 1970.

CHAPTER 2

General Issues in a Single-case Approach

2.1. INTRODUCTION

Two issues basic to any science are variability and generality of findings. These issues are handled somewhat differently from one area of science to another, depending on the subject matter. The first section of this chapter concerns variability.

In applied research, where individual behavior is the primary concern, it is our contention that the search for sources of variability in individuals must occur if we are to develop a truly effective clinical science of human behavior change. After a brief discussion of basic assumptions concerning sources of variability in behavior, specific techniques and procedures for dealing with behavioral variability in individuals are outlined. Chief among these are repeated measurement procedures that allow careful monitoring of day-to-day variability in individual behavior, and rapidly changing, improvised experimental designs that facilitate an immediate search for sources of variability in an individual. Several examples of the use of this procedure to track down sources of inter-subject or intra-subject variability are presented.

The second section of the chapter deals with generality of findings. Historically, this has been a thorny issue in applied research. The seeming limitations in establishing wide generality from results in a single case are obvious, yet establishment of generality from results in large groups has also proved elusive. After a discussion of important types of generality of findings, the shortcomings of attempting to generalize from group results in applied research are discussed. Traditionally, the major problems have been an inability to draw a truly random sample from human behavior disorders and the difficulty in generalizing from groups to an individual. Applied researchers attempted to solve the problem by making groups as homogeneous as possible so that results would be applicable to an individual who showed the characteristics of the homogeneous group. An alternative method of establishing generality of findings is the replication of single case experiments. The relative merits of establishing generality of findings from homogeneous groups and replication of single case experiments is discussed at the end of this section.

Finally, research questions that cannot be answered through experimentation on single cases are outlined and an appropriate combination of within-subject and between-subject research strategies is suggested.

2.2. VARIABILITY

The notion that behavior is a function of a multiplicity of factors finds wide agreement among scientists and professional investigators. Most scientists also agree that as one moves up the phylogenetic scale, the sources of variability in behavior become greater. In response to this, many scientists choose to work with lower life forms in the hope that laws of behavior will emerge more readily and be generalizable to the infinitely more complex area of human behavior. Applied researchers do not have this luxury. The task of the investigator in the area of human behavior disorders is to discover functional relations among treatments and specific behavior disorders over and above the welter of environmental and biological variables impinging on the patient at any given time. Given these complexities, it is small wonder that most treatments, when tested, produce small effects or, in Bergin and Strupp's terms, "weak" results (Bergin and Strupp, 1972).

Variability in basic research

Even in basic research, behavioral variability is enormous. In attempting to deal with this problem, many experimental psychologists assumed that variability was intrinsic to the organism rather than imposed by experimental or environmental factors (Sidman, 1960). If variability were an intrinsic component of behavior, then procedures had to be found to deal with this issue before meaningful research could be conducted. The solution involved experimental designs and confidence level statistics that would elucidate functional relations among independent and dependent variables over and above the intrinsic variability. Sidman (1960) notes that this is not the case in some other sciences, such as physics. Physics assumes that variability is imposed by error of measurement or other identifiable factors. Experimental efforts are then directed to discovering and eliminating as many sources of variability as possible so that functional relations can be determined with more precision. Sidman proposes that basic researchers in psychology also adopt this strategy. Rather than assuming that variability is intrinsic to the organism, one should make every effort to discover sources of behavioral variability among organisms such that laws of behavior could be studied with the precision and specificity found in physics. This precision, of course, would require close attention to the behavior of the individual organism. If one rat behaves differently than three other rats in an experimental condition, the proper tactic is to find out why. If the experimenter succeeds, the factors that produce that variability can be eliminated and a "cleaner" test of the effects of the original independent variable can be made. Sidman recognized that behavioral variability may never be entirely eliminated, but that isolation of as many sources of variability as possible will enable an investigator to estimate how much variability actually is intrinsic.

Variability in applied research

Applied researchers, by and large, have not been concerned with this argument. Every clinician is aware of multiple social or biological factors that are imposed on his data. If asked, many investigators might also assume some intrinsic variability in clients attributable to capriciousness in nature; but most are more concerned with the effect of uncontrollable but potentially observable events in the environment. For example, the sudden appearance of a significant relative or the loss of a job during treatment of depression may affect the course of depression to a far greater degree than the particular intervention procedure. Menstruation may cause marked changes in behavioral measures of "anxiety." Even more disturbing are the multiple unidentifiable sources of variability that cause broad fluctuation in a patient's clinical course. Most applied researchers assume this variability is imposed rather than intrinsic, but they may not know where to begin to factor out the sources.

The solution, as in basic research, has been to accept broad variability as an unavoidable evil, to employ experimental design and statistics that hopefully control variability, and to look for functional relations that supersede the "error."

As Sidman observes when discussing these tactics in basic research:

> The rationale for statistical immobilization of unwanted variables is based on the assumed random nature of such variables. In a large group of subjects, the reasoning goes, the uncontrolled factor will change the behavior of some subjects in one direction and will affect the remaining subjects in the opposite way. When the data are averaged over all the subjects, the effect of the uncontrolled variables are presumed to add algebraically to zero. The composite data are then regarded as though they were representative of one ideal subject who had never been exposed to the uncontrolled variables at all. (Sidman, 1960, p. 162)

Although one may question this strategy in basic research, as Sidman has, the amount of control an experimenter has over the behavioral history and current environmental variables impinging on the laboratory animal makes this strategy at least feasible. In applied research, when control over behavioral histories or even current environmental events is limited or non-existent, there is far less probability of discovering a treatment that is effective over and above these uncontrolled variables. This, of course, was the major cause of the inability of early group comparison studies to demonstrate that the treatment under consideration was effective. As noted in Chapter 1, some clients were improving while others were worsening, despite the presence of the treatment. Presumably, this variability was not intrinsic but due to current life circumstances of the clients.

Clinical vs. statistical significance

The experimental designs and statistics gleaned from the laboratories of experimental psychology have an added disadvantage in applied research. The purpose of research in any basic science is to discover functional relations among dependent and independent variables. Once discovered, these functional relationships become principles that add to our knowledge of behavior. In applied research, however, the discovery of functional relations is not sufficient. The purpose of applied research is to effect *meaningful* clinical or socially relevant behavioral changes. For example, if depression were reliably measureable on a 0-100 scale, with 100 representing severe depression, a treatment that improved *each patient* in a group of depressives from 80 to 75 would be statistically significant if all depressives in the control group remained at 80. This statistical significance, however, would be of little use to the practicing clinician since a score of 75 could still be in the suicidal range. An improvement of 40 or 50 points might be necessary before the clinician would consider the change clinically important. Elsewhere, we have referred to the issue as statistical vs. clinical significance (Barlow and Hersen, 1973). In this simplified example, statisticians might observe that this issue is easily correctable by setting a different criterion level for "effectiveness." In the jungle of applied research, however, when any effect superseding the enormous "error" or variance in a group of heterogeneous clients is remarkable, the clinician and even the researcher will often overlook this issue and consider a treatment that is statistically significant to be also clinically effective.

As Chassan (1960) points out, statistical significance can underestimate clinical effectiveness as well as overestimate it. This unfortunate circumstance occurs when a treatment is quite effective with a few members of the experimental group while the remaining members do not improve or deteriorate somewhat. Statistically, then, the experimental group does not differ from the control group whose members are relatively unchanged. When broad divergence such as this occurs among clients in response to an intervention, statistical treatments will average out the clinical effects along with changes due to unwanted sources of variability. In fact, this type of inter-subject variability is the rule rather than the exception. Bergin (1966) clearly illustrated the years that were lost to applied research because clinical investigators overlooked the marked effectiveness of these treatments on *some* clients. The issue of clinical vs. statistical significance is, of course, not restricted to between-group comparisons, but is something applied researchers must consider whenever statistical tests are applied to clinical data (see Chapter 8).

Nevertheless, the advantages of attempting to eliminate the enormous inter-subject variability in applied research through statistical methods has intuitive appeal to both researchers and clinicians who want quick answers to pressing clinical or social questions. In fact, to the clinician who might observe one severely depressive patient inexplicably get better while another equally depressed patient commits suicide, this variability may well seem to be intrinsic to the nature of the disorder rather than imposed by definable social or biological factors.

Highlighting variability in the individual

In any case, whether variability in applied research is intrinsic to some degree or not, the alternative to the treatment of inter-subject variability by statistical means is to highlight variability and begin the arduous task of determining sources of variability in the individual. To the applied researcher, this task is staggering. In realistic terms he must look at each individual who differs from other clients in terms of response to treatment and attempt to determine why. Since the complexities of human environments, both external and internal, are enormous, the possible causes of these differences number in the millions.

With the complexities involved in this search, one may legitimately question where to begin. Since inter-subject variability begins with one client differing in his response from some other clients, a logical starting point is the individual. If one is to concentrate on individual variability, however, the manner in which one observes this variability must also change. If one depressed patient deteriorates during treatment while others improve or remain stable, it is difficult to speculate on reasons for this deterioration if the only data available are observations before and after treatment. It would be much to the advantage of the clinical researcher to have followed this one patient's course *during* treatment so that the beginning of deterioration could be pinpointed. In this hypothetical case the patient may have begun to improve until a point midway in treatment when deterioration began. Perhaps a disruption in his family life occurred or he missed a treatment session, while other patients whose improvement continued did not experience these events. It would then be possible to speculate on these or other factors that were correlated with such change. In single case research the investigator could accommodate himself to the variability with immediate alteration in experimental design to test out hypothesized sources of these changes.

Repeated measures

The basis of this search for sources of variability is repeated measurement of the dependent variable or problem behavior. If this tactic has a familiar ring to clinicians, it is no accident since this is precisely the strategy every clinician uses in his daily practice. It is no secret to clinicians or other behavior change agents in applied settings that behavioral improvement from an initial observation to some endpoint sandwiches marked variability in the behavior between these points. A major activity of clinicians is observing this variability and making appropriate changes in treatment strategies or environmental circumstances, where possible, to eliminate these fluctuations from a general improving trend. Since measures in the clinic seldom go beyond gross observation and treatment consists of a combination of factors, it is difficult for the clinician to pinpoint potential sources of variability, but he speculates; with increased clinical experience, the effective clinician may guess right more often than wrong. In some cases, weekly observation may go on for years. As Chassan (1967) points out, "The existence of variability as a basic phenomenon in the study of individual psychopathology implies that a single observation of a patient state, in general, can offer only a minimum of information about the patient state. While such information is literally better than no information, it provides no more data than does any other statistical sample of one" (p. 182). He then quotes Wolstein (1954) from a psychoanalytic point of view, who comments on diagnostic categories: "These terms are 'ad hoc' definitions which move the focus of inquiry away from repetitive patterns with observable frequencies to fixed momentary states. But this notion of the momentary present is specious and deceptive; it is neither fixed nor momentary nor immediately present, but an inferred condition" (p. 39).

The relation of this strategy to process research, described in Chapter 1, is obvious. But the search for sources of individual variability cannot be restricted to repeated measures of one small segment of a client's course somewhere between the beginning and end of treatment as in process research. With the multitude of events impinging on the organism, significant behavior fluctuation may occur at any time—from the beginning of an intervention until well after completion of treatment. The necessity of repeated, frequent measures to begin the search for sources of individual variability is apparent. Procedures for repeated measures of a variety of behavior problems are described in Chapter 4.

Rapidly changing designs

If one is committed to determining sources of variability in individuals, repeated measurement alone is insufficient. In a typical case, no one event is clearly associated with behavioral fluctuation, and repeated observation will permit only a temporal correlation of several events with the behavioral fluctuation. In the clinic this temporal correlation provides differing degrees of evidence on an intuitive level concerning their causality. For instance, if a claustrophobic became trapped in an elevator on his way to the therapist's office and suddenly worsens, the clinician can make a reasonable inference that this event caused the fluctuation. Usually, of course, sources of variability are not so clear and the applied researcher must guess from among several correlated events. However, it would add little to science if an investigator merely reported at the end of an experiment that fluctuation in behaviors were observed and were correlated with several events. The task confronting the applied researcher at this point is to devise experimental designs to isolate the cause of the change, or lack of change. One advantage of single case experimental designs is that the investigator can begin an immediate search for the cause of an experimental behavior trend by altering his experimental design on the spot. This feature, when properly employed, can provide immediate information on hypothesized sources of variability. In Skinner's words, "A prior design in which variables are distributed, for example, in a Latin square, may be a severe handicap. When effects on behavior can be immediately observed, it is most efficient to explore relevant variables by manipulating them in an improvised and rapidly changing design. Similar practices have been responsible for the greater part of modern science" (Honig, 1966, p. 21).

2.3. EXPERIMENTAL ANALYSIS OF SOURCES OF VARIABILITY THROUGH IMPROVISED DESIGNS

In single case designs there are at least three patterns of variability highlighted by repeated measurement. In the first pattern a subject may not respond to a treatment previously demonstrated as effective with other subjects. In a second pattern a subject may improve when no treatment is in effect as in a baseline phase. This "spontaneous" improvement is often considered to be the result of "placebo" effects. These two patterns of inter-subject variability are quite common in applied research. In a third pattern the variability is intra-subject in that marked cyclical patterns emerge in the measures that supersede the effect of any independent variable. Using improvised and rapidly changing designs, it is possible to follow Skinner's suggestion and begin an immediate search for sources of this variability. Examples of these efforts are provided below.

Subject fails to improve

One experiment from our laboratories illustrates the use of an "improvised and rapidly changing design" to determine why one subject did not improve with a treatment that had been successful with other subjects. The purpose of this experiment was to explore the effects of a classical conditioning procedure on increasing heterosexual arousal in homosexuals (Herman, Barlow, and Agras, 1974). In this study, heterosexual arousal as measured by penile circumference change to slides of nude females was the major dependent variable. Measures of homosexual arousal and reports of heterosexual urges and fantasies were also recorded. The design is a basic A-B-A-B with a baseline procedure, making it technically an A-B-C-B-C, where A is baseline, B is a control phase, backward conditioning, and C is the treatment phase or classical conditioning. In classical conditioning the client viewed two slides for 1 minute each. One slide depicted a female, which became the CS. A male slide, to which the client became aroused routinely, became the UCS. During classical conditioning, the client viewed the CS (female slide) for 1 minute, followed immediately by the UCS (male slide) for 1 minute in the typical classical conditioning paradigm. During the B or control phase, however, the order of presentation was reversed (UCS-CS), resulting in a backward conditioning paradigm which, of course, should not produce any learning.

During experiment one (see Fig. 2-1), no increases in heterosexual arousal were noted during baseline or backward conditioning. A sharp rise occurred, however, during classical conditioning. This was followed by a downward trend in heterosexual arousal during a return to the backward conditioning control phase, and further increases in arousal during a second classical conditioning phase, suggesting that the classical conditioning procedure was producing the observed increase.

In attempting to replicate this finding on a second client, some variation in responding was noted. Again, no increase in heterosexual arousal occurred during baseline or backward conditioning phases; but none occurred during the first classical conditioning phase either, even though the number of UCS slides was increased from one to three. At this point, it was noted that his response latency to the male slide was approximately 30 seconds. Thus, the classical conditioning procedure was adjusted slightly, such that 30 seconds of viewing the female slide alone was followed by 30 seconds of viewing both the male and female slides simultaneously (side by side), followed by 30 seconds of the male slide alone. This adjustment (labeled simultaneous presentation) produced increases in heterosexual arousal in the separate measurement sessions, which reversed during a return to the original classical conditioning procedure and increased once again

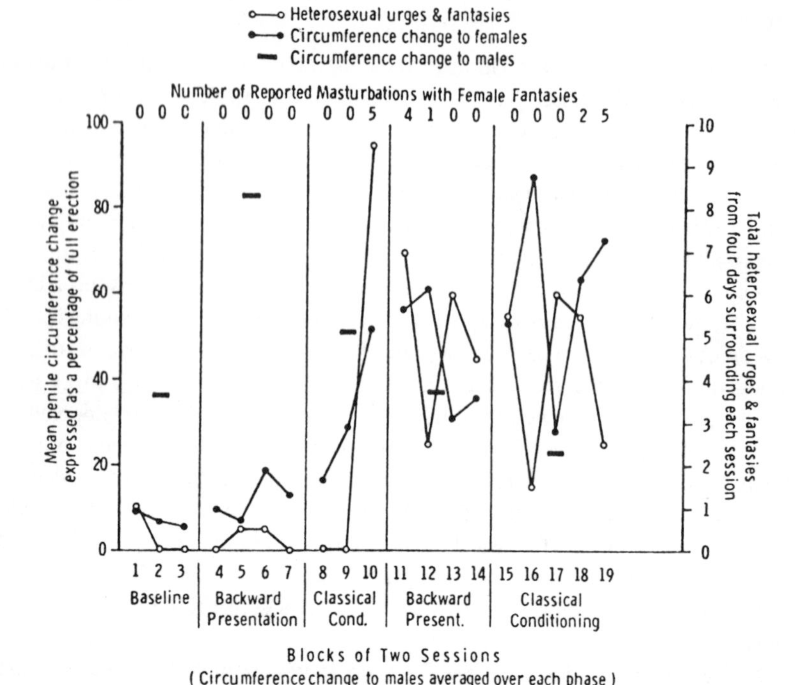

Fig. 2-1. Mean penile circumference change to male and female slides expressed as a percentage of full erection and total heterosexual urges and fantasies collected from 4 days surrounding each session. Data are presented in blocks of two sessions (circumference change to males averaged over each phase). Reported incidence of masturbation accompanied by female fantasy is indicated for each blocked point. (Fig. 1, p. 36, from: Herman, S. H., Barlow, D. H., and Agras, W. S. An experimental analysis of classical conditioning as a method of increasing heterosexual arousal in homosexuals. *Behavior Therapy*, 1974, 5, 33-47. Reproduced by permission.)

during the second phase in which the slides were presented simultaneously. The experiment suggested that classical conditioning was also effective with this client but only after a sensitive temporal adjustment was made.

Merely observing the "outcome" of the two subjects at the end of a fixed point in time would have produced the type of inter-subject variability so common in outcome studies of therapeutic techniques. That is, one subject would have improved with the initial classical conditioning procedure while one subject remained unchanged. If this pattern continued over additional subjects, the result would be the typical "weak" effect (Bergin and Strupp, 1972) with large inter-subject variability. Highlighting the variability through repeated measure-

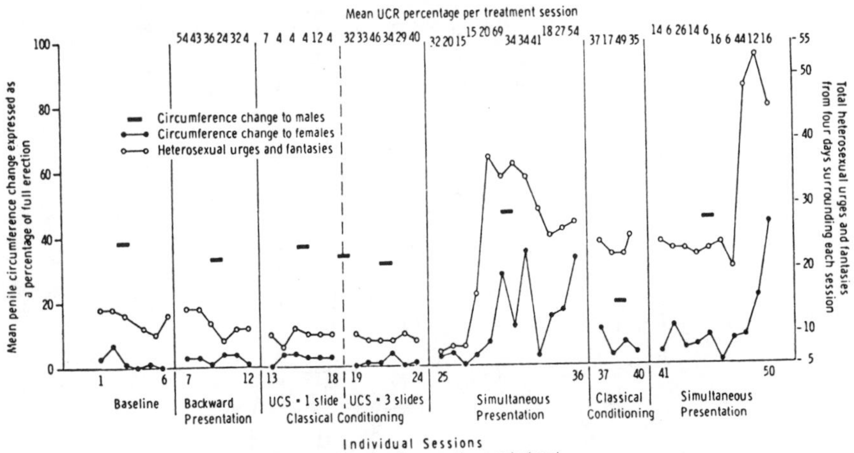

Fig. 2-2. Mean penile circumference change to male and female slides expressed as a percentage of full erection and total heterosexual urges and fantasies collected from 4 days surrounding each session. Data are presented for individual sessions with circumference change to males averaged over each phase. Mean UCR percentage is indicated for each treatment session. (Fig. 2, p. 40, from: Herman, S. H., Barlow, D. H., and Agras, W. S. An experimental analysis of classical conditioning as a method of increasing heterosexual arousal in homosexuals. *Behavior Therapy*, 1974, 5, 33-47. Reproduced by permission.)

ment in the individual and improvising a new experimental design as soon as a variation in response was noted (in this case no response) allowed an immediate search for the cause of this unresponsiveness. It should also be noted that this research tactic resulted in immediate clinical benefit to the patient, providing a practical illustration of the merging of scientist and practitioner roles in the applied researcher.

Subject improves "spontaneously"

A second source of variability quite common in single case research is the presence of "spontaneous" improvement in the absence of the therapeutic variable to be tested. This effect is illustrated in a second experiment on increasing heterosexual arousal in homosexuals (Herman, Barlow, and Agras, 1974).

In this study, the original purpose was to determine the effectiveness of orgasmic reconditioning, or pairing masturbation with heterosexual cues, in producing heterosexual arousal. The heterosexual cues chosen were movies of a female assuming provocative sexual positions. The initial phase consisted of measurements

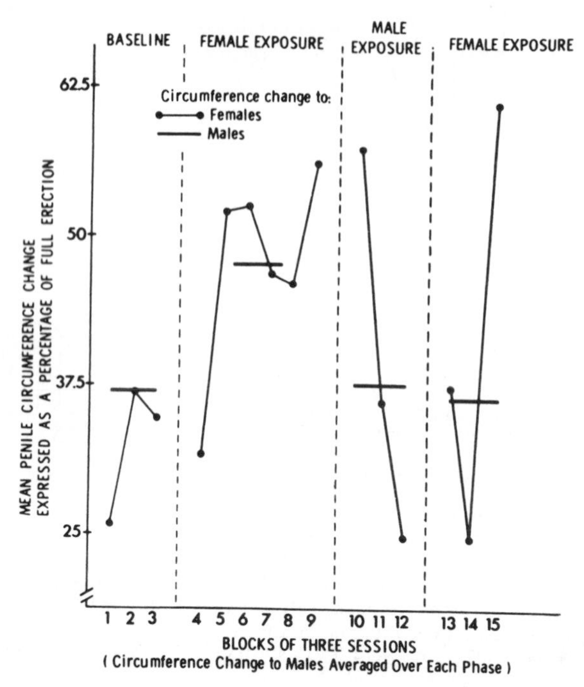

Fig. 2-3. Mean penile circumference change expressed as a percentage of full erection to nude female (averaged over blocks of three sessions) and nude male (averaged over each phase) slides. (Fig. 1, p. 338, from: Herman, S. H., Barlow, D. H., and Agras, W. S. An experimental analysis of exposure to explicit heterosexual stimuli as an effective variable in changing arousal patterns of homosexuals. *Behaviour Research and Therapy*, 1974, 12, 335-345. Reproduced by permission.)

of arousal patterns without any "treatment," which served as a baseline of sexual arousal. Before pairing masturbation with this movie, a control phase was administered where all elements of the treatment were present with the exception of masturbation. That is, the subject was instructed that this was "treatment" and that looking at movies would help him learn heterosexual arousal. Although no increase in heterosexual arousal was expected during this phase, this procedure was experimentally necessary to isolate the pairing of masturbation with the cues in the next phase as the effective treatment. The effects of masturbation were never tested in this experiment, however, since the first subject demonstrated unexpected but substantial increases in heterosexual arousal during the "control" phase in which he simply viewed the erotic movie (see Fig. 2-3). Once again it became necessary to "improvise" a new experimental design at the end of this control phase in an attempt to determine the cause of this unexpected increase. On the hunch that the erotic heterosexual movie was responsible for these gains rather than other therapeutic variables such as expectancy, etc., a second erotic movie without heterosexual content was introduced, in this case a homosexual movie. Heterosexual arousal dropped in this condition and increased once again when the heterosexual movie was introduced. This experiment, and subsequent replication, demonstrated that the erotic heterosexual movie was responsible for improvement. Determination of the effects of masturbation was delayed for future experimentation.

Subject displays cyclical variability

A third pattern of variability, highlighted by repeated measurement in individual cases, is observed when behavior varies in a cyclical pattern. The behavior may follow a regular pattern (i.e., weekly) or may be irregular. A common temporal pattern, of course, is the behavioral or emotional fluctuation noted during menstruation. Of more concern to the clinician is the marked fluctuation occurring in most behavioral disorders over a period of time. In most instances the fluctuation cannot be readily correlated with specific, observable environmental or psychological events due to the extent of the behavioral or emotional fluctuation and the number of potential variables that may be affecting the behavior. As noted in the beginning of this chapter, experimental clinicians can often make educated guesses, but the technique of repeated measurement can illustrate relationships that might not be readily observable.

A good example of this method is found in an early case of severe, daily asthmatic attacks reported by Metcalfe (1956). In the course of assessment, Metcalfe had the patient record in diary form asthmatic attacks as well as all activities during the day, such as games, shopping expeditions, meetings with her mother, and other social visits. These daily recordings revealed that asthmatic attacks most often followed meetings with the patient's mother, particularly if these meetings occurred in the home of the mother. After this relationship was demonstrated, the patient experienced a change in her life circumstances which resulted in moving some distance away from her mother. During the ensuing 20 months, only nine attacks were recorded despite the fact that these attacks had occurred daily for a period of 2 years prior to intervention. What is more remarkable is that eight of the attacks followed her now infrequent visits to her mother.

One again, the procedure of repeated measurement highlighted individual fluctuation, allowing a search for correlated events that bore potential causal relationships to the behavior disorder. It should be noted that no experimental analysis was undertaken in this case to isolate the mother as the cause of asthmatic attacks. However, the dramatic reduction of high-frequency attacks after decreased contact with the mother provides reasonably strong evidence on the contributory effects of visits to the mother in an A-B fashion. What is more convincing, however, is the reoccurrence of the attacks at widely spaced intervals after visits to the mother during the 20-month follow-up. This series of naturally occurring events approximates a contrived A-B-A-B. . . design and effectively isolates the mother's role in the patient's asthmatic attacks (see Chapter 5).

Searching for "hidden" sources of variability

In the preceding case functional relations become obvious without experimental investigation due to the overriding effects of one variable on the behavior in question and a series of fortuitous events (from an experimental point of view) during follow-up. Seldom in applied research is one variable so predominant. The more usual case is one where marked fluctuations in behavior occur that cannot be correlated with any one variable. In these cases, close examination of repeated measures of the target behavior and correlated internal or external events does not produce an obvious relationship. Most likely, many events may be correlated at one time or another with deterioration or improvement in a client. At this point, it becomes necessary to employ sophisticated experimental designs if one is to search for the source of variability. The experienced applied researcher must first choose the most likely variables for investigation from among the many impinging on the client at any one time. In the case described above, not only visits to the mother but visits to other relatives as well as stressful situations at work might all have contributed to the variance. The task of the clinical investigator is to tease out the relevant variables by manipulating one variable, such as visits to mother, while holding other variables constant. Once the contribution of visits to mother to behavioral fluctuation has been determined, the investigator must go on to the next variable, and so on.

In many cases, behavior is a function of an interaction of events. These events may be naturally occurring environmental variables or perhaps a combination of treatment variables which, when combined, affect behavior differently than each variable in isolation. For example, when testing out a variety of treatments for anorexia nervosa (Agras, Barlow, Chapin, Abel, and Leitenberg, 1974), it was discovered that size of meals served to the patients seemed related to caloric intake. An improvised design at this point in the experiment demonstrated that size of meals was related to caloric intake only if feedback and reinforcement were present. This discovery led to inclusion of this procedure in a recommended treatment package for anorexia nervosa. Experimental designs to determine the effects of combinations of variables will be discussed in Section 6.5 of Chapter 6.

2.4. BEHAVIOR TRENDS AND INTRA-SUBJECT AVERAGING

When testing the effects of specific interventions on behavior disorders, the investigator is less interested in small day-to-day fluctuations that are a part of so much behavior. In these cases the investigator must make a judgment on how much behavioral variability to ignore when looking for functional relations among overall trends in behavior and treatment in question. To the investigator interested in determining all sources of variability in individual behavior, this is a very difficult choice. For applied researchers, the choice is often determined by the practical considerations of discovering a therapeutic variable that "works" for a specific behavior problem in an individual. The necessity of determining the effects of a given treatment may constrain the applied researcher from improvising designs in mid-experiment to search for a source of each and every fluctuation that may appear.

In correlational designs, where one simply introduces a variable and observes the "trend," statistics have been devised to determine the significance of the trend over and above the behavioral fluctuation (Campbell and Stanley, 1966; see also Chapter 8 by Kazdin). In experimental designs such as A-B-A-B, where one is looking for cause-effect relationships, investigators will occasionally resort to averaging two or more data points within phases. This intra-subject averaging, which is sometimes called "blocking," will usually make trends in behavior more visible so that the clinician can judge the magnitude and clinical relevance of the effect. This procedure is dangerous, however, if the investigator is under some illusion that the variability has somehow disappeared or is unimportant to an understanding of the controlling effects of the behavior in question. This method is simply a procedure to make large and clinically significant changes as a result of introduction and withdrawal of treatment more apparent. To illustrate the procedure, the original data on caloric intake in a subject with anorexia nervosa will be presented for comparison with published data (Agras, Barlow, Chapin, Abel, and Leitenberg, 1974). The data as published are presented in Fig. 2-4. After the baseline phase, material reinforcers such as cigarettes, etc., were administered contingent on weight gain in a phase labeled "reinforcement." In the next phase, informational "feedback" was added to reinforcement. Feedback consisted of presenting the subject with daily weight counts of caloric intake after each meal and counts of number of mouthfuls eaten. The data indicate that caloric intake was relatively stable during the reinforcement phase but increased sharply when feedback was added to reinforcement. Six data points are presented in each of the reinforcement and reinforcement-feedback phases. Each data point represents the mean of 2 days. With this method of data presentation, caloric intake during reinforement looks quite stable.

In fact, there was a good deal of day-to-day variability in caloric intake during

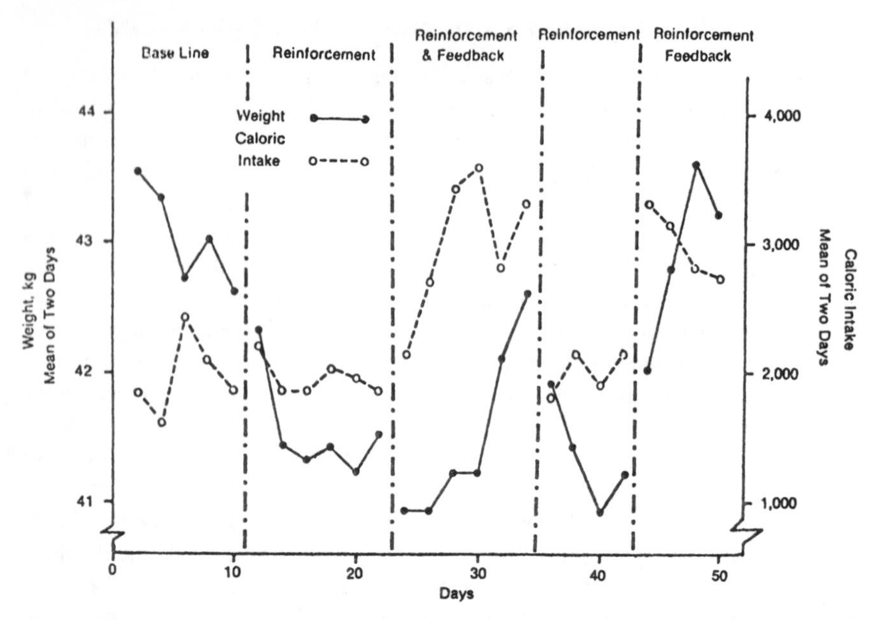

Fig. 2-4. Data from an experiment examining the effect of feedback on the eating behavior of a patient with anorexia nervosa (Patient 4). (Fig. 3, p. 283, from: Agras, W. S., Barlow, D. H., Chapin, H. N., Abel, G. G., and Leitenberg, H. Behavior modification of anorexia nervosa. *Archives of General Psychiatry*, 1974, **30**, 279-286. Reproduced by permission.)

this phase. If one examines the day-to-day data, caloric intake ranged from 1450 to 3150 over the 12-day phase. Since the variability assumed a pattern of roughly one day of high caloric intake followed by a day of low intake, the average of 2 days presents a stable pattern. When feedback was added during the next 12-day phase, the day-to-day variability remained but the range was displaced upward from 2150 to 3800 calories per day. Once again, this pattern of variability was approximately one day of high caloric intake followed by a low value. In fact, this pattern obtained throughout the experiment.

In this experiment, feedback is clearly a potent therapeutic procedure over and above the variability whether one examines the data day-by-day or in blocks of 2 days. The averaged data, however, present a clear picture of the effect of the variable over time. Since the major purpose of the experiment was to demonstrate the effects of various therapeutic variables with anorexics, we chose to present the data in this way. It was not our intention, however, to ignore the daily variability. The fairly regular pattern of change suggests several environmental or metabolic factors that may account for these changes. If one were interested in more basic research on eating patterns in anorexics, one would have

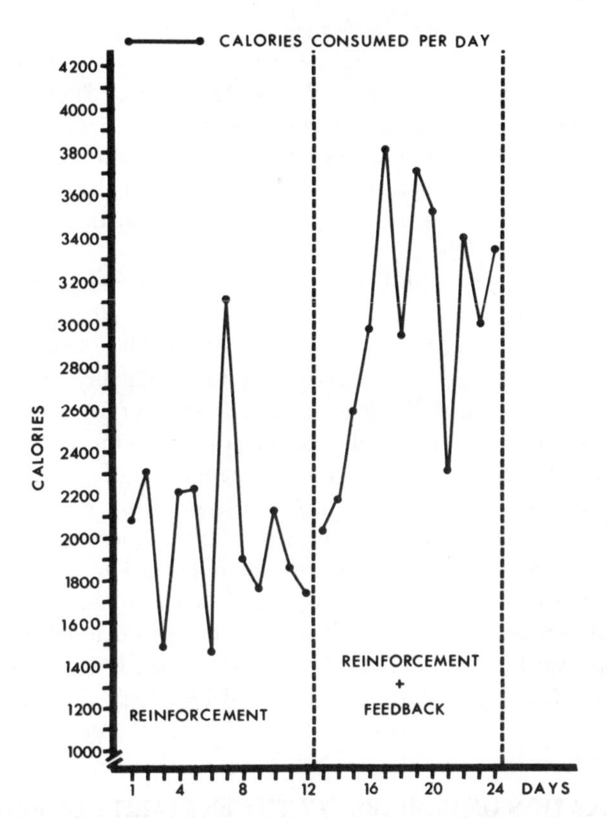

Fig. 2-5. Caloric intake presented on a daily basis during reinforcement and reinforcement and feedback phases for the patient whose data is presented in Fig. 2-4. (Replotted from Fig. 3, p. 283, from: Agras, W. S., Barlow, D. H., Chapin, H. N., Abel, G. G., and Leitenberg, H. Behavior modification of anorexia nervosa. *Archives of General Psychiatry*, 1974, **30**, 279-286.)

to explore possible sources of this variability in a finer analysis than we chose to undertake here.

It is possible, of course, that feedback might not have produced the clear and clinically relevant increase noted in these data. If feedback resulted in a small increase in caloric intake that was clearly visible only when data were averaged, one would have to resort to statistical tests to determine if the increase could be attributed to the therapeutic variable over and above the day-to-day variability (see Chapter 8). Once again, however, one may question the clinical relevance of the therapeutic procedure if the improvement in behavior is so small that the investigator must use statistics to determine if change actually occurred. If this situation were obtained, the preferred strategy might be to improvise on the experimental design and augment the therapeutic procedure such that more relevant and substantial changes are produced. The issue of clinical vs. statistical significance, which was discussed in some detail above, is a recurring one in single case research. In the last analysis, however, this is always reduced to judgments by therapists, educators, etc. on the magnitude of change that is relevant to the setting. In most cases these magnitudes are greater than changes that are merely statistically significant.

The above example notwithstanding, the conservative and preferred approach of data presentation in single case research is to present all of the data so that other investigators may examine the intra-subject variability first-hand and draw their own conclusions on the relevance of this variability to the problem.

Large intra-subject variability is a common feature during repeated measurements of target behaviors in a single case, particularly in the beginning of an experiment when the subject may be accommodating to intrusive measures. How much variability the researcher is willing to tolerate before introducing an independent variable (therapeutic procedure) is largely a question of judgment on the part of the investigator. Similar procedural problems arise when introduction of the independent variable itself results in increased variability. Here the experimenter must consider alteration in length of phases to determine if variability will decrease over time (as it often does), clarifying the effects of the independent variable. These procedural questions will be discussed in some detail in Chapter 3.

2.5. RELATION OF VARIABILITY TO GENERALITY OF FINDINGS

The search for sources of variability within individuals and the use of improvised and fast-changing experimental designs appear to be contrary to one of the most cherished goals of any science—the establishment of generality of findings. Studying the idiosyncrasies of one subject would seem, on the surface, to confirm Underwood's (1957) observation that intensive study of individuals will lead to discovery of laws that are applicable only to that individual. In fact, the identification of sources of variability in this manner lead to increases in generality of findings.

If one assumes that behavior is lawful, then identifying sources of variability in one subject should give us important leads in sources of variability in other similar subjects undergoing the same treatments. As Sidman (1960) points out,

> Tracking down sources of variability is then a primary technique for establishing generality. Generality and variability are basically antithetical concepts. If there are major undiscovered sources of variability in a given set of data, any attempt to achieve subject or principle generality is likely to fail. Every time we discover and achieve

> control of a factor that contributes to variability, we increase the likelihood that our data will be reproducible with new subjects and in different situations. Experience has taught us that precision of control leads to more extensive generalization of data. (p. 152)

And again,

> It is unrealistic to expect that a given variable will have the same effects upon all subjects under all conditions. As we identify and control a greater number of the conditions that determine the effects of a given experimental operation, in effect we decrease the variability that may be expected as a consequence of the operation. It then becomes possible to produce the same results in a greater number of subjects. Such generality could never be achieved if we simply accepted inter-subject variability and gave equal status to all deviant subjects in an investigation. (p. 190)

In applied research, when inter-subject and intra-subject variability is enormous and putative sources of the variability are difficult to control, the establishment of generality is a difficult task indeed. But the establishment of a science of human behavior change depends heavily on procedures to establish generality of findings. This important issue will be discussed in the next section.

2.6. GENERALITY OF FINDINGS

Types of generality

Generalization means many things. In applied research, generalization usually refers to the process in which behavioral or attitudinal changes in the treatment setting "generalize" to other aspects of the client's life. In educational research this can mean generalization of behavioral changes from the classroom to the home. Generalization of this type can be determined by observing behavioral changes outside of the treatment setting.

There are at least three additional types of generality in behavior change research, however, that are more relevant to the present discussion. The first is generality of findings across subjects or clients; that is, if a treatment effects certain behavior changes in one subject, will the same treatment also work in other subjects with similar characteristics? As we shall see below, this is a large question since subjects can be "similar" in many different ways. For instance, subjects may be similar in that they have the same diagnostic labels or behavioral disorders (e.g., schizophrenia or phobia). In addition, subjects may be of similar age (e.g., between 14 and 16) or come from similar socioeconomic backgrounds.

Generality across behavior change agents is a second type. For instance, will a therapeutic technique that is effective when applied by one behavior change agent also be effective when applied to the same problem by different agents? A common example is the classroom. If a young, attractive, female teacher successfully uses reinforcement principles to control disruptive behavior in her classroom, will an older teacher who is more stern also be able to apply successfully the same principles to similar problems in her class? Will an experienced therapist be able to treat a middle-aged claustrophobic more effectively than a naive therapist who used exactly the same procedure?

A third type of generality concerns the variety of settings in which clients are found. The question here is will a given treatment or intervention applied by the same or similar therapist, to similar clients, work as well in one setting as another? For example, would reinforcement principles that work in the classroom also work in a summer camp setting or would desensitization of an agoraphobic in an urban office building be more difficult than in a rural setting?

These questions are very important to clinicians who are concerned with which treatments are most effective with a given client in a given setting. Typically, clinicians have looked to the applied researcher to answer these questions.

Problems in generalizing from a single case

The most obvious limitation of studying a single case is that one does not know if the results from this case would be relevant to other cases. Even if one isolates the active therapeutic variable in a given client through a rigorous single case experimental design, critics (Kiesler, 1971; Underwood, 1957) note that there is little basis for inferring that this therapeutic procedure would be equally effective when applied to clients with similar behavior disorders (client generality) or that different therapists using this technique would achieve the same results (therapist generality). Finally, one does not know if the technique would work in a different setting (setting generality). This issue, more than any other, has retarded the development of single case methodology in applied research and has caused many authorities on research to deny the utility of studying a single case for any other purpose than the generation of hypotheses (e.g., Kiesler, 1971). Conversely, in the search for generality of applied research findings, the group comparison approach appeared to be the logical answer (Underwood, 1957).

In the specific area of individual human behavior change, however, there are issues that limit the usefulness of a group approach in establishing generality of findings. On the other hand, the newly developing procedures of direct and systematic replication offer an alternative, in some instances, for establishing generality of findings relevant to individuals. The purpose of this section is to outline the major issues, assumptions, and goals of generality of findings as related to behavior change in an individual and describe the advantages and disadvantages of the various procedures to establish generality of findings.

2.7. LIMITATION OF GROUP DESIGN IN ESTABLISHING GENERALITY OF FINDINGS

In Chapter 1, Section 1.5, several limitations of group designs in applied research noted by Bergin and Strupp (1972) were outlined. One of the limitations referred to difficulties in generalizing results from a group to an individual. In this category, two problems stand out. The first is inferring that results from a relatively homogeneous group are representative of a given population. The second is generalizing from the average response of a heterogeneous group to a particular individual. These two problems will be discussed in turn.

Random sampling and inference in applied research

After the brilliant work of Fisher, early applied researchers were most concerned with drawing a truly random sample of a given population so that results would be generalizable to this population. For instance, if one wished to draw some conclusion on the effects of a given treatment for schizophrenia, one would have to draw a random sample of all schizophrenics.

In reference to the three types of generality mentioned above, this means that the clients under study (e.g., schizophrenics) must be a random sample of all schizophrenics, not only for behavioral components of the disorder, such as loose associations or withdrawn behavior, but also for other patient characteristics such as age, sex, socioeconomic status, etc. These conditions must be fulfilled before one can infer that a treatment that demonstrates a statistically significant effect would also be effective for other schizophrenics outside of the study. As Edgington (1967) points out, "In the absence of random samples hypothesis testing is still possible, but the significance statements are restricted to the effect of the experimental treatments on the subjects actually used in the experiment, generalization to other individuals being based on logical non-statistical considerations" (p. 195). If one wishes to make statements about effectiveness of a treatment across therapists or settings, random samples of therapists and settings must also be included in the study.

Random sampling of characteristics in the animal laboratories of experimental psychology is feasible, at least across subjects, since most relevant characteristics such as genetic and environmental determinants of individual behavior can be controlled. In clinical or educational research, however, it is extremely difficult to sample adequately the population of a particular syndrome. One reason for this is the vagueness of many diagnostic categories (e.g., schizophrenia). In order to sample the population of schizophrenics one must be able to pinpoint the various behavioral characteristics that make up this diagnosis and ensure that any sample adequately represents these behaviors. The notorious unreliability of this diagnostic category, however, makes it very difficult to determine the adequacy of a given sample. In one center "schizophrenia" may mean bizarre behavior and hallucinations. In another center a thought disorder may be a *sine qua non* of the diagnosis.

A second problem that arises when one is attempting an adequate sample of a population is the availability of clients who have the needed behavior or characteristics to fill out the sample (see Chapter 1, Section 1.5). In laboratory animal research this is not a problem since subjects with specified characteristics or genetic backgrounds can be ordered or produced in the laboratories. In applied research, however, one must study what is available and this may result in a heavy weighting on certain client characteristics and inadequate sampling of other characteristics. Results of a treatment applied to this sample cannot be generalized to the population. For example, techniques to control disruptive behavior in the classroom will be less than generalizable if they are tested in a class where students are from predominantly middle-class suburbs with inner-city students underrepresented.

Even in the great snake phobic epidemic of the 1960s, where the behavior in question was circumscribed and clearly defined, the clients to whom various treatments were applied were almost uniformly female college sophomores whose fear was neither too great (they could not finish the experiment on time) nor too little (they would finish it too quickly). Most investigators admitted that the purpose of these experiments was not to generalize treatment results to clinical populations, but to test theoretical assumptions and generate hypotheses. The fact remains however, that these results cannot even be generalized beyond female college sophomores to the population of snake fearers, where age, sex, and amount of fear would all be relevant.

It should be noted that all examples above refer to generality of findings across clients with similar behavior and background characteristics. Most studies at least consider the importance of generality of findings along this dimension although few have been successful. What is perhaps more important is the failure of most studies to consider the generality problem in the other two dimensions—namely, setting generality and behavior change agent (therapist) generality. Several investigators (e.g., Kazdin, 1973; McNamara and MacDonough, 1972) have recently suggested that this information may be more important than client generality. For example, Paul (1969) notes after a survey of group studies that the results of systematic desensitization seemed to be a function of the qualifications of the therapist rather than differences among clients. Furthermore, in regard to setting generality, Brunswick (1956) suggests that, "In fact, proper sampling of situations and problems may be in the end more important than proper sampling of subjects considering the fact that individuals are probably on the whole much more alike than are situations among one another" (p. 39).

The failure to be able to make statistically inferential statements, even about populations of clients based on most clinical research studies, does not mean that no statements about generality can be made. As Edgington (1966) points out, one can make statements at least on generality of findings to similar clients based on logical non-statistical considerations. Edgington refers to this as "logical generalization" and this issue along with generality to settings and therapists will be discussed below in relation to the establishment of generality of findings from a single case.

Problems in generalizing from the group to the individual

The above discussion might be construed as a plea for more adequate sampling procedures involving larger numbers of clients seen in many different settings by a variety of therapists–in other words, the notion of the "grand collaborative study," which emerged from the conferences on research in psychotherapy in the 1960s (e.g., Bergin and Strupp, 1972; Strupp and Luborsky, 1962). On the contrary, one of the pitfalls of a truly random sample in applied research is that the more adequate the sample, in that all relevant population characteristics are represented, the less relevance will this finding have for a specific individual. The major issue here is that the better the sample, the more heterogeneous the group. The average response of this group, then, will be less likely to represent a given individual in the group. Thus, if one were establishing a random sample of severe depressives, one should include clients of various age, racial, and socioeconomic backgrounds. In addition, clients with various combinations of the behavior and thinking or perceptual disorder associated with severe depression must be included. It would be desirable to include some patients with severe agitation, others demonstrating psychomotor retardation, still others with varying degrees and types of depressive delusions, and somatic correlates such as terminal sleep disturbance. As this sample becomes truly more random and representative, the group becomes more heterogeneous. The specific effects of a given treatment on an individual with a certain combination of problems becomes lost in the group average. For instance, a certain treatment might alleviate severe agitation and terminal sleep disturbance but have a deleterious effect on psychomotor retardation and depressive delusions. If one were to analyze the results, one could infer that the treatment, on the average, is better than no treatment for the population of patients with severe depression. For the individual clinician, this finding is not very helpful and could actually be dangerous if the clinician's patient had psychomotor retardation and depressive delusions.

Most studies, however, do not pretend to draw a truly random sample of patients with a given diagnosis or behavior disorder. Most studies choose clients or patients on the basis of availability and then randomly assign these subjects into two groups that are matched on relevant characteristics. Typically, the treatment is administered to one group while the other group becomes the no-treatment control. This arrangement, which has characterized much clinical and educational research, suffers for two reasons: to the extent that the "available" clients are not a random sample, one cannot generalize to the population; but to the extent that the group is heterogeneous on any of a number of characteristics one cannot make statements about the individual. The only statement that can be made concerns the average response of a group with that particular makeup which, unfortunately, is unlikely to be duplicated again. As Bergin (1966) noted, it is even difficult to say anything important about individuals within the group based on the average response since his analysis demonstrated that some were improving and some deteriorating. The result, as Chassan (1967) eloquently pointed out, was that the behavior change agent did not know which treatment or aspect of treatment was effective with which clients and, furthermore, ran the risk of employing a treatment that was statistically better than no treatment but actually might make his particular patient worse.

Improving generality of findings to the individual through homogeneous groups

What Bergin and Strupp (1972) and others (e.g., Kiesler, 1971; Paul, 1967) recognized was that if anything important was going to be said about the individual, after experimenting with a group, then the group would have to be homogeneous for relevant client characteristics. For example, in studing a group of agoraphobics, they should all be in one age group with a relatively homogeneous amount of fear and approximately equal background (personality) variables. Naturally, clients in the control group must also be homogeneous for these characteristics.

Although this approach sacrifices random sampling and the ability to make inferential statements about the population of agoraphobics, one can begin to say something about agoraphobics with the same or similar characteristics as those in the study through the process of logical generalization (Edgington, 1967). That is, if a study shows that a given treatment is successful with a homogeneous group of 20- to 30-year-old female agoraphobics with certain personality characteristics, then a clinician can be relatively confident that a 25-year-old female agoraphobic with those personality characteristics will respond well to that same treatment.

The process of logical generalization depends on similarities between the patients in the homogeneous group and the individual in question in the clinician's office. But if one can generalize in logical fashion from a patient whose results or characteristics are well specified as part of a homogeneous group, then one can also logically generalize from a single individual whose response and biographical characteristics are specified. In fact, the rationale has enabled applied researchers to generalize the results of single case experiments for years (Dukes, 1965; Shontz, 1965). To increase the base for generalization from a single case experiment, one simply repeats the same experiment several times on similar patients, thereby providing the clinician with results from a number of patients.

2.8. HOMOGENEOUS GROUPS vs. REPLICATIONS OF A SINGLE-CASE EXPERIMENT

Because the issue of generalization from single-case experiments in applied research is a major source of controversy (Underwood, 1957), the sections to follow will describe our views of the relative merits of replication studies vs. generalization from homogeneous groups.

As a basis for comparison, it is useful to compare the single-case approach with Paul's (1967, 1969) incisive analysis of the power of various experimental designs using groups of clients. Within the context of the power of these various designs to establish cause-effect relationships, Paul reviews the several procedures commonly used in applied research. These procedures range from case studies with and without measurement from which cause-effect relationships can seldom if ever be extracted, through series of cases typically reporting percentage of success with no control group. Finally, Paul cites the two major experimental designs capable of establishing functional relationships between treatments and the average response of clients in the group. The first is what Paul refers to as the nonfactorial design with no-treatment control, in other words the comparison of an experimental (treatment) group with a no-treatment control group. The second design is the powerful factorial design, which not only establishes cause-effect relations between treatments and clients but specifies what type of clients under what conditions improve with a given treatment. The single case replication strategy paralleling the nonfactorial design with no-treatment control is direct replication. The replication strategy paralleling the factorial design is called systematic replication.

Direct replication and treatment/no-treatment control group design

When Paul's article was written (1967), applied research employing single case designs, usually of the A-B-A variety, was just beginning to appear (e.g., Ullmann and Krasner, 1965). Paul quickly recognized the validity or power of this design, noting that "The level of product for this design approaches that of the nonfactorial group design with no-treatment controls" (p. 117). When Paul speaks of level of product here he is referring, in Campbell and Stanley's (1966) terms, to internal validity—i.e., the power of the design to isolate the independent variable (treatment) as responsible for experimental effects—and to external validity or the ability to generalize findings across relevant domains such as client, therapist, and setting. We would agree with Paul's notions that the level of product of a single case experimental design only "approaches" that of treatment/no-treatment group designs, but for somewhat different reasons. It is our contention that the single case A-B-A design "approaches" rather than equals the nonfactorial group design with no-treatment controls only because the number of clients is considerably less in a single case design (N=1) than in a group design, where eight, ten, or more clients are not uncommon. It is our further contention that in terms of external validity or generality of findings, a series of single case designs in similar clients in which the original experiment is directly replicated three or four times can far surpass the experimental group/no-treatment control group design. Some of the reasons for this assertion are outlined below.

Results generated from an experimental group/no-treatment control group study as well as a direct replication series of single case experimental designs yield some information on generality of findings across clients but cannot address the question of generality across different therapists or settings. Typically, the group study employs one therapist in one setting who applies a given treatment to a group of clients. Measures are taken on a pre-post basis. Pre- and post-measures are also taken from a matched group of clients in the control group who do not receive the intervening treatment. For example, ten depressive patients homogeneous on behavioral and emotional aspects of their depression as well as personality characteristics, would be compared to a matched group of patients who did not receive treatment. Logical generalization to other clients (but not to other therapists or settings) would depend on the degree of homogeneity among the depressives in both groups. As noted above, the less homogeneous the depression in the experiment, the greater the difficulty for the practicing clinician in determining if that treatment is effective for his particular patient. A solution to this problem would be to specify in some detail the characteristics of each patient in the treatment group and present individual data on each patient. The clinician could then observe those patients that are most like his particular client and determine if these experimental patients improved more than the average response in the control group. For example, after describing in detail the case history and presenting symptomatology of ten depressives, one could administer a pre-test measuring severity of depression to the ten depressives and a matched control group of ten depressives. After treating the ten depressives in the experimental group, the post-test would be administered. When results are presented, the improvement (or lack of improvement) of each patient in the treatment group could be presented either graphically or in numerical form along with the means and standard deviations for the control group. After the usual procedure to determine statistical significance, the clinician could examine the *amount* of improvement of each patient in the experimental group to determine (1) if the improvement were *clinically* relevant, and (2) if the improvement exceeded any drift toward improvement in the control group. To the extent that some patients in the treatment group are similar to the clinician's patient, the clinician could begin to determine, through logical generalization, whether the treatment might be effective with his patient.

However, a series of single case designs where the original experiment is replicated on a number of patients also enables one to determine generality of findings across patients (but not across therapists or settings). For example, in the same hypothetical group of depressives, the treatment could be administered in an A-B-A-B design, where A represents baseline measurement and B represents the treatment. The comparison here is still between treatment and no treatment. As results accumulate across patients, generality of findings is established and the results are readily translatable to the practicing clinician, since he can quickly determine which patient with which characteristics improved and which patient did not improve. To the extent that therapist and treatment are alike across patients, this is the clinical prototype of a direct replication series (Sidman, 1960), and it represents the most common replication tactic in the experimental single case approach to date.

Given these results, other attributes of the single case design provide added strength in generalizing results to other clients. The first attribute is flexibility (noted in Section 2.3 above). If a particular procedure works well in one case but works less well or fails when attempts are made to replicate this in a second or third case, slight alterations in the procedure can be made immediately. In many cases, reasons for the inability to replicate the findings can be ascertained immediately, assuming that procedural deficiencies were, in fact, responsible for the lack of generality. An example of this result was outlined in Section 2.3 above, describing inter-subject variability. In this example, one patient improved with treatment but a second did not. Use of an improvised experimental design at this point allowed identification of the reason for failure. This finding should increase generality of findings by enabling immediate application of the altered procedure to another patient with a similar response pattern. This is an example of Sidman's (1960) assertion that "tracking down sources of variability is then a primary technique of establishing generality" (see also Kazdin, 1973; Leitenberg, 1973; Skinner, 1966). If alterations in the procedure do not produce clinical improvement, either differences in background, *personality* characteristics, or differences within the behavior disorder itself can be noted, suggesting further hypotheses on procedural changes that can be tested on this type of client at a later date.

Finally, using the client as his own control in successive replication provides an added degree of strength in generalizing the effect of treatment across differing clients. In group or single case designs employing no-treatment controls or attention-placebo controls, it is possible and even quite likely that certain environmental events in a no-treatment control group or phase will produce considerable improvement (e.g., placebo effects). In a nonfactorial group design, where treated clients show more improvement than clients in a no-treatment control, one can conclude that the treatment is effective and then proceed in generalizing results to other clients in clinical situations. However, the *degree* of the contribution of nonspecific environmental factors to the improvement of *each individual* client is difficult to judge. In a single case design (for example, the A-B-A-B or true withdrawal design), the influence of environmental factors on each individual client can be estimated by observing the degree of deterioration when treatment is withdrawn. If environmental or other factors are operating during treatment, improvement will continue during the withdrawal phase, perhaps at a slower rate, necessitating further experimental inquiry. Even in a nonfactorial group design with powerful effects, the contribution of this factor to *individual* clients is difficult to ascertain.

Systematic replication and factorial designs

Direct replication series and nonfactorial designs with no-treatment controls come to grips with only one aspect of generality of findings—generality across clients. These designs are not capable of simultaneously answering questions on generality of findings across therapists, settings, or clients that differ in some substantial degree from the original homogeneous group. For example, one might ask, if the treatment works for 25-year-old female agoraphobics with certain personality characteristics, will it also work for a 40-year-old female agoraphobic with different personality characteristics?

In the therapist domain, the obvious question concerns the effectiveness of treatment as related to that particular therapist. If the therapist in the hypothetical study were an older, more experienced therapist, would the treatment work as well with a young therapist? Finally, even if several therapists in one setting were successful, could therapists in another setting and geographical area attain similar results?

To answer all of these questions would require literally hundreds of experimental group/no-treatment control group studies where each of the factors relevant to generalization was varied one at a time (e.g., type of therapist, type of client). Even if this were feasible, however, the results could not always be attributed to the factor in question as replication after replication ensued, since other sources of variance due to faulty random assignment of clients to the group could appear.

In reviewing the status and goals of psychotherapy research, both Paul (1967, 1969) and Kiesler (1971) proposed the application of one of the most sophisticated experimental designs in the armamentarium of the psychological researcher–the factorial design–as an answer to the above problem. In this design, relevant factors in all three areas of generality of concern to the clinician can be examined. The power of this design is in the specificity of the conclusions. For example, the effects of two anti-depressant pharmacological agents and a placebo might be evaluated in two different settings (the inpatient ward of a general hospital and an outpatient community mental health center) on two groups of depressives (one group with moderate to severe depression and a second group with mild depression). A therapist in the psychiatric ward setting would administer each treatment to one half of each group of depressives–the moderate to severe group and the mild group. All depressives would be matched as closely as possible on background variables such as age, sex, personality characteristics, etc. The same therapist could then travel to the community mental health center and carry out the same procedure. Thus we have a 2x2x2 factorial design. Possible conclusions from this study are numerous, but results might be so specific as to indicate that anti-depressants do work but only with moderate to severe depressives and only if hospitalized on a psychiatric ward. It would not be possible to draw conclusions on the importance of a particular type of therapist since this factor was not systematically varied. Of course, the usual shortcomings of group designs are also present here since results would be presented in terms of group averages and inter-subject variability. However, to the extent that subjects in each experimental cell were homogeneous and to the extent that improvement was large and clinically important rather than merely statistically significant, then results would certainly be a valuable contribution. The clinical practitioner would be able to examine the characteristics of those subjects in the improved group and conclude that under similar conditions (i.e., an inpatient psychiatric unit) his moderate to severe depressive patient would be likely to improve, assuming, of course, that his patient resembled those in the study. Here again, the process of logical generalization rather than statistical inference from a sample to a population is the active mechanism.

Thus, while the factorial design can be effective in specifying generality of findings across all important domains in applied research (within the limits discussed above), one major problem remains: applied researchers seldomly do this kind of study. As noted in Chapter 1, Section 1.5, the major reasons for this are practical. The enormous investment of money and time necessary to collect large numbers of homogeneous patients has severely inhibited this type of endeavor. And often, even in several different settings, the necessary number of patients to complete a study is just not available unless one is willing to wait years. Added to this are procedural difficulties in recruiting and paying therapists, ensuring adequate experimental controls such as double-blind procedures within a large setting, and overcoming resistance to assigning a large number of patients to placebo or control conditions, as well as coping with the laborious task of recording and analyzing large amounts of data (Barlow and Hersen, 1973; Bergin and Strupp, 1972).

In addition, the arguments raised in the last section on inflexibility of the group design are also applicable here. If one patient does not improve or reacts in an unusual way to the therapeutic procedure, administration of the procedure must continue for the specified number of sessions. The unsuccessful or aberrant results are then, of course, averaged into the group results from that experimental cell, thus precluding an immediate analysis of the inter-subject variability which will lead to increased generality.

Systematic replication procedures involve exploring the effects of different settings, therapists, or clients on a procedure previously demonstrated as successful in a direct replication series. In other words, to borrow the example from the factorial design, a single case design may demonstrate that a treatment for severe depression works on an inpatient unit. Several direct replications then establish generality among homogeneous patients. The next task is to replicate the procedure once again, in different settings with different therapists or with patients with different background characteristics. Thus, the goals of systematic replication in terms of generality of findings are similar to those of the factorial study.

At first glance, it does not appear as if replication techniques within single case methodology would prove any more practical in answering questions concerning generality of findings across therapists, settings, and types of behavior disorder. While direct replication can begin to provide answers to questions on generality of findings across similar clients, the large questions of setting and therapist generality would also seem to require significant collaboration among diverse investigators, long-range planning, and a large investment of money and time—the very factors that were noted by Bergin and Strupp (1972) to preclude these important replication effects. The surprising fact concerning this particular method of replication, however, is that these issues are not interfering with the establishment of generality of findings, since systematic replication is in progress in a number of areas of applied research. In view of the fact that systematic replication has the same advantages of logical generalization as direct replication, the information yielded by the procedure has direct applicability to the clinic. Examples from these ongoing systematic replication series as well as procedures and guidelines for systematic replication will be described in Chapter 9.

2.9. APPLIED RESEARCH QUESTIONS REQUIRING GROUP DESIGNS

It was observed in Chapter 1 that applied researchers during the 1950s and 1960s often considered single case vs. between-group comparison research as an "either-or" proposition. Most investigators in this period chose one methodology or the other and eschewed the alternative. Much of this polemic characterized the idiographic-nomethetic dichotomy in the 1950s (Allport, 1961). This type of argument, of course, prevented many investigators from asking the obvious question: Under what condition is one type of design more appropriate than another? As single case designs became more sophisticated, the number of questions that could be answered by this strategy increased. But, there are many instances in which single case designs either cannot answer the relevant applied research question or are at least inappropriate to the problem at hand. The purpose of this book, of course, is to make a case for the relevance of single case experimental designs and to cover those issues, areas, and examples where a single case approach is appropriate and important. We would be remiss, however, in ignoring those areas where experimental designs involving groups offer a better answer.

Technique building vs. technique testing

Leitenberg (1973) says that the single case approach is a good way to start an investigation of therapeutic procedures but may not be the best way to end it. In addition to practical considerations noted in Chapter 1, the major advantage of the single case approach in beginning an investigation is the ability to isolate mechanisms of therapeutic action in a global treatment. Isolation of these mechanisms of action then make it possible to combine various treatment variables in a more powerful treatment "package." There is little question that a single case approach, with its flexibility, can determine individual sources of variability and quickly bring the investigator to the point where he is ready to construct a global treatment package. At this point, however, it may be desirable to test this "package" in a large group outcome study where one is less concerned with individual variability since most sources of variability would have been explored in previous single case designs. This strategy, in which single case designs are used to isolate relevant therapeutic variables before testing a "treatment package," has been called "technique building" by Bergin and Strupp (1972). Since most factors influencing generality of findings would also have been determined through direct and systematic replication, one could proceed with an appropriate test of this powerful combination of therapeutic variables in a large group comparison, "technique testing" approach. This is one example of an effective combination of experimental strategies in achieving an applied research goal.

Actuarial questions

There are several related questions or issues that require experimental strategies involving groups. Baer (1971) refers to one as actuarial, although he might have said political. The fact is, after a package has been put together, society wants to know the magnitude of its effects. This information is often best conveyed in terms of percentage of people who improved compared to an untreated group. If one can say that a treatment works in 75 out of 100 cases where only 15 out of 100 would improve without treatment, this is the kind of information that is readily understood by society. In a systematic replication series, the results would be stated differently. Here the investigator would say that under certain conditions the treatment works, while under other conditions it does not work and other therapeutic variables must be added. While this statement might be adequate for the practicing clinician or educator, little information on the magnitude of effect is conveyed. Since society supports research and, ultimately, benefits from it, this actuarial approach is not trivial. As Baer (1971) points out, this problem ". . . is similar to that of any insurance company, we merely need to know how often a behavioral analysis changes the relevant behavior of society toward the behavior, just as the insurance company needs to know how often age predicts death rates" (p. 366). It should be noted, however, that a study such as this cannot answer why a treatment works; it is simply capable of communicating the size of the effect. But if the treatment package is the result of a series of single case designs, then one should already know why it works and demonstration of the magnitude of effect is all that is needed.

Several cautions should be noted when utilizing the group design in this manner. Firstly, the cost and practical limitation of running a large group study do not allow unlimited replication of this effort, if it can be done at all. Thus, one should have a well-developed treatment package that has been thoroughly tested in single case experimental designs and replications before embarking on this effort. Preferably, the investigator should be well into a systematic replication series so that he has some idea of the client, setting, or therapeutic variables that predict success. Groups can then be constructed in a homogeneous fashion. Premature application of the group comparison design, where a treatment or the conditions under which it is effective have not been adequately worked out, can only produce the characteristic "weak" effect with large inter-subject variability that is so prevalent in group comparison studies to date (Bergin and Strupp, 1972).

Modification of group behavior

A related issue on the appropriateness of group design arises when the applied researcher is not concerned with the fate of the individual but rather with the effectiveness of a given procedure on a well-defined group. A particularly good example is the classroom. If the problem is a mild but annoying one, such as disruptive behavior in the classroom, the researcher and school administrator may be more interested in quickly determining what procedure is effective in remedying this problem for the classroom as a whole. The goal in this case is changing behavior of a well-defined group rather than individuals within that group. It may not be important that two or three children remain somewhat out of order if the classroom is substantially more quiet. A particularly good example is an experiment on the modification of classroom noise reported in Chapter 7, Fig. 7-7 (Wilson and Hopkins, 1973). A similar approach might be desirable with any coexisting group of people, such as a ward in a state hospital where the control of disruptive behavior would allow more efficient execution of individual therapeutic programs (see Chapter 5, Fig. 5-11) (Ayllon and Azrin, 1965). This stands in obvious contrast to a series of patients with severe clinical problems who do not coexist in some geographical location but are seen sequentially and assigned to a group only for experimental consideration. In this case, the applied researcher would be ill-disposed to ignore the significant human suffering of those individuals who did not improve or perhaps deteriorated.

When group behavior is the target, however, and a comparison of treated and untreated classrooms, etc., is desirable, one is not limited to between-subject designs in these instances since within-subject designs are also feasible. There are many examples where A-B-A or multiple baseline designs have been used in classroom research with repeated measures of the average behavior of the group (e.g., Wolf and Risley, 1971).

Once again, it is a good idea to have a treatment that has been adequately worked out on individuals before attempting to modify behavior of a group. If not, the investigator will encounter intolerable inter-subject variability that will weaken the effects of his intervention.

Comparison of two treatment packages

A final instance in which a group comparison design is desirable occurs when one wishes to compare two treatment packages for the same behavior disorder, with preliminary evidence that each treatment is likely to be effective. "Treatment" in this case refers to a relatively complex package of therapeutic variables that differ considerably in administration. Often these treatments may have different theoretical underpinnings. This strategy was relatively popular in earlier decades when various psychotherapies from different schools were pitted against one another. In recent years, the treatments and target behaviors have become more specific, but the strategy remains. Within behavior therapy, a good example is the recent spate of studies comparing systematic desensitization and flooding or implosion for phobic behavior (e.g., DeMoor, 1970; Mealiea and Nawas, 1971). It is not possible to answer a question of the relative effectiveness of two treatment packages with a single case analysis. The appropriate experimental design is a between-group comparison. This problem is quite different from analyzing the relative effectiveness and interaction of two different but well-defined individual therapeutic variables. In this case it is possible, albeit difficult, to compare one variable with another and also to examine the effects of combining these two variables. An example of this type of investigation is the series examining the effects of two variables–feedback and praise–on phobic behavior (see Chapter 6, Section 6.5). This procedure is termed an "interaction" design. Obviously it is not possible to alternate or combine fixed treatment packages such as systematic desensitization and implosion since each contains numerous ingredients.

One may question, however, the strategy and timing of this type of analysis, if not the goal. For example, both systematic desensitization and flooding or implosion emanated from experimental work with animals in the laboratory and/or theoretical assumptions of the etiology and maintenance of phobic behavior in humans. It is possible that a comparison of the two treatments was premature. An alternative strategy would be to experimentally analyze, in single cases, therapeutic mechanisms of action in both of these treatments, discard the superfluous procedural components, and combine the active ingredients in an improved treatment package for phobias. In fact, the studies comparing the two treatments to date have yielded equivocal results (Morganstern, 1973), while other investigators have been searching for mechanisms of action in these treatments (e.g., Agras *et al.*, 1971). This search has suggested that these treatments and other procedures to reduce phobic behavior share a common mechanism of action–the facilitation of exposure to phobic stimuli (Davison and Wilson, 1973). If this is confirmed by future research, the early studies comparing these two treatments would be premature. It is possible, of course, that these treatments or any two treatments would not be reducible to common mechanisms of action, in which case a large group comparison study would be quite appropriate to determine the superior procedure. However, this should occur only after considerable preliminary work on the effective ingredients in each treatment. Once again, the general guideline would be to develop the strongest treatment and to determine the conditions under which it is effective or ineffective. Only at this point should the investigator consider a large group comparison approach.

References

Agras, W. S., Barlow, D. H., Chapin, H. N., Abel, G. G., and Leitenberg, H. Behavior modification of anorexia nervosa. *Archives of General Psychiatry*, 1974, **30**, 279-286.

Agras, W. S., Leitenberg, H., Barlow, D. H., Curtis, N. A., Edwards, J., and Wright, D. Relaxation in systematic desensitization. *Archives of General Psychiatry*, 1971, **25**, 511-514.

Allport, G. W. *Pattern and growth in personality.* New York: Holt, Rinehart and Winston, 1961.

Ayllon, T., and Azrin, N. H. The measurement and reinforcement of behavior of psychotics. *Journal of the Experimental Analysis of Behavior*, 1965, 8, 357-383.

Baer, D. M. Behavior modification: You shouldn't. In E. Ramp and B. L. Hopkins (Eds.), *A new direction for education: Behavior analysis.* Lawrence, Kansas: Lawrence University Press, 1971.

Barlow, D. H., and Hersen, M. Single-case experimental designs: Uses in applied clinical research. *Archives of General Psychiatry*, 1973, **29**, 319-325.

Bergin, A. E. Some implications of psychotherapy research for therapeutic practice. *Journal of Abnormal Psychology*, 1966, **71**, 235-246.

Bergin, A. E., and Strupp, H. H. *Changing frontiers in the science of psychotherapy.* New York: Aldine-Atherton, 1972.

Brunswick, E. *Perception and the representative design of psychological experiments.* Berkeley: University of California Press, 1956.

Campbell, D. T., and Stanley, J. C. *Experimental and quasi-experimental designs for research.* Chicago: Rand-McNally, 1966.

Chassan, J. B. Statistical inference and the single case in clinical design. *Psychiatry*, 1960, **23**, 173-184.

Chassan, J. B. *Research design in clinical psychology and psychiatry.* New York: Appleton-Century-Crofts, 1967.

Davison, G. C., and Wilson, G. T. Processes of fear-reduction in systematic desensitization: Cognitive and social reinforcement factors in humans. *Behavior Therapy*, 1973, **4**, 1-21.

DeMoor, W. Systematic desensitization versus prolonged high intensity stimulation (flooding). *Journal of Behavior Therapy and Experimental Psychiatry*, 1970, **1**, 45-52.

Dukes, W. F. N=1. *Psychological Bulletin*, 1965, **64**, 74-79.

Edgington, E. S. Statistical inference and nonrandom samples. *Psychological Bulletin*, 1966, **66**, 485-487.

Edgington, E. S. Statistical inference from N=1 experiments. *The Journal of Psychology*, 1967, **65**, 195-199.

Herman, S. H., Barlow, D. H., and Agras, W. S. An experimental analysis of classical conditioning as a method of increasing heterosexual arousal in homosexuals. *Behavior Therapy*, 1974, **5**, 33-47.

Herman, S. H., Barlow, D. H., and Agras, W. S. An experimental analysis of exposure to explicit heterosexual stimuli as an effective variable in changing arousal patterns of homosexuals. *Behaviour Research and Therapy*, 1974, **12**, 335-345.

Honig, W. K. (Ed.) *Operant behavior: Areas of research and application.* New York: Appleton-Century-Crofts, 1966.

Kazdin, A. E. Methodological and assessment considerations in evaluating reinforcement programs in appled settings. *Journal of Applied Behavior Analysis*, 1973, **6**, 517-531.

Kiesler, D. J. Experimental designs in psychotherapy research. In A. E. Bergin and S. L. Garfield (Eds.), *Handbook of psychotherapy and behavior change: An empirical analysis.* Pp. 36-74. New York: Wiley, 1971.

Leitenberg, H. The use of single-case methodology in psychotherapy research. *Journal of Abnormal Psychology*, 1973, **82**, 87-101.

McNamara, J. R., and MacDonough, T. S. Some methodological considerations in the design and implementation of behavior therapy research. *Behavior Therapy*, 1972, **3**, 361-378.

Mealiea, W. L., and Nawas, M. M. The comparative effectiveness of systematic desensitization and implosive therapy in the treatment of snake phobia. *Journal of Behavior Therapy and Experimental Psychiatry*, 1971, **2**, 85-94.

Metcalfe, M. Demonstration of a psychosomatic relationship. *British Journal of Medical Psychology*, 1956, **29**, 63-66.

Morganstern, K. Implosive therapy and flooding procedures: A critical review. *Psychological Bulletin*, 1973, **79**, 318-334.

Paul, G. L. Strategy of outcome research in psychotherapy. *Journal of Consulting Psychology*, 1967, **31**, 104-118.

Paul, G. L. Behavior modification research: Design and tactics. In C. M. Franks (Ed.), *Behavior therapy: Appraisal and status.* Pp. 29-62. New York: McGraw-Hill, 1969.

Shontz, F. C. *Research methods in personality.* New York: Appleton-Century-Crofts, 1965.

Sidman, M. *Tactics of scientific research: Evaluating experimental data in psychology.* New York: Basic Books, 1960.

Skinner, B. F. Operant behavior. In W. K. Honig (Ed.), *Operant behavior: Areas of research and application.* Pp. 12-32. New York: Appleton-Century-Crofts, 1966.

Strupp, H. H., and Luborsky, L. (Eds.), *Research in psychotherapy*, Vol. 2. Washington, D. C.: American Psychological Association, 1962.

Ullmann, L. P., and Krasner, L. (Eds.), *Case studies in behavior modification.* New York: Holt, Rinehart and Winston, 1965.

Underwood, B. J. *Psychological research.* New York: Appleton-Century-Crofts, 1957.

Wilson, C. W., and Hopkins, B. L. The effects of contingent music on the intensity of noise in junior high home economics classes. *Journal of Applied Behavior Analysis*, 1973, **6**, 269-275.

Wolf, M. M., and Risley, T. R. Reinforcement: Applied research. In R. Glaser (Ed.), *The nature of reinforcement.* Pp. 310-325. New York: Academic Press, 1971.

Wolstein, B. *Transference, its meaning and function in psychoanalytic therapy.* New York: Grune and Stratton, 1954.

CHAPTER 3

General Procedures in Single-case Research

3.1. INTRODUCTION

Advantages of the experimental single case design and general issues involved in this type of research were briefly outlined in Chapter 2. In the present chapter a more detailed analysis of general procedures characteristic of all experimental single case research will be undertaken. Although previous discussion of these procedures has appeared periodically in the psychological and psychiatric literatures (Baer, Wolf, and Risley, 1968; Barlow and Hersen, 1973; Browning and Stover, 1971; Hersen, 1973a; Kazdin, 1973; Leitenberg, 1973; Risley and Wolf, 1972; Wolf and Risley, 1971), a comprehensive analysis, both from a theoretical and applied framework, is very much needed.

A review of the literature on applied clinical research for the last 5 to 10 years shows that there is a substantial increase in the number of articles reporting the use of the experimental single case design strategy. These papers have appeared in a wide variety of educational, psychological, and psychiatric journals. However many researchers have proceeded without the benefit of carefully thought-out guidelines and, as a consequence, needless errors in design and practice have resulted. Even in the *Journal of Applied Behavior Analysis*, which is primarily devoted to the experimental analysis model of research, errors in procedure and practice are not uncommon in reported investigations.

In the succeeding sections of this chapter, theoretical and practical applications of repeated measurement, methods for choosing an appropriate baseline, changing one independent variable at a time, reversals and withdrawals, length of phases, and techniques for evaluating effects of "irreversible" procedures will be considered. For heuristic purposes, both correct and incorrect applications of the aforementioned will be examined. Illustrations of actual and hypothetical cases will be provided.

3.2. REPEATED MEASUREMENT

Aspects of repeated measurement techniques have already been discussed in Chapter 2. However, in this section we will examine some of the issues in greater detail. In the typical psychotherapy outcome study (e.g., Truax, Wargo, Carkhuff, Kodman, and Moles, 1966), in which the randomly assigned or matched-group design is used, dependent measures (e.g., Q sort scores filled out by patients) are obtained only on a pre- and post-therapy basis. Thus, possible fluctuations, including upward and downward trends and curvilinear relationships, occurring throughout the course of therapy are omitted from the analysis. However, whether one espouses a behavioral, client-centered, existential, or psychoanalytic position, the experienced clinician is undoubtedly cognizant that changes unfortunately *do not* follow a smooth linear function from the beginning of treatment to its ultimate conclusion.

Practical implications and limitations

There are a number of important practical implications and limitations in applying repeated measurement techniques when conducting experimental single case research (see Chapter 2, Section 2.2, for general discussion). First of all, the operations involved in obtaining such measurements (whether they be motoric, physiological, or attitudinal) must be clearly specified, observable, public, and replicable in all respects. When measurement techniques require the use of human observers, independent reliability checks must be established (see Chapter 4, Section 4.3, for specific details). Secondly, measurements taken repeatedly, especially over extended periods of time, must be done under exacting and totally standardized conditions with respect to measurement devices used, personnel involved, time or times of day measurements recorded, instructions to the subject, and the specific environmental conditions (e.g., location) where the measurement sessions occur.

Deviations from any of the aforementioned conditions may well lead to *spurious* effects in the data and might result in erroneous conclusions. This is of particular import at the point when the prevailing condition is experimentally altered (e.g., change from baseline to reinforcement conditions). In the event that an adventitious change in measurement conditions were to coincide with a modification in experimental procedure, resulting differences in the data could not be scientifically attributed to the experimental manipulation, inasmuch as a correlative change may have taken place. Under these circumstances, the conscientious experimenter would either have to renew his efforts or experimentally manipulate and evaluate the change in measurement technique.

The importance of maintaining standard measurement conditions bears some illustration. Elkin, Hersen, Eisler, and Williams (1973) recently examined the separate and combined effects of feedback, reinforcement, and increased food presentation in a male anorexia nervosa patient. With respect to measurement, two dependent variables–caloric intake and weight–were examined daily. Caloric intake was monitored throughout the 42-day study without the subject's knowledge. Three daily meals (each at a specified time) were served to the subject while he dined alone in his room for a 30-minute period. At the conclusion of each of the three daily meals, unknown to the subject, the caloric value of the food remaining on his tray was subtracted from the standard amount presented. Also, the subject was weighed daily at approximately 2:00 p.m., in the same room, on the same scale, with his back turned toward the dial, and, for the most part, by the same experimenter. In this study consistency of the experimenter was not considered crucial to maintaining accuracy and freedom from bias in measurement. However, maintaining consistency of the time of day weighed was absolutely essential, particularly in terms of the number of meals (two) consumed until that point.

There are certain instances when a change in the experimenter will seriously affect the subject's responses over time. Indeed, this was empirically evaluated by Agras, Leitenberg, Barlow, and Thomson (1969), in a multiple schedule design (see Chapter 7). However, in most single case research, unless explicitly planned, such change may mar the results obtained. For example, when employing the Behavioral Assertiveness Test (Eisler, Miller, and Hersen, 1973) over time repeatedly as a standard behavioral measure of assertiveness, it is clear that the use of different role models to promote responding might result in unexpected interaction with the experimental condition (e.g., feedback or instructions) being manipulated. Even when using more objective measurement techniques, such as the mechanical strain gauge for recording penile circumference change (Barlow, Becker, Leitenberg, and Agras, 1970) in sexual deviates, extreme care should be exercised with respect to instructions given and to the role of the examiner (male research assistant) involved in the measurement session. A substitute for the original male experimenter, particularly in the case of a homosexual in the early stages of his experimental treatment, could conceivably result in spurious correlated changes in penile circumference data.

There are several other important issues to be considered when using repeated measurement techniques in applied clinical research. For example, frequency of measurements obtained per unit time should be given most careful attention. The experimenter obviously must ensure that a sufficient number of measurements are recorded so that a representative sample is obtained. On the other hand, the experimenter must exercise caution to avoid taking too many measurements in a given period of time as fatigue on the part of the subject may result. This is of paramount importance when taking measurements that require an active response on the subject's part (e.g., number of erections to sexual stimuli over a specified time period, or repeated modeling of responses during the course of a session in assertive training).

A unique problem related to measurement traditionally faced by investigators working in institutional settings (state hospitals, training schools for the retarded, etc.) involves the major environmental changes that take place at night and on weekends. The astute observer who has worked in these settings is quite familiar with the distinction that is made between the "day" and "night" hospital and the "work week" and the "weekend" hospital. Unless the investigator is in the favored position to exert considerable control over his environment (as were Ayllon and Azrin (1968) in their studies on token economy), careful attention should be paid to such differences. One possible solution would be to restrict the taking of measurements across similar conditions (e.g., measurements taken only during the day). A second solution would involve plotting separate data for day and night measurements.

A totally different measurement problem is faced by the experimenter who is intent on using self-report data on a repetitive basis. When using this type of assessment technique, the possibility always exists, even in clinical subjects, that the subject's natural responsivity will not be tapped, but that data in conformity to "experimental demand" (Orne, 1962) are being recorded. The use of alternate forms and the correlation of self-report (attitudinal) measures with motoric and physiological indices of behavior are some of the methods to ensure validity of responses. This is of particular utility when measures obtained from the different response systems correlate both highly and positively. Discrepancies in verbal and motoric indices of behavior have been a subject of considerable speculation and study in the behavioral literature, and the reader is referred to the following for a more complete discussion of those issues: Begelman and Hersen (1973), Hersen (1973b), Hersen, Eisler, and Miller (1973), Hersen, Eisler, Miller, Johnson, and Pinkston (1973), and Lang (1968).

A final issue, related to repeated measurement, involves the problem of extreme daily variability of a target behavior under study. For example, repetitive time sampling on a random basis within specified time limits is a most useful technique for a variable subject to extreme fluctuations and responsivity to environmental events (see Hersen, Eisler, Alford, and Agras, 1973; Williams, Barlow, and Agras, 1972). Similar problems in measurement include the area of cyclic variation, an excellent example being the effect of the female's estrus cycle on behavior. Issues related to cyclic variation in terms of extended measurement sessions will be discussed more specifically in Section 3.6 of this chapter.

3.3. CHOOSING A BASELINE

In most experimental single case designs (the exception is the B-A-B design) the initial period of observation involves the repeated measurement of the natural frequency of occurrence of the target behaviors under study. This initial period is defined as the baseline, and it is most frequently designated as the A-phase of study (Barlow and Hersen, 1973; Kazdin and Bootzin, 1972; Krasner, 1971; Leitenberg, 1973; Risley and Wolf, 1972). It should be noted that this phase was earlier labeled $O_1O_2O_3O_4$ by Campbell and Stanley (1966) in their analysis of quasi-experimental designs for research (time series analysis).

The primary purpose of baseline measurement is to have a standard by which the subsequent efficacy of an experimental intervention may be evaluated. In addition, Risley and Wolf (1972) point out that, from a statistical framework, the baseline period functions as a predictor for the level of the target behavior attained in the future. A number of statistical techniques for analyzing time series data have appeared in the literature (Gentile, Roden, and Klein, 1972; Gottman, 1973; Gottman, McFall, and Barnett, 1969; Revusky, 1967); the use of these methods will be discussed in Chapter 8.

Baseline stability

When selecting a baseline, its stability and range of variability must be carefully examined. McNamara and MacDonough (1972) have raised an issue that is continuously faced by all of those involved in applied clinical research. They specifically pose the following question: "How long is long enough for a baseline?" (p. 364). Unfortunately, there is no simple response or formula that can be applied to this question, but a number of suggestions have been made. Baer, Wolf, and Risley (1968) recommend that baseline measurement be continued over time "until its stability is clear" (p. 94). McNamara and MacDonough concur with Wolf and Risley's (1971) recommendation that repeated measurement be applied until a stable pattern emerges. However, there are some practical and ethical limitations to extending initial measurement beyond certain limits. The first involves a problem of logistics. For the experimenter working in an institutional setting (unless in an extended care facility), the subject under study will have to be discharged within a designated period of time, whether upon self-demand, familial pressure, or exhaustion of insurance company compensation. Secondly, even in a facility giving extended care to its patients, there is an obvious ethical question as to how long the applied clinical researcher can withhold a treatment application. This assumes even greater magnitude when the target behavior under study results in serious discomfort either to the subject or to others in his environment (see Johnston, 1972, p. 1036). Finally, although

McNamara and MacDonough (1972) argue that "The use of an extended baseline is a most easily implemented procedure which may help to identify regularities in the behavior under study" (p. 361), unexpected effects on behavior may be found as a result of extended measurement through self-recording procedures (McFall, 1970; Simkins, 1971). Such effects have been found when subjects were asked to record their behaviors under repeated measurement conditions. For example, McFall (1970) found that when he asked smokers to monitor their rate of smoking, increases in their actual smoking behavior occurred. By contrast, smokers asked to monitor rate of resistance to smoking did not show parallel changes in their behavior. The problem of self-recorded and self-reported data will be discussed in more detail in Chapter 4.

In the context of basic animal research, where the behavioral history of the organism can be determined and controlled, Sidman (1960) has recommended that, for stability, rates of behavior should be within a 5 percent range of variability. Indeed, the "basic science" researcher is in position to create his baseline data through a variety of interval and ratio scheduling effects. However, even in animal research, where scheduling effects are programed to ensure stability of baseline conditions, there are instances where unexpected variations take place as a consequence of extrinsic variables. When such variability is presumed to be "extrinsic" rather than "intrinsic," Sidman (1960) encourages the researcher to first examine the source of variability through the method of experimental analysis. Then, extrinsic sources of variation can be systematically eliminated and controlled.

Sidman acknowledges, however, that the applied clinical researcher, by virtue of his subject matter, where control over the behavioral history is nearly impossible, is at a distinct disadvantage. He notes that "The behavioral engineer must continuously take variability as he finds it, and deal with it as an unavoidable fact of life" (Sidman, 1960, p. 192). He also acknowledges that "The behavioral engineer seldom has the facilities or the time that would be required to eliminate variability he encounters in a given problem" (p. 193). When variability in baseline measurements is extensive in applied clinical research, it might be useful to apply statistical techniques for purposes of comparing one phase to the next. This would certainly appear to be the case when such variability exceeds a 50 percent level. The use of statistics under these circumstances would then meet the kind of criticism that has been leveled at the applied clinical researcher who uses single case methodology. For example, Bandura (1969) argues that there is no difficulty in interpreting performance changes when differences between phases are large (e.g., the absence of overlapping distributions) and when such differences can be replicated across subjects (see Chapter 9). However, he underscores the difficulties in reaching valid conclusions when there is "considerable variability during baseline conditions" (Bandura, 1969, p. 243).

Examples of baselines

With the exception of a brief discussion in Barlow and Hersen's (1973) paper, which was primarily directed toward a psychiatric readership, the different varieties of baselines commonly encountered in applied clinical research have neither been examined nor presented in logical sequence in the experimental literature. Thus, the primary function of this section is to provide and familiarize the interested applied researcher with examples of baseline patterns. For the sake of convenience, hypothetical examples, based on actual patterns reported in the literature, will be illustrated and described. Methods for dealing with each pattern are to be outlined and an attempt to formulate some specific rules (à la cookbook style) will be undertaken.

The issue concerning the ultimate length of the baseline measurement phase was previously discussed in some detail. However, it should be pointed out here that "A minimum of three separate observation points, plotted on the graph, during this baseline phase are required to establish a trend in the data" (Barlow and Hersen, 1973, p. 320). Thus, three successively increasing or decreasing points would constitute establishment of either an upward or downward trend in the data. Obviously, in two sets of data in which the same trend is exhibited, differences in the slope of the line will indicate the extent or power of the trend. By contrast, a pattern in which only minor variation is seen would indicate the recording of a stable baseline pattern. An example of such a stable baseline pattern is depicted in Fig. 3-1. Mean number of facial tics averaged over three daily 15-minute videotaped sessions are presented for a 6-day period. Visual inspection of these data reveal no apparent upward or downward trend. Indeed, data points are essentially parallel to the abscissa, while variability remains at a minimum. This kind of baseline pattern, which shows a constant rate of behavior, represents the most desirable trend, as it permits an unequivocal departure for analyzing the subsequent efficacy of a treatment intervention. Thus, the beneficial or detrimental effects of the following intervention should be clear. In addition, should there be an absence of effects following introduction of a treatment, it will also be apparent. Absence of such effects, then, would graphically appear as a continuation of the steady trend first established during the baseline measurement phase.

A second type of baseline trend that frequently is encountered in applied clinical research is such that the subject's condition under study appears to be worsening (known as the "deteriorating baseline"–Barlow and Hersen, 1973). Once again, using our hypothetical data on facial tics, an example of this kind of baseline trend is presented in Fig. 3-2. Examination of this figure shows a steadily increasing linear function, with the number of tics observed augmenting

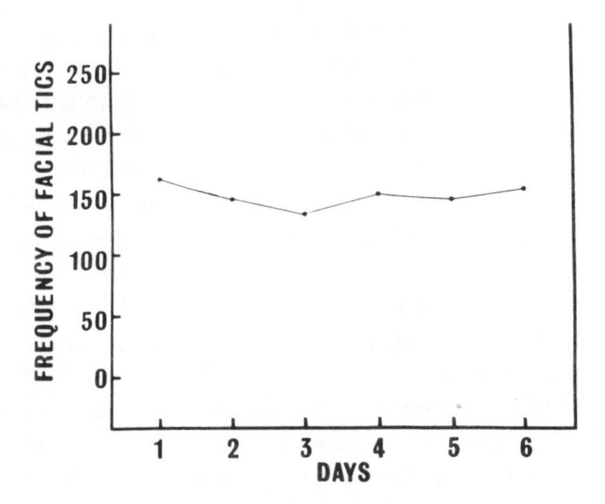

Fig. 3-1. The stable baseline. Hypothetical data for mean number of facial tics averaged over three daily 15-minute videotaped sessions.

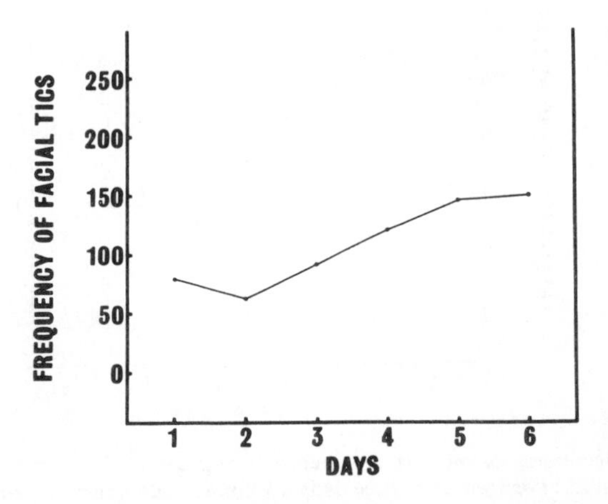

Fig. 3-2. The increasing baseline (target behavior deteriorating). Hypothetical data for mean number of facial tics averaged over three daily 15-minute videotaped sessions.

over days. The "deteriorating baseline" is an acceptable pattern inasmuch as the subsequent application of a successful treatment intervention should lead to a reversed trend in the data (i.e., a decreasing linear function over days). However, should the treatment be ineffective, no change in the slope of the curve would be noted. If, on the other hand, the treatment application leads to further deterioration (i.e., if the treatment is actually detrimental to the patient—see Bergin, 1966), it would be most difficult to assess its effects using the "deteriorating baseline." In other words, a differential analysis as to whether trend in the data were simply a continuation of the baseline pattern or whether application of a detrimental treatment specifically led to its continuation could not be made. Only if there appeared to be a pronounced change in the slope of the curve following introduction of a detrimental treatment could some kind of valid conclusion be reached on the basis of visual inspection. Even then, the withdrawal and reintroduction of the treatment would be required to establish its controlling effects, But, from both clinical and ethical considerations, this procedure would be clearly unwarranted.

A baseline pattern that provides difficulty for the applied clinical researcher is one that reflects steady improvement in the subject's condition during the course of initial observation. An example of this kind of pattern appears in Fig. 3-3. Inspection of this figure shows a linear decrease in tic frequency over a

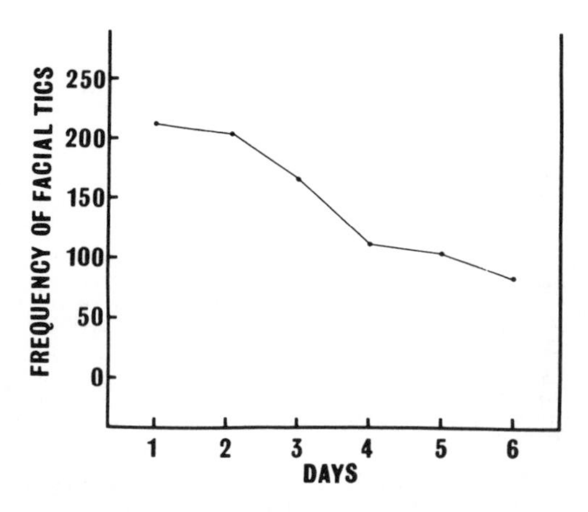

Fig. 3-3. The decreasing baseline (target behavior improving). Hypothetical data for mean number of facial tics averaged over three daily 15-minute videotaped sessions.

6-day period. The major problem posed by this pattern, from a research standpoint, is that application of a treatment strategy while improvement is already taking place will not allow for an adequate assessment of the intervention. Secondly, should improvement be maintained following initiation of the treatment intervention, the experimenter would be unable to attribute such continued improvement to the treatment unless a marked change in the slope of the curve were to occur. Moreover, removal of the treatment and its subsequent reinstatement would be required to show any controlling effects.

An alternative (and possibly a more desirable) strategy involves the continuation of baseline measurement with the expectation that a plateau will be reached. At that point, a steady pattern will emerge and the effects of treatment could then be easily evaluated. It is also possible that improvement seen during baseline assessment is merely a function of some "extrinsic" variable (Sidman, 1960) of which the experimenter is currently unaware. Following Sidman's (1960) recommendations, it then behooves the methodical experimenter, assuming that time limitations and clinical and ethical considerations permit, to evaluate empirically, through experimental analysis, the possible source (e.g., "placebo" effects) of covariation. The results of this kind of analysis could indeed lead to sime interesting hunches, which then might be subjected to further verification through the experimental analysis method (see Chapter 2, Section 2.3).

The extremely variable baseline presents yet another problem for the clinical researcher. Unfortunately, this kind of baseline pattern is frequently obtained during the course of applied clinical research, and various strategies for dealing with it are required. An example of the variable baseline is presented in Fig. 3-4.

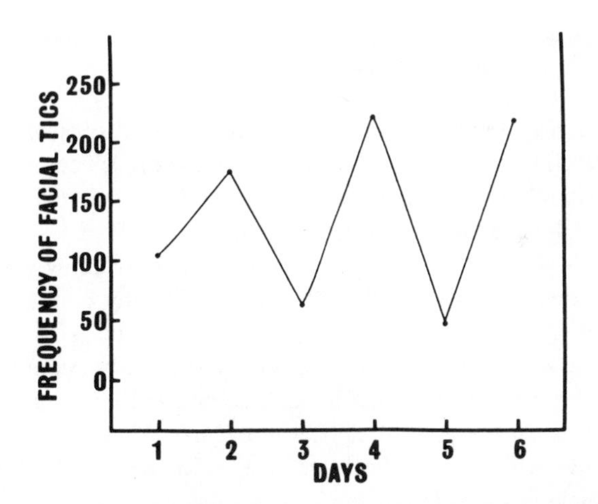

Fig. 3-4. The variable baseline. Hypothetical data for mean number of facial tics averaged over three 15-minute videotaped sessions.

An examination of these data indicate a tic frequency of about 24 to 255 tics per day, with no discernible upward or downward trend clearly in evidence. However, a distinct pattern of alternating low and high points is present. One possibility (previously discussed in dealing with extreme initial variability) is to simply extend the baseline observation until some semblance of stability is attained, an example of which appears in Fig. 3-5.

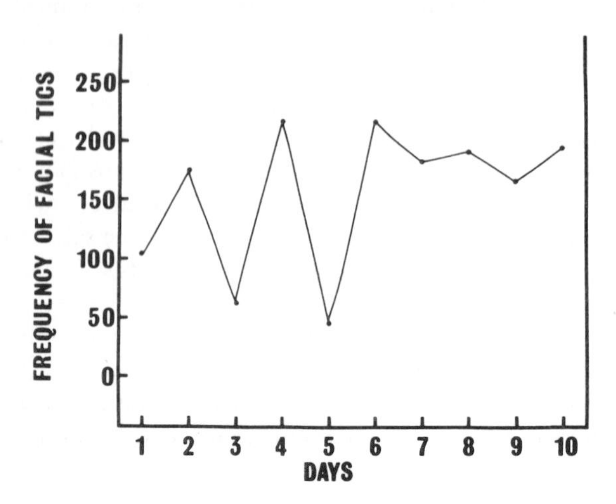

Fig. 3-5. The variable-stable baseline. Hypothetical data for mean number of facial tics averaged over three daily 15-minute videotaped sessions.

A second strategy involves the use of inferential statistics when comparing baseline and treatment phases, particularly where there is considerable overlap between succeeding distributions. However, if overlap is that extensive, the statistical model will be equally ineffective in finding differences as appropriate probability levels will not be reached. Further details regarding graphic presentation and statistical analyses of data will appear in Chapter 8.

A final strategy for dealing with the variable baseline is to systematically assess the sources of variability. However, as pointed out by Sidman (1960), the amount of work and time involved in such an analysis is better suited for the "basic scientist" than the applied clinical researcher. There are times when the clinical researcher will have to "learn to live" with such variability or he will have to select measures that fluctuate to a lesser degree.

Another possible baseline pattern is one in which there is an initial period of deterioration, which is then followed by a trend toward improvement (see Fig. 3-6). This type of baseline (increasing-decreasing) poses a number of problems for the experimenter. First, when time and conditions permit, an empirical examination of the covariants leading to reversed trends would be of heuristic value. Second, while the trend toward improvement is continued in the latter half of the baseline period of observation, application of a treatment will lead to

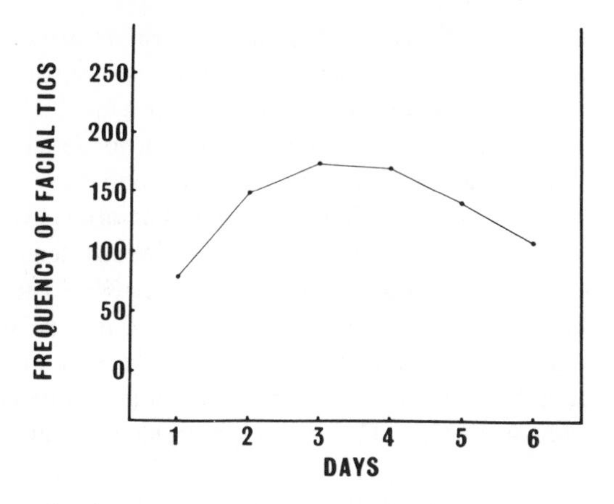

Fig. 3-6. The increasing-decreasing baseline. Hypothetical data for mean number of facial tics averaged over three daily 15-minute videotaped sessions.

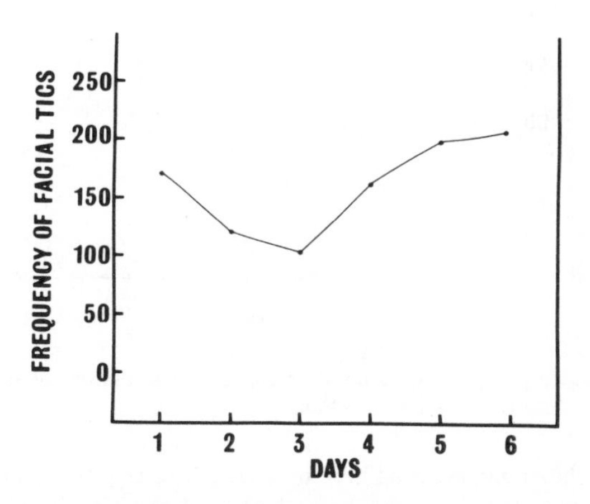

Fig. 3-7. The decreasing-increasing baseline. Hypothetical data for mean number of facial tics averaged over three daily 15-minute videotaped sessions.

the same difficulties in interpretation that are present in the "improving" baseline, previously discussed. Therefore, the most useful course of action to pursue involves continuation of measurement procedures until a stable and steady pattern emerges.

Very similar to the increasing-decreasing pattern is its reciprocal, the decreasing-increasing type of baseline (see Fig. 3-7). This kind of baseline pattern often reflects the "placebo" effects of initially being part of an experiment or being monitored (either self or observed). Although "placebo" effects are always of interest to the clinical researcher, when faced with time pressures, the preferred course of action is to continue measurement procedures until a steady pattern in the data is clear. If extended baseline measurement is not feasible, introduction of the treatment, following the worsening of the target behavior under study, is an acceptable procedure, particularly if the controlling effects of the procedure are subsequently demonstrated via its withdrawal and reinstatement.

A final baseline trend, the unstable baseline, also causes difficulty for the applied clinical researcher. A hypothetical example of this type of baseline, obtained under extended measurement conditions, appears in Fig. 3-8. Examination of these data not only reveals extreme variability, but also reveals the absence of a particular pattern. Therefore, the problems found in the variable

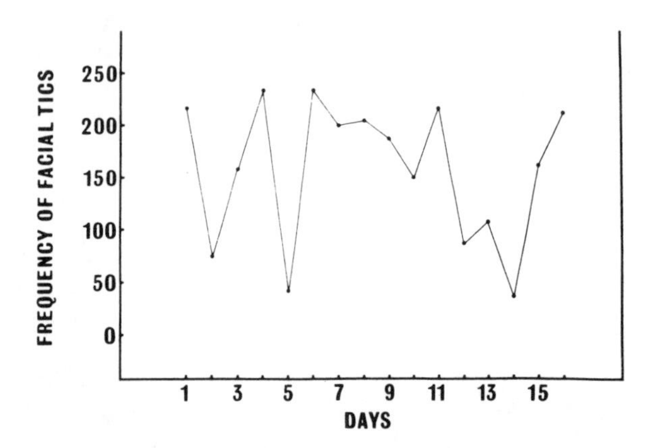

Fig. 3-8. The unstable baseline. Hypothetical data for mean number of facial tics averaged over three daily 15-minute videotaped sessions.

baseline are further compounded here by the lack of any trend in the data. This, of course, heightens the difficulty in evaluating these data through the method of experimental analysis. Even the procedure of blocking data usually fails to eliminate all instability on the basis of visual analysis. To date, no completely satisfactory strategy for dealing with this type of baseline pattern has appeared; at best, the kinds of strategies for dealing with the variable baseline are also recommended here.

3.4. CHANGING ONE VARIABLE AT A TIME

A cardinal rule of experimental single case research is to change *one* variable at a time when proceeding from one phase to the next (Barlow and Hersen,

1973). Barlow and Hersen (1973) point out that when two variables are simultaneously manipulated, the experimental analysis does not permit conclusions as to which of the two components (or how much of each) contributes to improvements in the target behavior. It should be underscored that the *one* variable rule holds, regardless of the particular phase (beginning, middle, or end) that is being evaluated. These strictures are most important when examining the interactive effects of treatment variables (Barlow and Hersen, 1973; Elkin, Hersen, Eisler, and Williams, 1973; Leitenberg, Agras, Thomson, and Wright, 1968). A more complete discussion of interaction designs appears in Chapter 6, Section 6.5.

Correct and incorrect applications

A frequently committed error during the course of experimental single case research involves the simultaneous manipulation of two variables so as to assess their presumed interactive effects. A review of the literature suggests that this type of error is often made in the latter phases of experimentation. In order to clarify the issues involved, selected examples of correct and incorrect applications are to be presented.

For illustrative purposes, let us assume that baseline measurement in a study consists of the number of social responses (operationally defined) emitted by a chronic schizophrenic during a specified period of observation. Let us further assume that subsequent introduction of a single treatment variable involves application of contingent (token) reinforcement following each social response that is observed on the ward. At this point in our hypothetical example, only one variable (token reinforcement) has been added across the two experimental phases (baseline to the first treatment phase). In accordance with design principles followed in the A-B-A-B design, the third phase would consist of a return to baseline conditions, again changing (removing) only one variable across the second and third phase. Finally, in the fourth phase token reinforcement would be reinstated (addition of one variable from Phase 3 to 4). Thus, we have a procedurally correct example of the A-B-A-B design (see Chapter 5) in which only one variable is altered at a time from phase to phase.

In the following example we will present an inaccurate application of single case methodology. Using our previously described measurement situation, let us assume that baseline assessment is now followed by a treatment combination comprised of token reinforcement and social reinforcement. At this point, the experiment is labeled A-BC. Phase 3 is a return to baseline conditions (A), while Phase 4 consists of social reinforcement alone (C). Here we have an example of an A-BC-A-C design, with A = baseline, BC = token and social reinforcement, A = baseline, and C = social reinforcement. In this experiment the researcher is hopeful of teasing the relative effects of token and social reinforcement. However, this is a *totally erroneous* assumption on his part. From the A-BC-A portion of this experiment, it is feasible only to assess the combined BC effect over baseline (A), assuming that the appropriate trends in the data appear. Evaluation of the individual effects of the two variables (social and token reinforcement) comprising the treatment package is not possible. Moreover, application of the C condition (social reinforcement alone) following the second baseline also does not permit firm conclusions, either with respect to the effects of social reinforcement alone or in contrast to the combined treatment of token and social reinforcement. The experimenter is not in a position to examine the interactive effects of the BC and C phases as they are not adjacent to one another.

If our experimenter were interested in accurately evaluating the interactive effects of token and social reinforcement, the following extended design would be considered appropriate (A-B-A-B-BC-B-BC). When this experimental strategy is used, the interactive effects of social and token reinforcement can be examined systematically by comparing differences in trends between the adjacent B (token reinforcement) and BC (token and social reinforcement) phases. The subsequent return to B and reintroduction of the combined BC would allow for analysis of the additive and controlling effects of social reinforcement, assuming expected trends in the data occur. The interaction design will be discussed in detail in Chapter 6, Section 6.5.

A published example of the correct manipulation of variables across phases appears in Fig. 3-9. In this study, Leitenberg, Agras, Thomson, and Wright (1968) examined the separate and combined effects of feedback and praise on the mean number of seconds a knife-phobic patient allowed himself to be exposed to a knife. An examination of the seven phases of study reveals the following progression of variables: (1) feedback, (2) feedback and praise, (3) feedback, (4) no feedback and no praise, (5) feedback, (6) feedback and praise, and (7) feedback. A comparison of adjacent phases shows that only one variable was manipulated (added or subtracted) at a time across phases. In a similar design, Elkin, Hersen, Eisler, and Williams (1973) assessed additive and subtractive effects of therapeutic variables in a case of anorexia nervosa. The following progression of variables was used in a six-phase experiment: (1) 3000 calories–*baseline*, (2) 3000 calories–*feedback*, (3) 3000 calories–*feedback and reinforcement*, (4) 45000 calories–*feedback and reinforcement*, (5) 3000 calories–*feedback and reinforcement*, (6) 4500 calories–*feedback and reinforcement.* Again, changes from one phase to the next (italicized) never involved more than the manipulation of a single variable.

A number of design options (e.g., A-B-A-C, A-B-A-BC) frequently used by applied clinical researchers also violate the principle of altering one variable at a

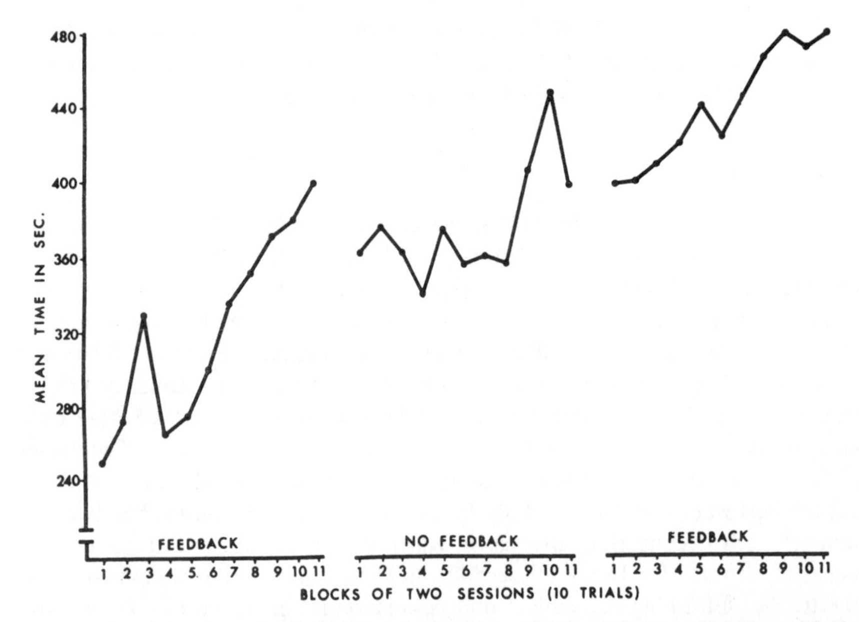

Fig. 3-9. Time in which a knife was kept exposed by a phobic patient as a function of feedback, feedback *plus* praise, and no feedback or praise conditions. (Fig. 2, p. 131, from: Leitenberg, H., Agras, W. S., Thomson, L., and Wright, D. E. Feedback in behavior modification: An experimental analysis in two phobic cases. *Journal of Applied Behavior Analysis*, 1968, 1, 131-137. Copyright by Society for the Experimental Analysis of Behavior, Inc., reproduced by permission.)

time. Let us consider a hypothetical example of the A-B-A-C design in which A = baseline, B = feedback, A = baseline, and C = reinforcement. The A-B-A phase of the experiment follows a stepwise sequential analysis, but the introduction of the C variable in Phase 4 does not permit any definitive conclusions for two reasons. First, without a return to baseline (A), the controlling effects of C cannot be determined. At this point only suggested trends in the data can be inferred. Second, even if the study were to be extended to an A-B-A-C-A design, the differential effects (if any) between B and C cannot be properly assessed. As previously noted, the A-B-A-B-BC-B-BC interaction design is required. In short, the experimental single case design strategy allows for a comparison of additive and sequential effects, but does not provide for analysis of comparative effects from nonadjacent phases.

As in the previous example, the last phase of the A-B-A-BC design does not result in unequivocal conclusions from the experimental analysis framework. Again, with A = baseline, B = feedback, and C = reinforcement, the combination of feedback and reinforcement in the BC phase does not answer the question as to the added effects of reinforcement. Only through an extension of the design (A-B-A-BC-B-BC) in which reinforcement is withdrawn and reintroduced (BC-B-BC) are its additive effects appropriately evaluated.

Exceptions to the rule

In a number of experimental single case studies (Barlow, Leitenberg, and Agras, 1969; Eisler, Hersen, and Agras, 1973; Pendergrass, 1972; Ramp, Ulrich, and Dulaney, 1971) legitimate exceptions to the rule of maintaining a consistent stepwise progression (additive or subtractive) across phases have appeared. In this section the exceptions will be discussed and examples of published data will be presented and analyzed. For example, Ramp, Ulrich, and Dulaney (1971) examined the effects of instructions and delayed timeout in a 9-year-old male elementary school student who proved to be a disciplinary problem. Two target behaviors (intervals out of seat without permission and intervals talking without permission) were selected for study in four separate phases. During baseline, the number of 10-second time intervals in which the subject was out of seat or talking were recorded for 15-minute sessions. In Phase 2 instructions simply involved the teacher informing the subject that permission for being out of seat and talking were required (raising his hand). The third phase consisted of a delayed timeout procedure. A red light, mounted on the subject's desk, was illuminated for a 1–3-second period immediately following an instance of out-of-seat or talking behavior. Number of illuminations recorded were cumulated each day, with each classroom violation resulting in a 5-minute detention period in a specially constructed timeout booth while other children participated in gym and recess activities. The results of this study appear in Fig. 3-10. Relabeling of the four experimental phases yields an A-B-C-A design. Inspection of the figure shows that the baseline (A) and instructions (B) phases do not differ significantly for either of the two target behaviors under study. Thus, although the independent variables differ across these phases, the resulting dependent measures are essentially alike. However, institution of the delayed timeout contingency (C) yielded a marked decrease in classroom violations. Subsequent removal of the timeout contingency in Phase 4 (A) led to a renewed increase in classroom violations.

Since the two initial phases (A and B) yield similar data (instructions did not appear to be effective), equivalence of the baseline and instructions phases are assumed. If one then collapses data across these two phases, an A-C-A design emerges, with some evidence demonstrated for the controlling effects of delayed timeout. In this case the A-C-A design follows the experimental analysis used in the case of the A-B-A design (see Chapter 5). However, further confirmation of the controlling effects would require a return to the C condition (delayed

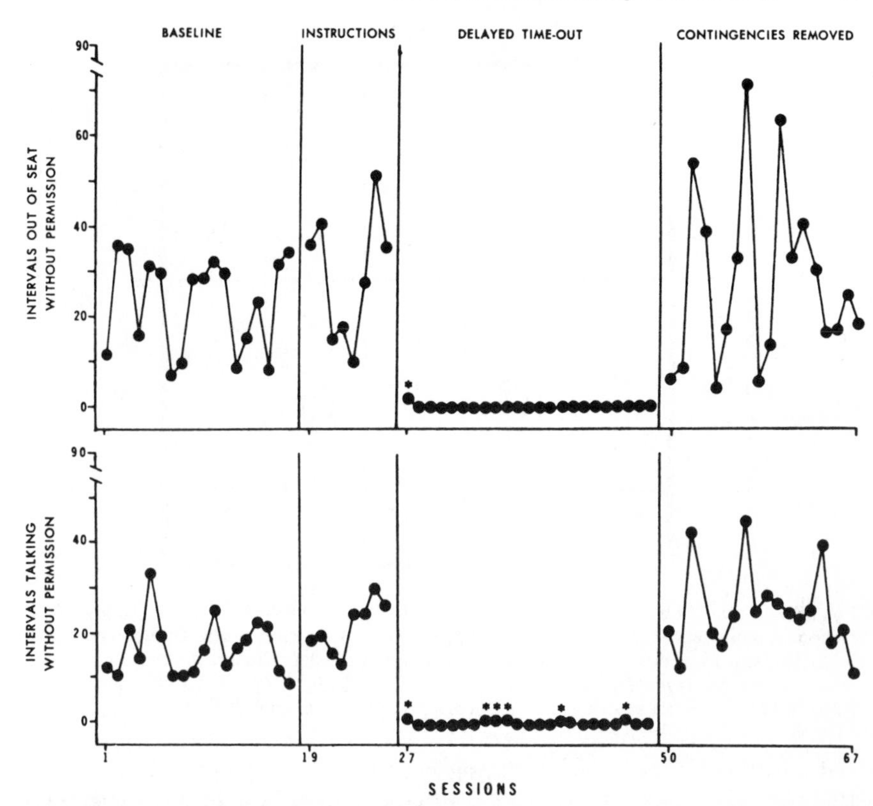

Fig. 3-10. Each point represents one session and indicates the number of intervals in which the subject was out of his seat (top) or talking without permission (bottom). A total of 90 such intervals was possible within a 15-minute session. Asterisks over points indicate sessions that resulted in time being spent in the booth. (Fig. 1, p. 237, from: Ramp, E., Ulrich, R., and Dulaney, S. Delayed timeout as a procedure for reducing disruptive classroom behavior: A case study. *Journal of Applied Behavior Analysis*, 1971, 4, 235-239. Copyright by Society for the Experimental Analysis of Behavior, Inc., reproduced by permission.)

timeout). This new design would then be labeled as follows: A=B-C-A-C. It should be noted that without the functional equivalence of the first two phases (A = B) this would essentially be an incorrect experimental procedure. The functional equivalence of different adjacent experimental phases warrants further illustration. An excellent example is provided by Pendergrass (1972), who used an A-B-A=C-B design strategy. In her study, Pendergrass evaluated the effects of timeout and observation of punishment being administered (timeout) to a co-subject in an 8-year-old retarded boy. Two negative high-frequency behaviors were selected as targets for study. They were: (1) banging objects on the floor

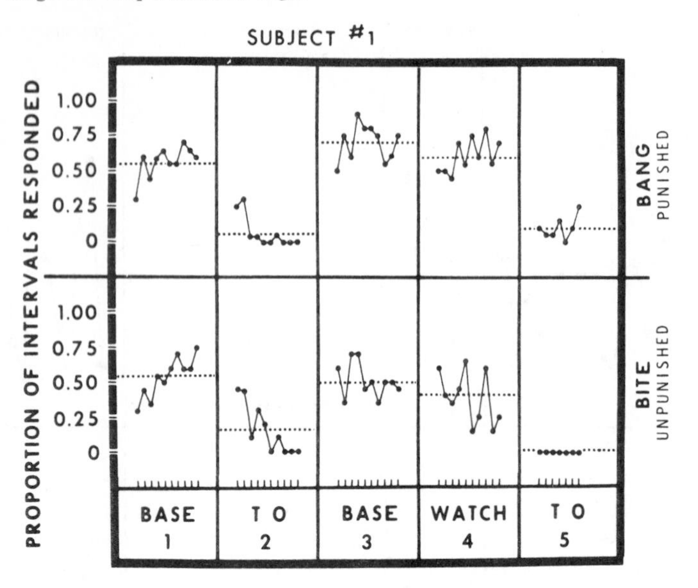

Fig. 3-11. Proportion of total intervals in which Bang (punished) and Bite (unpunished) responses were recorded for S1 in 47 free play periods. (Fig. 1, p. 88, from: Pendergrass, V. E. Timeout from positive reinforcement following persistent, high-rate behavior in retardates. *Journal of Applied Behavior Analysis*, 1972, 5, 85-91. Copyright by Society for the Experimental Analysis of Behavior, Inc., reproduced by permission.)

and on others (bang), and (2) the subject's biting of his lips and hand (bite). However, only one of the two target behaviors (bang) was directly subjected to treatment effects, but generalization and side effects of treatment on the second behavior (bite) were examined concurrently. Results of the study are presented in Fig. 3-11. Timeout following baseline assessment led to a significant decrease in both the punished and unpunished behaviors. A return to baseline conditions in Phase 3 resulted in high levels of both target behaviors. Institution of the "watch" condition (observation of punishment) did not lead to an appreciable decrease, hence the functional equivalence of Phases 3 (A) and 4 (C). In Phase 5 the reinstatement of timeout led to renewed improvement in target behaviors.

In this study the ineffectiveness of the "watch" condition is functionally equivalent to the continuation of the baseline phase (A), despite obvious differences in procedure. With respect to labeling of this design, it is most appropriately designated as follows: A-B-A=C-B (the equal sign between A and C represents their functional equivalence insofar as dependent measures are concerned).

A further exception to the basic rule occurs when the experimenter is interested in the total impact of a treatment package containing two or more

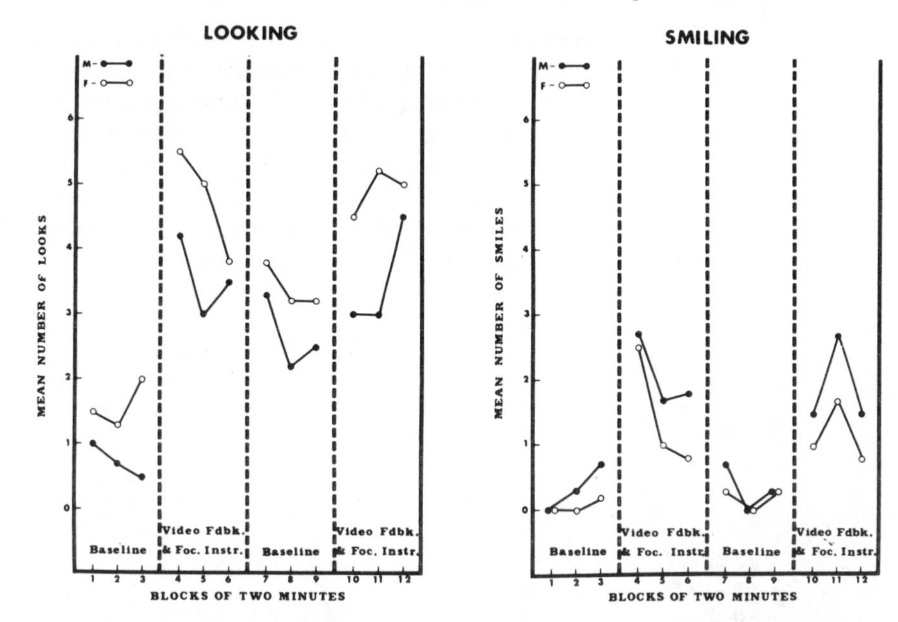

Fig. 3-12. Mean number of looks and smiles for three couples in 10-second intervals plotted in blocks of 2 minutes for the Videotape feedback plus Focused Instructions Design. (Fig. 3, p. 556, from: Eisler, R. M., Hersen, M., and Agras, W. S. Effects of videotape and instructional feedback on nonverbal marital interaction: An analog study. *Behavior Therapy*, 1973, 4, 551-558. Reproduced by permission.)

components (e.g., instructions, feedback, and reinforcement). In this case, more than one variable is manipulated at a time across adjacent experimental phases. An example of this type of design appeared in a series of analogue studies reported by Eisler, Hersen, and Agras (1973). In one of their studies the combined effects of videotape feedback and focused instructions were examined in an A-BC-A-BC design, with A = baseline and BC = videotape feedback and focused instructions. As is apparent from inspection of Fig. 3-12, analysis of these data follow the A-B-A-B design pattern, with the exception that the B phase is represented by a compound treatment variable (BC). However, it should be pointed out that despite the fact that improvements over baseline appear for both target behaviors (looking and smiling) during videotape feedback and focused instructions conditions, this type of design will obviously allow for no conclusions as to the relative contribution of each treatment component.

A final exception to the one variable rule appears in a study by Barlow, Leitenberg, and Agras (1969), in which the controlling effects of the noxious scene in covert sensitization were examined in two sexual deviates (a case of pedophilia and one of homosexulaity). In each case an A-BC-B-BC experimental design was used (Barlow and Hersen, 1973). In both cases the four experimental phases were as follows: (1) A = baseline, (2) BC = covert sensitization treatment (verbal description of deviant sexual activity and introduction of the "nauseous" scene), (3) B = verbal description of deviant sexual activity but *no* introduction of the "nauseous" scene, and (4) BC = covert sensitization (verbal description of sexual activity and introduction of the "nauseous" scene). For purposes of illustration, data from the pedophilic case appear in Fig. 3-13.

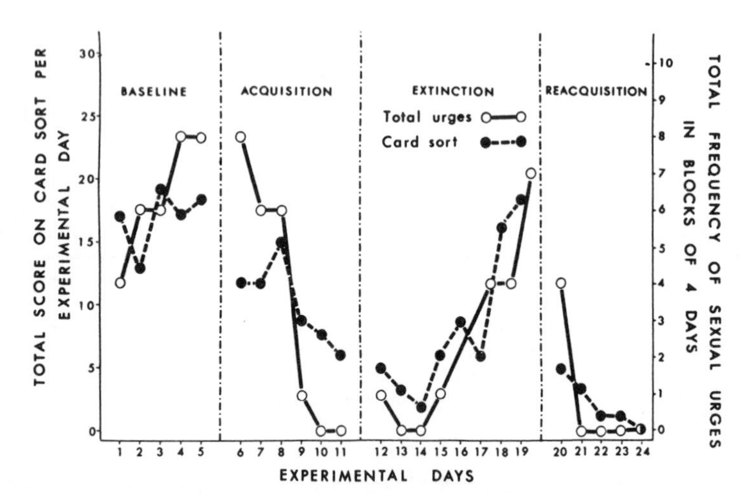

Fig. 3-13. Total score on card sort per experimental day and total frequency of pedophilic sexual urges in blocks of 4 days surrounding each experimental day. (Lower scores indicate less sexual arousal.) (Fig. 1, p. 599, from: Barlow, D. H., Leitenberg, H., and Agras, W. S. Experimental control of sexual deviation through manipulation of the noxious scene in covert sensitization. *Journal of Abnormal Psychology*, 1969, **74**, 596-601. Copyright 1969 by the American Psychological Association, reproduced by permission.)

Examination of the design strategy reveals that covert sensitization treatment (BC) required instigation of both components. Thus, initial differences between baseline (A) and acquisition (BC) only suggests efficacy of the total treatment package. When the "nauseous" scene is removed during extinction (B), the resulting increase in deviant urges and card sort scores similarly ***suggests*** the controlling effects of the "nauseous" scene. In reacquisition (BC), where the "nauseous" scene is reinstated, renewed decreases in the data confirm its controlling effects. Therefore, despite an initial exception to changing one variable at a time across adjacent phases, a stepwise subtractive and additive progression is maintained in the last two phases, with valid conclusions derived from the ensuing experimental analysis.

Issues in drug evaluation

Issues discussed in the previous section that pertain to changing of variables across adjacent experimental phases and the functional equivalence in data following procedurally different operations are identical when analyzing the effects of drugs on behavior. It is of some interest that experimenters with both a behavior modification bias (e.g., Liberman, Davis, Moon, and Moore, 1973) and those adhering to the psychoanalytic tradition (e.g., Bellak and Chassan, 1964) have used remarkably similar design strategies when investigating drug effects on behavior, either alone or in combination with psychotherapeutic procedures.

Keeping in mind the one variable rule, the following sequence of experimental phases has appeared in a number of studies: (1) no drug, (2) placebo, (3) active drug, (4) placebo, and (5) active drug. This kind of design, in which a stepwise application of variables appears, permits conclusions with respect to possible placebo effects (no drug to placebo phase) and those with respect to the controlling influences of active drugs (placebo, active drug, placebo, active drug). Within the experimental analysis framework, Liberman, Davis, Moon, and Moore (1973) have labeled this sequence the A-A_1-B-A_1-B design. More specifically, they examined the effects of stelazine on a number of asocial responses emitted by a withdrawn schizophrenic patient. The particular sequence used was as follows: (A) no drug, (A_1) placebo, (B) stelazine, (A_1) placebo, and (B) stelazine. Similarly, within the psychoanalytic framework, Bellak and Chassan (1964) assessed the effects of chlordiazepoxide on variables (primary process, anxiety, confusion, hostility, "sexual flooding," depersonalization, ability to communicate) rated by a therapist during the course of ten weekly interviews. A double-blind procedure was used in which neither the patient nor the therapist was informed about changes in placebo and active medication conditions. In this study, an A-A_1-B-A_1-B design was employed with the following sequential pattern: (A) no drug, (A_1) placebo, (B) chlordiazepoxide, (A_1) placebo, and (B) chlordiazepoxide.

Once again, pursuing the one variable rule, Liberman, Davis, Moon, and Moore (1973) have shown how the combined effects of drugs and behavioral manipulations can be evaluated. Maintaining a constant level of medication (600 mg. of chlorpromazine per day), the controlling effects of timeout on delusional behavior (operationally defined) were examined as follows: (1) baseline plus 600 mg. of chlorpromazine, (2) timeout plus 600 mg. of chlorpromazine, and (3) removal of timeout plus 600 mg. of chlorpromazine. In this study (AB-CB-AB) the only variable manipulated across phases was the time contingency.

There are several other important issues related to the investigation of drug effects in single case experimental designs that merit careful analysis. They include the double-blind evaluation of results, long-term carryover effects of phenothiazines, and length of phases. These will be discussed in some detail in Section 3.6 of this chapter and in Chapter 6.

3.5. REVERSAL AND WITHDRAWAL

In their survey of the methodological aspects of applied behavior analysis, Baer, Wolf, and Risley (1968) state that there are two types of experimental designs that can be used to show the controlling effects of treatment variables in individuals. These two basic types are commonly referred to as the "reversal" and "multiple baseline" design strategies. In this section we will concern ourselves only with the "reversal" design. The prototypic A-B-A design and all of its numerous extension and permutations (see Chapter 5 for details) are usually placed in this category (Barlow and Hersen, 1973; Johnston, 1972; Kazdin, 1973; Kazdin and Bootzin, 1972; Krasner, 1971; Liberman, Davis, Moon, and Moore, 1973; McNamara and MacDonough, 1972; Wolf and Risley, 1971).

When speaking of a reversal, one typically refers to the removal (withdrawal) of the treatment variable that is applied after baseline measurement has been concluded. In practice, the reversal involves a withdrawal of the B phase (in the A-B-A design) after behavioral change has been successfully demonstrated. If the treatment (B phase) indeed exerts control over the targeted behavior under study, a decreased or increased trend (depending on which direction indicates deterioration) in the data should follow its removal.

In describing their experimental efforts when using A-B-A designs, applied clinical researchers frequently have referred to both their procedures and resulting data as reversals. This, then, represents a terminological confusion between the independent variable and the dependent variable. However, from either a semantic, logical, or scientific standpoint, it is untenable that both a cause and an effect should be given an identical label. A careful analysis reveals that a reversal involves a *specific technical operation*, and that its resultant (changes in the target behavior(s)) is simply examined in terms of rates of the data (increased, decreased, or no change) in relation to patterns seen in the previous experimental phase. To summarize, a reversal is an active procedure; the obtained data may or may not reflect a particular trend.

The reversal design

A still finer distinction regarding reversals is made by Leitenberg (1973) in his examination of experimental single case design strategies. He contends that the "reversal" design (e.g., A-B-A-B design) is inappropriately labeled, and that the term withdrawal (i.e., withdrawal of treatment in the second A phase) is a more accurate description of the actual technical operation. Indeed, a distinction between a withdrawal and a reversal is made, and Leitenberg shows how the latter refers to a specific kind of experimental strategy. It should be underscored that, although ". . . this distinction. . . is typically not made in the behavior modification literature" (Leitenberg, 1973), the point is well taken and should be considered by applied clinical researchers.

To illustrate and clarify this distinction, an excellent example of the "reversal" design, selected from the child behavior modification literature, will be presented. Allen, Hart, Buell, Harris, and Wolf (1964) were concerned with the contingent effects of reinforcement on the play behavior of a 4½-year-old girl who evidenced social withdrawal with peers in a preschool nursery setting. Two target behaviors were selected for study: (1) percentage of interaction with adults, and (2) percentage of interaction with children. Observations were recorded daily during 2-hour morning sessions. As can be seen in Fig. 3-14, baseline data show that about 15 percent of the child's time was spent interacting with children, whereas approximately 45 percent of the time was spent in interactions with adults. The remaining 40 percent involved "isolate" play. Inasmuch as the authors hypothesized that teacher attention fostered interactions with adults, in the second phase of experimentation an effort was made to demonstrate that the same teacher attention, when presented contingently in the form of praise following the child's interaction with other children, would lead to an increase of such interactions. Conversely, "isolate" play and approaches to adults were ignored. Inspection of Fig. 3-14 reveals that contingent reinforcement (praise) increased the percentage of interaction with children and led to a concomitant decrease in interactions with adults. In the third phase a "true" *reversal* of contingencies was put into effect. That is to say, contingent reinforcement (praise) was now administered when the child approached adults but interaction with other children was ignored. Examination of Phase 3 data reflects the reversal in contingencies. Percentage of time spent with children decreased substantially while percentage of time spent with adults showed a marked increase. Phase 2 contingencies were then reinstated in Phase 4, and the remaining points on the graph are concerned with follow-up measures.

Reversal and withdrawal designs compared

A major difference in the "reversal" and withdrawal design is that in the third phase of the reversal design, following instigation of the therapeutic procedure, the same procedure is now applied to an alternative but incompatible behavior.

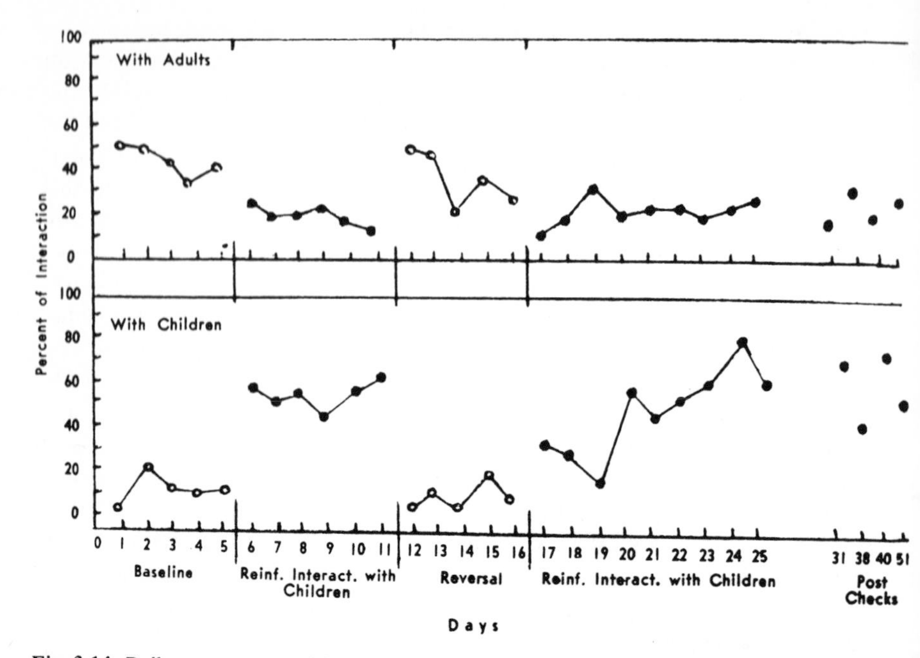

Fig. 3-14. Daily percentages of time spent in social interaction with adults and with children during approximately 2 hours of each morning session. (Fig. 2, p. 515, from: Allen, K. E., Hart, B., Buell, J. S., Harris, F. R., and Wolf, M. M. Effects of social reinforcement on isolate behavior of a nursery school child. *Child Development*, 1964, 35, 511-518. Reproduced by permission of The Society For Research in Child Development, Inc.)

By contrast, in the withdrawal design, the A phase following introduction of the treatment variable (e.g., token reinforcement) simply involves its removal and a return to baseline conditions. Leitenberg (1973) argues that "Actually, the reversal design although it can be quite dramatic is somewhat more cumbersome . . ." (pp. 90-91) than the more frequently employed withdrawal design. Moreover, the withdrawal design is much better suited for investigations that do not emanate from the operant (reinforcement) framework (e.g., the investigation of drugs, examination of non-behavioral therapies).

Withdrawal of treatment

The specific point at which the experimenter removes his treatment variable (second A phase in the A-B-A design) in the withdrawal design is multidetermined. Among the factors to be considered are time limitations imposed by the treatment setting, staff cooperation when working in institutions (Johnston,

1972), and ethical considerations when removal of treatment can possibly lead to some harm to the subject (e.g., head banging in a retardate) or others in his environment (e.g., physical assaults towards wardmates in disturbed inpatients). Assuming that these important environmental considerations can be dealt with adequately and judiciously, a variety of parametric issues must be taken into account before instituting withdrawal of the treatment variable. One of these issues involves the overall length of adjacent treatment phases; this will be examined in Section 3.6 of this chapter.

In this section we will consider the implementation of treatment withdrawal in relation to data trends appearing in the first two phases (A and B) of study. We will illustrate both correct and incorrect applications using hypothetical data. Let us consider an example in which A refers to baseline measurement of the frequency of social responses emitted by a withdrawn schizophrenic. The subsequent treatment phase (B) involves contingent reinforcement in the form of praise, while the third phase (A) represents the withdrawal of treatment and a return to "original" baseline conditions. For purposes of illustration, we will assume stability of "initial" baseline conditions for each of the following examples.

In our first example (see Fig. 3-15) data during contingent reinforcement show a clear upward trend. Therefore, institution of withdrawal procedures at the conclusion of this phase will allow for analysis of the controlling effects of reinforcement, particularly if the return to baseline results in a downward trend

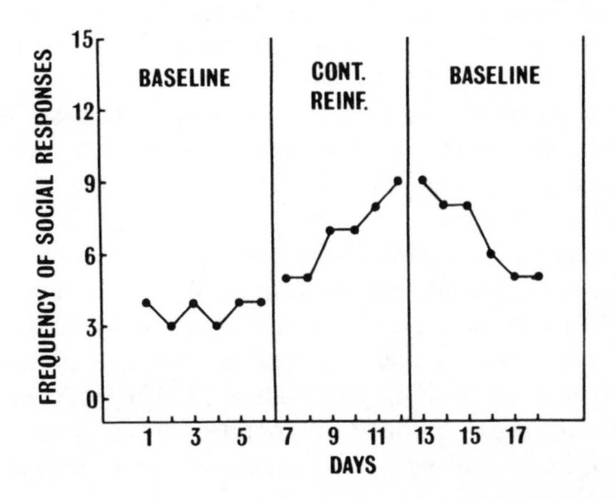

Fig. 3-15. Increasing treatment phase followed by decreasing baseline. Hypothetical data for frequency of social responses in a schizophrenic patient per 2-hour period of observation.

in the data. Equally acceptable is a baseline pattern (second A phase) in which there is an immediate loss of treatment effectiveness, which is then maintained at a low-level stable rate (this pattern is the same as the initial baseline phase).

In our second example (see Fig. 3-16) data during contingent reinforcement show the immediate effects of treatment and are maintained throughout the phase. After these initial effects, there is no evidence of an increased rate of responding. However, the withdrawal of contingent reinforcement at the conclusion of the phase *does* permit analysis of its controlling effects. Data in the

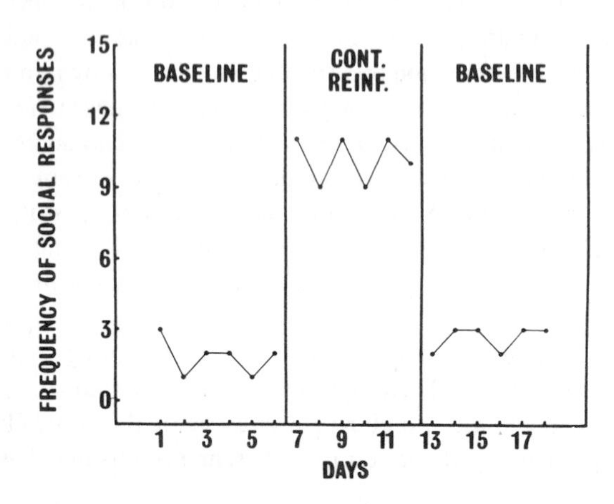

Fig. 3-16. High-level treatment phase followed by low-level baseline. Hypothetical data for frequency of social responses in a schizophrenic patient per 2-hour period of observation.

second baseline show no overlap with contingent reinforcement as there is a return to the stable but low rate of responding seen in the first baseline (as in Fig. 3-16). Equally acceptable would be a downward trend in the data as depicted in the second baseline in Fig. 3-14.

In our third example of a correct withdrawal procedure, examination of Fig. 3-17 indicates that contingent reinforcement resulted in an immediate increase in rate followed by a linear decrease and then a renewed increase in rate which then stabilized. Although it would be advisable to analyze contributing factors to the decrease and subsequent increase (Sidman, 1960), institution of the withdrawal procedure at the conclusion of the contingent reinforcement phase allows for an analysis of its controlling effects, particularly as a decreased rate was observed in the second baseline.

An example of the incorrect application of treatment withdrawal appears in Fig. 3-18. Inspection of the figure reveals that after a stable pattern is obtained in baseline, introduction of contingent reinforcement leads to an immediate and

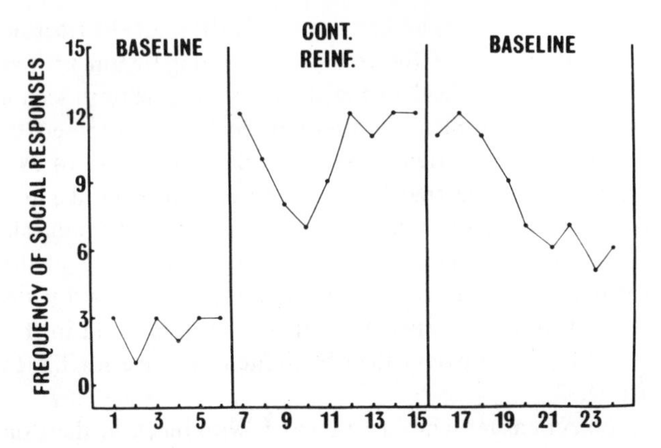

Fig. 3-17. Decreasing-increasing-stable treatment phase followed by decreasing baseline. Hypothetical data for frequency of social responsies in a schizophrenic per 2-hour period of observation.

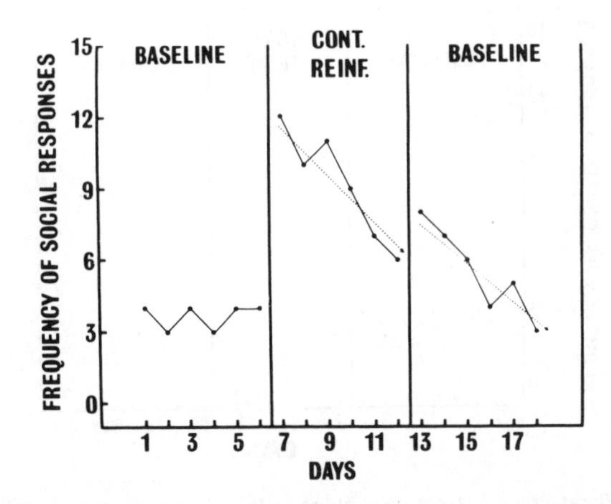

Fig. 3-18. High-level decreasing treatment phase followed by decreasing baseline. Hypothetical data for frequency of social responses in a schizophrenic patient per 2-hour period of observation.

dramatic improvement, which is then followed by a marked decreasing linear function. This trend is in evidence despite the fact that the last data point in contingent reinforcement is clearly above the highest point achieved in baseline. Removal of treatment and a return to baseline conditions on Day 13 similarly result in a decreasing trend in the data. Therefore, no conclusions as to the controlling effects of contingent reinforcement are possible as it is not clear whether the decreasing trend in the second baseline is a function of the treatment's withdrawal or mere continuation of the trend begun during treatment. Even if withdrawal of treatment were to lead to the stable low-level pattern seen in the first baseline period, the same problems in interpretation would be posed.

When the aforementioned trend appears during the course of experimental treatment, it is recommended that the phase be continued until a more consistent pattern emerges. However, if this strategy is pursued, the equivalent length of adjacent phases is altered (see Section 3.6). A second strategy, although admittedly somewhat weak, is to reintroduce treatment in Phase 4 (thus, we have an A-B-A-B design), with the expectation that a reversed trend in the data will reflect improvement. There would then be limited evidence for the treatment's controlling effects.

A similar problem ensues when treatment is withdrawn in the example that appears in Fig. 3-19. In spite of an initial upward trend in the data when contingent reinforcement is first introduced (B), the decreasing trend in the latter half

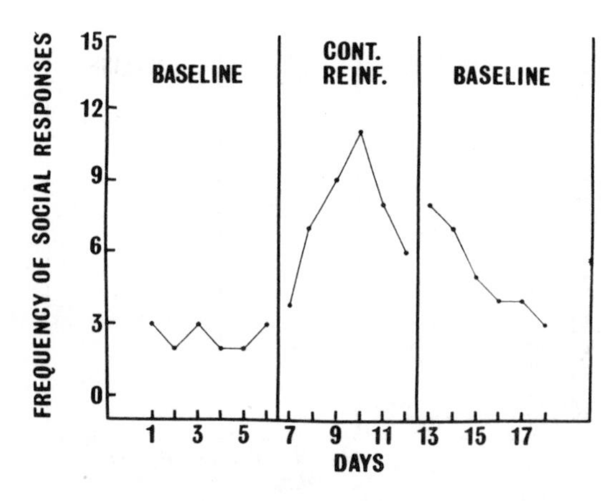

Fig. 3-19. Increasing-decreasing treatment phase followed by decreasing baseline. Hypothetical data for frequency of social responses in a schizophrenic patient per 2-hour period of observation.

of the phase, which is then followed by a similar decline during the second baseline (A), prevents an analysis of the treatment's controlling effects. Therefore, the same recommendations made in the case of Fig. 3-18 apply here.

Limitations and problems

As mentioned earlier, the applied clinical researcher faces some unique problems when he is intent on pursuing his experimental analysis by withdrawing a particular treatment technique. These problems are heightened in settings where he exerts relatively little control, either with respect to staff cooperation or in terms of other important environmental contingencies (e.g., when dealing with individual problems in the classroom situation, responses of other children throughout the varying stages of experimentation may spuriously affect the results). Although these concerns have been articulated elsewhere in the behavioral literature (Baer, Wolf, and Risley, 1968; Bijou, Peterson, Harris, Allen, and Johnston, 1969; Kazdin and Bootzin, 1972; Leitenberg, 1973), a brief summary of the issues at stake might be useful at this point.

A frequent criticism leveled at researchers using single case methodology is that removal of the treatment will lead to the subject's irreversible deterioration (at least in terms of the behavior under study). However, as Leitenberg (1973) points out, this is a weak argument as there is no supporting evidence to be found in the experimental literature. If the technique shows initial beneficial effects and it exerts control over the targeted behavior being examined, then when reinstated its controlling effects will be reestablished. To the contrary, Krasner (1971) reports that recovery of initially low levels of baseline performance often fails to occur in extended applications of the A-B-A design where multiple withdrawals and reinstatements of the treatment technique are instituted (e.g., A-B-A-B-A-B-A-B). Indeed, the possible carryover effects across phases and concomitant environmental events leading to improved conditions contribute to the researcher's difficulties in carrying out scientifically acceptable studies.

A less subtle problem encountered is one of staff resistance. Usually, the researcher working in an applied setting (be it at school, state institution for the retarded, or psychiatric hospital) is consulting with house staff on difficult problems. In his efforts to remediate the problem, the experimenter encourages staff to apply treatment strategies that are likely to achieve beneficial results. When staff members are subsequently asked to temporarily withdraw treatment procedures, some may openly rebel. "What teacher, seeing Johnny for the first time quietly seated for most of the day, would like to experience another week or two of bedlam just to satisfy the perverted whim of a psychologist?" (Johnston, 1972, p. 1035). In other cases the staff member or parent (when establishing parental retraining programs) may be unable to revert to his original manner of functioning (i.e., how he previously responded to certain classes of behavior). Indeed, this happened in a study reported by Hawkins, Peterson, Schweid, and Bijou (1966). Leitenberg (1973) argues that "In such cases, where the therapeutic procedure cannot be introduced and withdrawn at will, sequential ABA designs are obviated" (p. 98). Under these circumstances, the use of alternative experimental strategies such as multiple baseline (Baer, Wolf, and Risley, 1968) or time-lagged control (Gottman, 1973) designs are obviously better suited.

To summarize, the researcher using the withdrawal design must ensure that (1) he has full staff or parental cooperation on an *a priori* basis, (2) the withdrawal of treatment will lead to minimal environmental disruptions (i.e., no injury to subject or others in his environment will result) (see Peterson and Peterson, 1968), (3) the withdrawal period will be relatively brief, (4) outside environmental influences will be minimized throughout baseline, treatment, and withdrawal phases, and (5) final reinstatement of treatment to its logical conclusion will be accomplished as soon as it is technically feasible.

3.6. LENGTH OF PHASES

Although there has been some intermittent discussion in the literature with regard to the length of phases when carrying out single case experimental research (Barlow and Hersen, 1973; Bijou, Peterson, Harris, Allen, and Johnston, 1969; Chassan, 1967; Johnston, 1972), a complete examination of the problems faced and the decision to be made by the researcher has yet to appear. Therefore, in this section the major issues involved including individual and relative length of phases, carryover effects, and cyclic variations are to be considered. In addition, these considerations will be examined as they apply to the study of drugs on behavior.

Individual and relative length

When considering the individual length of phases independently of other factors (e.g., time limitations, ethical considerations, relative length of phases), most experimenters would agree that baseline and experimental conditions should be continued until some semblance of stability in the data is apparent. Johnston (1972) has examined these issues with regard to the study of punishment. He states that "It is necessary that each phase be sufficiently long to demonstrate stability (lack of trend and a constant range of variability) and to dispel any doubts of the reader that the data shown are sensitive to and representative of what was happening under the described conditions" (Johnston, 1972, p. 1036). He notes further ". . . that if there is indication of an increasing or decreasing trend in the data or widely variable rates from day to day (even with no trend) then the present condition should be maintained until the instability disappears or is shown to be representative of the current conditions" (p. 1036).

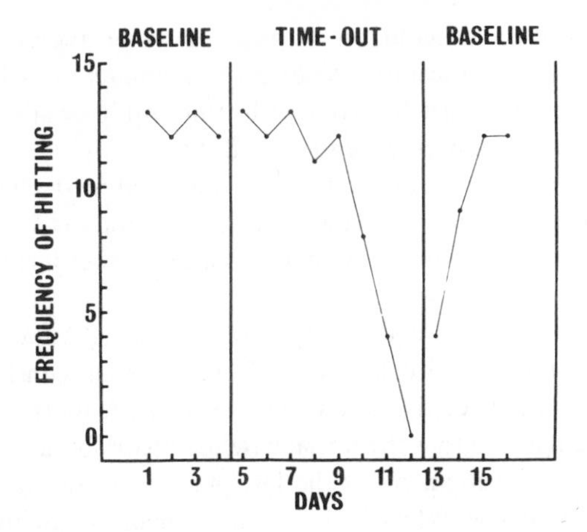

Fig. 3-20. Extension of the treatment phase in an attempt to show its effects. Hypothetical data in which the effects of timeout on daily frequency of hitting other children (based on a 2-hour free play situation) in a 3-year-old male child are examined.

The aforementioned recommendations reflect the ideal and apply best when each experimental phase is considered individually and independently of adjacent phases. If one were to fully carry out these recommendations, the possibility exists that widely disparate lengths in phases will result. The strategic difficulties inherent in unequal phases have been noted elsewhere by Barlow and Hersen (1973). Indeed, Barlow and Hersen (1973) cite the advantages of obtaining a relatively equal number of data points for each phase.

Let us illustrate the importance of their suggestions by considering the following hypothetical example, in which the effects of timeout on frequency of hitting other children during a free-play situation are assessed in a 3-year-old child. Examination of Fig. 3-20 shows a stable baseline pattern, with a high frequency of hitting behavior exhibited. Data for Days 5-7, when treatment (timeout) is first instigated, show no effects, but on Day 8 a slight decline in frequency appears. If the experimenter were to terminate treatment at this point, it is obvious that few statements about its efficacy could be made. Thus, he continues treatment for an additional 4 days (9-12) and an appreciable decrease of hitting is obtained. However, by extending (doubling) the length of the treatment phase, it is not clear whether additional treatment in itself leads to changes, whether some correlated variable (e.g., increased teacher attention to incompatible positive behaviors emitted by the child) results in changes, or whether the mere passage of time (maturational changes) accounts for the decelerated trend. Of course, the withdrawal of treatment on Days 13-16 (second baseline) leads to a marked increase in hitting behavior, thus suggesting the controlling effects of the timeout contingency. However, the careful investigator would reinstate timeout procedures to dispel any doubts as to its possible controlling effects over the target behavior of hitting. Additionally, once the treatment (timeout) phase has been extended to 8 days, it would be appropriate to maintain equivalence in subsequent baseline and treatment phases by also collecting approximately 8 days of data in each condition. Then, questions as to whether treatment effects are due to maturational or other uncontrollable influences will be satisfactorily answered.

As previously noted, the actual length of phases (as opposed to the ideal length) is often determined by factors aside from design considerations. However, where possible, the relative equivalence of phase lengths is desirable. If exceptions are to be made, either the initial baseline phase should be lengthened to achieve stability in measurement or the last phase (e.g., second B phase in the A-B-A-B design) should be extended to insure permanence of the treatment effects. In fact, with respect to this latter point, investigators should make an effort to follow their experimental treatments with a full clinical application of the most successful techniques available.

An example of the ideal length of alternating baseline and treatment phases appears in Miller's (1973) analysis of the use of Retention Control Training in a "secondary enuretic" child (see Fig. 3-21). Two target behaviors, number of enuretic episodes and mean frequency of daily urination, were selected for study in an A-B-A-B experimental design. During baseline, the child recorded the natural frequency of target behaviors and received counseling from the experimenter on general issues relating to home and school. Following baseline, the first week of RCT involved teaching the child to postpone urination for a 10-minute period after experiencing each urge. Delay of urination was increased to 20 and 30 minutes in the next 2 weeks. During Weeks 7-9, RCT was withdrawn, but was reinstated in Weeks 10-14.

Examination of Fig. 3-21 indicates that each of the first three phases consisted of 3 weeks, with data reflecting the controlling effects of RCT on both target behaviors. Reinstatement of RCT in the final phase led to renewed control and the treatment was extended to 5 weeks to ensure maintenance of gains.

It might be noted that phase and data patterns do not often follow the ideal sequence depicted in the Miller (1973) study. And, as a consequence, experimenters frequently are required to make accommodations either for ethical, procedural, or parametric reasons. Moreover, when working in an unexplored area where the issues are of social significance, deviations from some of our proposed rules during the earlier stages of investigation are acceptable. However, once technical procedures and major parametric concerns have been dealt with satisfactorily, a more vigorous pursuit of scientific rigor would be expected. In

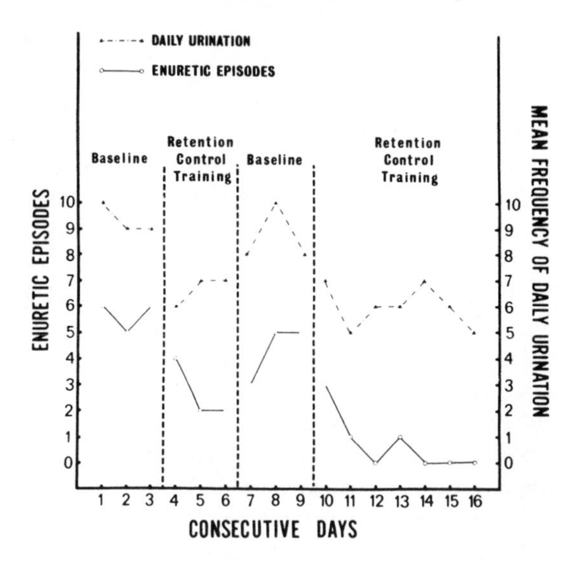

Fig. 3-21. Number of enuretic episodes per week and mean number of daily urinations per week for Subject 1. (Fig. 1, p. 291, from: Miller, P. M. An experimental analysis of retention control training in the treatment of nocturnal enuresis in two institutionalized adolescents. *Behavior Therapy*, 1973, 4, 288-294. Reproduced by permission.)

short, as in any scientific endeavor, as increased knowledge accrues the level of experimental sophistication should reflect its concurrent growth.

Carryover effects

A parametric issue that is very much related to the comparative lengths of adjacent baseline and treatment phases is one of overlapping (carryover) effects. Carryover effects in behavioral (as distinct from drug) studies usually appear in the second baseline phase of the A-B-A-B type design, and are characterized by the experimenter's inability to retrieve original levels of baseline responding. Not only is the original baseline rate not recoverable in some cases (e.g., Ault, Peterson, and Bijou, 1968; Hawkins, Peterson, Schweid, and Bijou, 1966), but on occasion (e.g., Zeilberger, Sampen, and Sloane, 1968) the behavior under study undergoes more rapid modification the second time the treatment variable is introduced.

Presence of carryover effects has been attributed to a variety of factors including changes in instructions across experimental conditions (Kazdin, 1973), the establishment of new conditioned reinforcers (Bijou, Peterson, Harris, Allen, and Johnston, 1969), the maintenance of new behavior through naturally occurring environmental contingencies (Krasner, 1971), and the differences in stimulus conditons across phases (Kazdin and Bootzin, 1972). Carryover effects in behavioral research are an obvious "clinical" advantage, but pose a problem experimentally as the controlling effects of procedures are then obfuscated.

Proponents of the group comparison approach (e.g., Bandura, 1969) contend that the presence of carryover effects in single case research is one of its major shortcomings as an experimental strategy. Both in terms of drug evaluation (Chassan, 1967) and with respect to behavioral research (Bijou, Peterson, Harris, Allen, and Johnston, 1969), short periods of experimentation (application of the treatment variable) are recommended to counteract these difficulties. Examining the problem from the operant framework, Bijou *et al.* (1969) argue that "In studies involving stimuli with reinforcing properties, relatively short experimental periods are advocated, since long ones might allow enough time for the establishment of new conditioned reinforcers" (p. 202).

A major difficulty in carrying out meaningful evaluations of drugs on behavior using single case methodology involves their carryover effects from one phase to the next. This is most problematic when withdrawing active drug treatment (B phase) and returning to the placebo (A_1 phase) condition in the A-A_1-B-A_1-B design. With respect to such effects, Chassan (1967) points out that "This, for instance, is thought likely to be the case in the use of monoaminoxidase inhibitors for the treatment of depression" (p. 204). Similarly, when using phenothiazine derivatives, the experimenter must exercise caution inasmuch as residuals of the drugs have been found to remain in body tissues for extended periods of time (as long as 6 months in some cases) following their discontinuance (Ban, 1969).

However, it is possible to examine the short-term effects of phenothiazines on designated target behaviors (Liberman, Davis, Moon, and Moore, 1973), but it behooves the experimenter to demonstrate, via blood and urine laboratory studies, that controlling effects of the drug are "truly" being demonstrated. That is to say, correlations (statistical and graphic data patterns) between behavioral changes and drug levels in body tissues should be demonstrated across experimental phases.

Despite the carryover difficulties encountered with the major tranquilizers and anti-depressants, the possibility of conducting *extended studies* in long-term facilities should be explored, assuming that high ethical and experimental standards prevail. In addition, study of the *short term* efficacy of the minor tranquilizers and amphetamines on selected target behaviors is quite feasible.

Cyclic variations

A most neglected issue in experimental single case research is that of cyclic variations (see Chapter 2, Sections 2.2 and 2.3, for a more general discussion of variability). Although the importance of cyclic variations is given attention by Sidman (1960) with respect to basic animal research, the virtual absence of similar discussions in the applied literature is striking. This issue is of paramount concern when using adult female subjects as their own controls in short-term (1 month or less) investigations. Despite the fact that the effects of the estrus cycle on behavior is given some consideration by Chassan (1967), he argues that "... a 4-week period (with random phasing) would tend to distribute menstrual weeks evenly between treatments" (p. 204). However, he does recognize that "The identification of such weeks in studies involving such patients would provide an added refinement for the statistical analysis of the data" (p. 204).

Whether examining drug effects or behavioral interventions, the implications of cyclic variation for single case methodology are enormous. Indeed, the psychiatric literature is replete with examples of the deleterious effects (leading to increased incidence of psychopathology) of the premenstrual and menstrual phases of the estrus cycle on a wide variety of target behaviors in pathological and nonpathological populations (e.g., Dalton, 1959, 1960a, 1960b, 1961; Glass, Heninger, Lansky, and Talan, 1971; Mandell and Mandell, 1967; Rees, 1953).

To illustrate, we will consider the following possibility. Let us assume that alternating placebo and active drug conditions are being evaluated (1 week each per phase) on the number of physical complaints issued daily in a young hospitalized female. Let us further assume that the first placebo condition coincides with the pre-menstrual and early part of the subject's menstrual cycle. Instigation of the active drug would then be confounded with cessation of the subject's menstrual phase. Assuming that resulting data suggest a decrease in somatic complaints, it is *entirely possible* that such change is primarily due to correlated factors (e.g., effects of the different portions of the subject's menstrual cycle). Of course, completion of the last two phases (A and B) of this A-B-A-B design might result in no change in data patterns across phases. However, interpretation of data would be complicated unless the experimenter were aware of the role placed by cyclic variation (i.e., the subject's menstrual cycle).

The use of extended measurement phases under these circumstances in addition to direct and systematic replications (see Chapter 9) across subjects is absolutely necessary in order to derive meaningful conclusions from the data.

3.7. EVALUATION OF IRREVERSIBLE PROCEDURES

There are certain kinds of procedures (e.g., surgical lesions, therapeutic instructions) that obviously cannot be withdrawn once they have been applied. Thus, when assessing these procedures in single case research, the use of reversal and withdrawal designs is generally precluded. The problem of irreversibility of behavior has attracted some attention and is viewed as a major limitation of single case design by some (e.g., Bandura, 1969). The notion here is that some therapeutic procedures produce results in "learning" that will not reverse when the procedure is withdrawn. Thus, one is unable to isolate that procedure as effective. In response to this, some have advocated withdrawing the procedure early in the treatment phase to effect a reversal. This strategy is based on the hypothesis that behavioral improvements may begin as a result of the therapeutic technique but are maintained at a later point by factors in the environment that the investigators cannot remove (see Kazdin, 1973; Leitenberg, 1973; Chapter 5). The most extreme cases of irreversibility may be a study of the effects of surgical lesions on behavior, or psychosurgery. Here the effect is clearly irreversible. This problem is easily solved, however, by turning to a multiple baseline design. In fact, the multiple baseline strategy is ideally suited for studying such variables in that withdrawals of treatment are not required to show the controlling effects of particular techniques (Baer, Wolf, and Risley, 1968; Barlow and Hersen, 1973; Johnston, 1972; Kazdin, 1973; Kazdin and Bootzin, 1972; Leitenberg, 1973). A complete discussion of issues related to the varieties of multiple baseline designs currently being employed by applied researchers appears in Chapter 7.

In this section, however, the limited use and evaluation of therapeutic instructions in withdrawal designs will be examined and illustrated. Let us consider the problems involved in "withdrawing" therapeutic instructions. In contrast to a typical reinforcement procedure, which can be introduced, removed, and reintroduced at will, an instructional set, after it has been given, technically cannot be withdrawn. Certainly, it can be stopped (e.g., Eisler, Hersen, and Agras, 1973) or changed (Agras, Leitenberg, Barlow, and Thomson 1969; Barlow, Agras, Leitenberg, Callahan, and Moore, 1972), but it is not possible to remove it in the same sense as one does in the case of reinforcement. Therefore, in light of these issues, when examining the interacting effects of instructions and other therapeutic variables (e.g., social reinforcement), instructions are typically maintained constant across treatment phases while the therapeutic variable is introduced, withdrawn, and reintroduced in sequence (Hersen, Gullick, Matherne, and Harbert, 1972).

Exceptions

There are some exceptions to the above that recently have appeared in the psychological literature. In two separate studies the short-term effects of instructions (Eisler, Hersen, and Agras, 1973) and the therapeutic value of instructional sets (Barlow, Leitenberg, Agras, Callahan, and Moore, 1972) were examined in withdrawal designs. In one of a series of analogue studies, Eisler, Hersen and Agras (1973) investigated the effects of focused instructions ("We would like you to pay attention as to how much you are looking at each other") on two nonverbal behaviors (looking and smiling) during the course of 24 minutes of free interaction in three married couples. An A-B-A-B design was used, with A consisting of 6 minutes of interaction videotaped between a husband and wife in a small television studio. The B phase also involved 6 minutes of videotaped interaction, but focused instructions on looking were administered three times at 2-minute intervals over a two-way intercom system by the experimenter from the adjoining control room. During the second A phase, instructions were discontinued, while in the second B they were renewed, thus completing 24 minutes of taped interaction.

Retrospective ratings of looking and smiling for husbands and wives (mean data for the three couples were used, as trends were similar in all cases) appear in Fig. 3-22. Looking duration in baseline for both spouses was moderate in frequency. In the next phase, focused instructions resulted in a substantial increase followed by a slightly decreasing trend. When instructions were discontinued in the second baseline, the downward trend was maintained. But reintroduction of instructions in the final phase led to an upward trend in looking. Thus, there was some evidence for the controlling effects of introducing, discontinuing, and reintroducing the instructional set. However, data for a second but "untreated" target behavior–smiling–showed almost no parallel effects.

Barlow, Leitenberg, Agras, Callahan, and Moore (1972) examined the effects of negative and positive instructional sets administered during the course of covert sensitization therapy for homosexual subjects. In a previous study (Barlow, Leitenberg, and Agras, 1969), pairing of the "nauseous" scene with deviant sexual imagery proved to be the controlling ingredient in covert sensitization. However, as the possibility was raised that therapeutic instructions or positive expectancy of subjects may have contributed to the treatment's overall efficacy, an additional study was conducted (Barlow, Leitenberg, Agras, Callahan, and Moore, 1972).

The dependent measure in the study by Barlow and his associates was mean percentage of penile circumference change to selected slides of nude males. Four homosexuals served as subjects in A-BC-A-BD single case designs. During A (baseline placebo), a positive instructional set was administered in that subjects

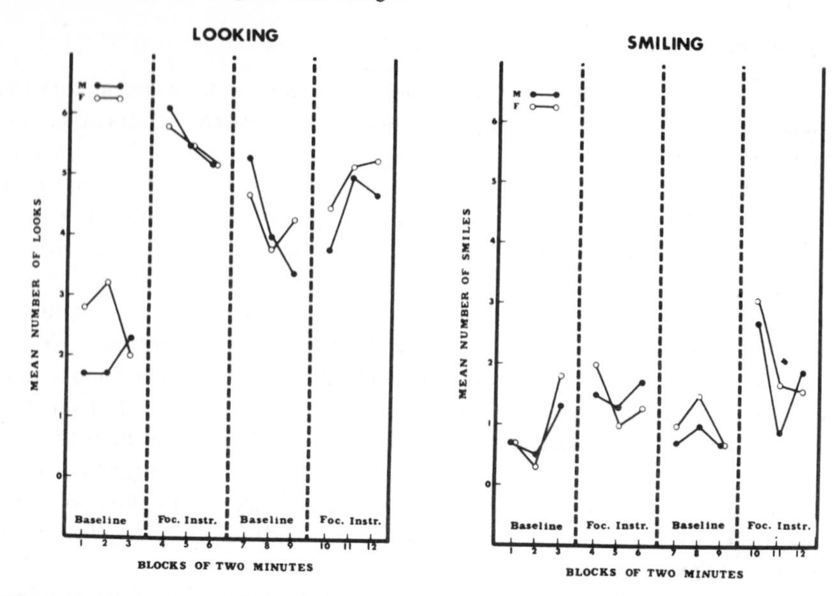

Fig. 3-22. Mean number of looks and smiles for three couples in 10-second intervals plotted in blocks of 2 minutes for the Focused Instructions Alone Design. (Fig. 4, p. 556, from: Eisler, R. M., Hersen, M., and Agras, W. S. Effects of videotape and instructional feedback on non-verbal marital interactions.: *Behavior Therapy*, 1973, **4**, 551-558. Reproduced by permission.)

were told that descriptions of homosexual scenes along with deep muscle relaxation would lead to improvement. In the BC phase standard covert sensitization treatment was paired with a negative instructional set (subjects were informed that increased sexual arousal would occur). In the next phase a return to baseline placebo conditions was instituted (A). In the final phase (BD) standard covert sensitization treatment was paired with a positive instructional set (subjects were informed that pairing of the "nauseous" scene with homosexual imagery, based on a review of their data, would lead to greatest improvement).

Mean data for the four subjects presented in blocks of two sessions appear in Fig. 3-23. Baseline data suggest that the positive set failed to effect a decreased trend. In the next phase (BC) a marked improvement was noted as a function of covert sensitization despite the instigation of a negative set. In the third phase (A) some deterioration was apparent although a positive set had been instituted. Finally, in the last phase (BD), covert sensitization coupled with positive expectation of treatment resulted in a renewed improvement.

In summary, data from this study show that covert sensitization treatment is the effective procedure and that therapeutic expectancy is definitely not the

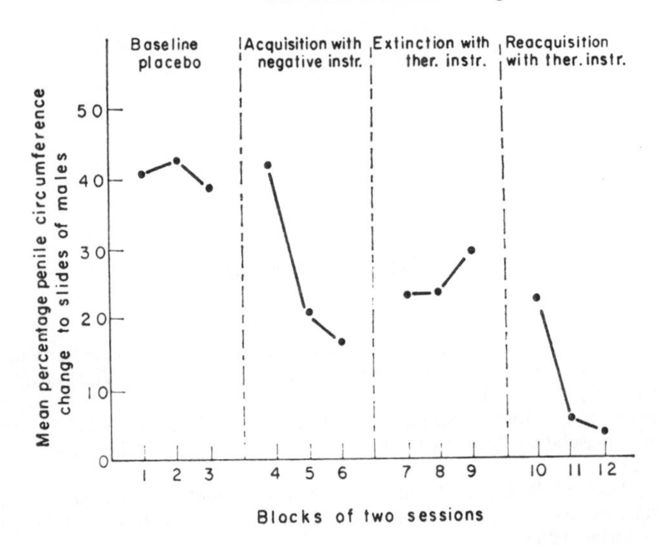

Fig. 3-23. Mean penile circumference changes to male slides for the four *S*s, expressed as a percentage of full erection. In each phase, data from the first, middle, and last pair of sessions are shown. (Fig. 1, p. 413, from: Barlow, D. H., Agras, W. S., Leitenberg, H., Callahan, E. J., and Moore, R. C. The contributions of therapeutic instruction to covert sensitization. *Behaviour Research and Therapy*, 1972, **10**, 411-415. Reproduced by permission.)

primary ingredient leading to success. To the contrary, a positive set paired with a placebo-relaxation condition in baseline did not yield improvement in the target behavior.

Although the design in this study permits conclusions as to the efficacy of positive and negative sets, a more direct method of assessing the problem could have been accomplished in the following design: (1) baseline placebo, (2) acquisition with positive instructions, (3) acquisition with negative instructions, and (4) acquisition with positive instructions. When labeled alphabetically, we then have an A-BC-BD-BC design. In the event that negative instructions were to exert a negative effect in the BD phase, a reversed trend in the data would appear. On the other hand, should negative instructions have no effect or a negligible effect, then a continued downward linear trend would appear across phases BC, BD and the return to BC.

References

Agras, W. S., Leitenberg, H., Barlow, D. H., and Thomson, L. E. Instructions and reinforcement in the modification of neurotic behavior. *American Journal of Psychiatry*, 1969, **125**, 1435-1439.

Allen, K. E., Hart, B., Buell, J. S., Harris, F. R., and Wolf, M. M. Effects of social reinforcement on isolate behavior of a nursery school child. *Child Development*, 1964, **35**, 511-518.

Ault, M. E., Peterson, R. F., and Bijou, S. W. The management of contingencies of reinforcement to enhance study behavior in a small group of young children. Unpublished manuscript, 1968.

Ayllon, T., and Azrin, N. H. *The token economy: A motivational system for therapy and rehabilitation.* New York: Appleton-Century-Crofts, 1968.

Baer, D. M., Wolf, M. M., and Risley, T. R. Some current dimensions of applied behavior analysis. *Journal of Applied Behavior Analysis*, 1968, **1**, 91-97.

Ban, T. *Psychopharmacology.* Baltimore: Williams and Wilkins, 1969.

Bandura, A. *Principles of behavior modification.* New York: Holt, Rinehart and Winston, 1969.

Barlow, D. H., Becker, R., Leitenberg, H., and Agras, W. S. A mechanical strain gauge for recording penile circumference change. *Journal of Applied Behavior Analysis*, 1970, **3**, 73-76.

Barlow, D. H., and Hersen, M. Single-case experimental designs: Uses in applied clinical research. *Archives of General Psychiatry*, 1973, **29**, 319-325.

Barlow, D. H., Leitenberg, H., and Agras, W. S. Experimental control of sexual deviation through manipulation of the noxious scene in covert sensitization. *Journal of Abnormal Psychology*, 1969, **74**, 596-601.

Barlow, D. H., Agras, W. S., Leitenberg, H., Callahan, E. J., and Moore, R. C. The contribution of therapeutic instruction to covert sensitization. *Behaviour Research and Therapy*, 1972, **10**, 411-415.

Begelman, D. A., and Hersen, M. An experimental analysis of the verbal-motor discrepancy in schizophrenia. *Journal of Clinical Psychology*, 1973, **31**, 175-179.

Bellak, L., and Chassan, J. B. An approach to the evaluation of drug effect during psychotherapy: A double-blind study of a single case. *Journal of Nervous and Mental Disease*, 1964, **139**, 20-30.

Bergin, A. E. Some implications of psychotherapy research for therapeutic practice. *Journal of Abnormal Psychology*, 1966, **71**, 235-246.

Bijou, S. W., Peterson, R. F., Harris, F. R., Allen, K. E., and Johnston, M. S. Methodology for experimental studies of young children in natural settings. *Psychological Record*, 1969, **19**, 177-210.

Browning, R. M., and Stover, D. O. *Behavior modification in child treatment: An experimental and clinical approach.* Chicago and New York: Aldine-Atherton, 1971.

Campbell, D. T., and Stanley, J. C. *Experimental and quasi-experimental designs for research.* Chicago: Rand McNally, 1966.

Chassan, J. B. *Research design in clinical psychology and psychiatry.* New York: Appleton-Century-Crofts, 1967.

Dalton, K. Menstruation and acute psychiatric illness. *British Medical Journal*, 1959, **1**, 148-149.

Dalton, K. Menstruation and accidents. *British Medical Journal*, 1960a, **2**, 1425-1426.

Dalton, K. School girl's behavior and menstruation. *British Medical Journal*, 1960b, **2**, 1647-1649.

Dalton, K. Menstruation and crime. *British Medical Journal*, 1961, **2**, 1752-1753.

Eisler, R. M., Hersen, M., and Agras, W. S. Effects of videotape and instructional feedback on nonverbal marital interaction: An analog study. *Behavior Therapy*, 1973, **4**, 551-558.

Eisler, R. M., Miller, P. M., and Hersen, M. Components of assertive behavior. *Journal of Clinical Psychology*, 1973, **29**, 295-299.

Elkin, T. E., Hersen, M., Eisler, R. M., and Williams, J. G. Modification of caloric intake in anorexia nervosa: An experimental analysis. *Psychological Reports*, 1973, **32**, 75-78.

Gentile, J. R., Roden, A. M., and Klein, R. D. An analysis-of-variance model for the intrasubject replication design. *Journal of Applied Behavior Analysis*, 1972, **5**, 193-198.

Glass, G. S., Heninger, G. R., Lansky, M., and Talan, K. Psychiatric emergency related to the menstrual cycle. *American Journal of Psychiatry*, 1971, **128**, 705-711.

Gottman, J. M. N-of-one and N-of-two research in psychotherapy. *Psychological Bulletin*, 1973, **80**, 93-105.

Gottman, J. M., McFall, R. M., and Barnett, J. T. Design and analysis of research using time series. *Psychological Bulletin*, 1969, **72**, 299-306.

Hawkins, R. P., Peterson, R. T., Schweid, E., and Bijou, S. W. Behavior therapy in the home. Amelioration of problem parent-child relations with the parent in a therapeutic role. *Journal of Experimental Child Psychology*, 1966, **4**, 99-107.

Hersen, M. Developments in behavior modification: An editorial. *Journal of Nervous and Mental Disease*, 1973a, **156**, 373-376.

Hersen, M. Self-assessment of fear. *Behavior Therapy*, 1973b, **4**, 241-257.

Hersen, M., Eisler, R. M., Alford, H. S., and Agras, W. S. Effects of token economy on neurotic depression: An experimental analysis. *Behavior Therapy*, 1973, **4**, 392-397.

Hersen, M., Eisler, R. M., and Miller, P. M. Development of assertive responses: Clinical, measurement, and research considerations. *Behaviour Research and Therapy*, 1973, **11**, 505-522.

Hersen, M., Eisler, R. M., Miller, P. M., Johnson, M. B., and Pinkston, S. G. Effects of practice, instructions, and modeling on components of assertive behavior. *Behaviour Research and Therapy*, 1973, **11**, 443-451.

Hersen, M., Gullick, E. L., Matherne, P. M., and Harbert, T. L. Instructions and reinforcement in the modification of a conversion reaction. *Psychological Reports*, 1972, **31**, 719-722.

Johnston, J. M. Punishment of human behavior. *American Psychologist*, 1972, **27**, 1033-1054.

Kazdin, A. E. Methodological and assessment considerations in evaluating reinforcement programs in applied settings. *Journal of Applied Behavior Analysis*, 1973, **6**, 517-531.

Kazdin, A. E., and Bootzin, R. R. The token economy: An evaluative review. *Journal of Applied Behavior Analysis*, 1972, **5**, 343-372.

Krasner, L. The operant approach in behavior therapy. In A. E. Bergin and S. L. Garfield (Eds.), *Handbook of psychotherapy and behavior change: An empirical analysis.* Pp. 612-652. New York: Wiley, 1971.

Lang, P. J. Fear reduction and fear behavior: Problems in treating a construct. In J. M. Schlien (Ed.), *Research in Psychotherapy.* Vol. 3. Pp. 90-101. Washington, D. C.: American Psychological Association, 1966.

Leitenberg, H. The use of single-case methodology in psychotherapy research. *Journal of Abnormal Psychology*, 1973, **82**, 87-101.

Leitenberg, H., Agras, W. S., Thomson, L., and Wright, D. E. Feedback in behavior modification. An experimetal analysis in two phobic cases. *Journal of Applied Behavior Analysis*, 1968, **1**, 131-137.

Liberman, R. P., Davis, J., Moon, W., and Moore, J. Research design for analyzing drug-environment-behavior interactions. *Journal of Nervous and Mental Disease*, 1973, **156**, 432-439.

Mandell, A. J., and Mandell, M. P. Suicide and the menstrual cycle. *Journal of the American Medical Association*, 1967, **200**, 792-793.

McFall, R. M. Effects of self-monitoring on normal smoking behavior. *Journal of Consulting and Clinical Psychology*, 1970, **35**, 135-142.

McNamara, J. R., and MacDonough, T. S. Some methodological considerations in the design and implementation of behavior therapy research. *Behavior Therapy*, 1972, **3**, 361-378.

Miller, P. M. An experimental analysis of retention control training in the treatment of nocturnal enuresis in two institutionalized adolescents. *Behavior Therapy*, 1973, **4**, 288-294.

Orne, M. T. On the social psychology of the psychological experiment: With particular reference to demand characteristics and their implications. *American Psychologist*, 1962, **17**, 776-783.

Pendergrass, V. E. Timeout from positive reinforcement following persistent, high-rate behavior in retardates. *Journal of Applied Behavior Analysis*, 1972, **5**, 85-91.

Peterson, R. F., and Peterson, L. The use of positive reinforcement in the self-control of self-destructive behavior in a retarded boy. *Journal of Experimental Child Psychology*, 1968, **6**, 351-360.

Ramp, E., Ulrich, R., and Dulaney, S. Delayed timeout as a procedure for reducing disruptive classroom behavior: A case study. *Journal of Applied Behavior Analysis*, 1971, **4**, 235-239.

Rees, L. Psychosomatic aspects of the premenstrual tension system. *Journal of Mental Science*, 1953, **99**, 62-73.

Revusky, S. H. Some statistical treatments compatible with individual organism methodology. *Journal of the Experimental Analysis of Behavior*, 1967, **10**, 319-330.

Risley, T. R., and Wolf, M. M. Strategies for analyzing behavioral change over time. In J. Nesselroade and H. Reese (Eds.), *Life-span developmental psychology: Methodological issues.* Pp. 175-183. New York: Academic Press. 1972.

Sidman, M. *Tactics of scientific research: Evaluating experimental data in psychology.* New York: Basic Books, 1960.

Simkins L. The reliability of self-recorded behaviors. *Behavior Therapy*, 1971, **2**, 83-87.

Truax, C. B., Wargo, D. G., Carkhuff, R. R., Kodman, F., and Moles, E. W. Changes in self-concepts during group psychotherapy as a function of alternate sessions and vicarious therapy pre-training in institutionalized mental patients and juvenile delinquents. *Journal of Consulting Psychology*, 1966, **30**, 309-314.

Williams, J. G., Barlow, D. H., and Agras, W. S. Behavioral measurement of severe depression. *Archives of General Psychiatry*, 1972, **27**, 330-333.

Wolf, M. M., and Risley, T. R. Reinforcement: Applied research. In R. Glaser (Ed.), *The nature of reinforcement.* Pp. 316-325. New York: Academic Press, 1971.

Zeilberger, J., Sampen, S. E., and Sloane, H. N., Jr. Modification of a child's problem behaviors in the home with the mother as therapist. *Journal of Applied Behavior Analysis*, 1968, **1**, 47-53.

CHAPTER 4

Repeated Measurement Techniques

4.1. INTRODUCTION

The use of repeated measurements is one of the most outstanding features of the experimental single case strategy. However, with the exception of the papers by Bijou and his colleagues (Bijou, Peterson, and Ault, 1968; Bijou, Peterson, Harris, Allen, and Johnston, 1969), relatively little attention has been given to measurement considerations in the surveys on single case methodology (e.g., Baer, Wolf, and Risley, 1968; Barlow and Hersen, 1973; Browning and Stover, 1971; Hersen, 1973a; Kazdin, 1973; Leitenberg, 1973; Risley and Wolf, 1972; Thoreson, 1972; Wolf and Risley, 1971). Nevertheless, it is clear that the applied clinical researcher using single case experimental strategies is frequently confronted with a variety of measurement problems. Following an initial period of assessment, the researcher must select relevant target behaviors. These target behaviors, then, constitute the criteria for improvement. At times, the researcher may be specifically concerned with behavioral (motoric) indices of improvement, subjective (self-report) indices of improvement, or physiological indices of improvement. Moreover, he may be concerned with improvements in varying degrees in all three response systems and the relationships and covariations among the three (Hersen, 1973b; Lang, 1968; Paul, 1967).

Regardless of the particular system that is being monitored, the researcher will have to determine how frequently measures are to be taken in order to be assured that an adequate sampling is obtained. In addition to such general considerations, each type of measurement system poses some unique problems for the investigator. For example, when taking measures of motoric activity (e.g., frequency of social initiations in withdrawn schizophrenics), the issue of interobserver reliability (agreement) must be dealt with satisfactorily if the response under study cannot be measured via mechanical means (i.e., automatic recording devices). On the other hand, when subjective (self-report) measures are used as criteria, the experimenter must be concerned with their reliability in terms of self-monitoring effects (Simkins, 1971) and their validity when examined in the light of the "demand characteristics" operating in a specific therapeutic situation (Orne, 1962). These two issues are of lesser significance when physiological measures are employed, but here mechanical and equipment failures are more likely to be encountered.

A major question also arises as to whether scores obtained on direct or indirect (e.g., responses to projective tests) asessment techniques are to be used as dependent measures (see Cautela, 1968; Goldfried and Kent, 1972; Goldfried and Pomeranz, 1968; Kanfer and Saslow, 1969; Miller, 1973; Mischel, 1972; Scott and Johnson, 1972). Although we certainly acknowledge the possibility of using validated portions of indirect assessment techniques as dependent measures in single case experimentation (see Greenspoon and Gersten, 1967), our own biases for using *direct measurement techniques* (motoric, self-report, and physiological) will be reflected in this chapter. These biases are well supported by research data indicating the superiority of the direct measurement approach in predicting behavior (Mischel, 1968, 1971).

We might note at this point that the problems of assessment in general, and the measurement in applied research in particular, in themselves, are sufficiently broad and sufficiently complex so that entire books could be readily written on each of those subjects alone. However, it is obvious that in one chapter we will not be able to cover all aspects of assessment and measurement in fine detail. But, we will outline the major points at issue in developing good measurement systems that can be used effectively in single case strategies. We will also briefly examine the basic assumptions underlying indirect and direct measurement in addition to summarizing some of the research findings in that area. Finally, we will describe some of the more standardized techniques for assessing neurotic and psychotic disorders, interpersonal problems, alcoholism, sexual deviations, and child behavior disorders. Illustrations of measurement systems (motoric, self-report, and physiological) will be presented and analyzed with respect to their inherent advantages and disadvantages in single case experimental designs.

4.2. INDIRECT AND DIRECT MEASUREMENT

Underlying assumptions

There are some primary underlying assumptions that clearly differentiate indirect and direct measurement. By indirect measurement, we not only refer to the large variety of projective tests (e.g., Rorschach, TAT, Figure Drawings, Sentence Completion Tests, etc.) available to the psychological examiner, but we also include objective personality inventories such as the Minnesota Multiphasic Personality Inventory and the California Psychological Inventory, which contain disguised properties.

When using indirect measurement techniques, it is assumed that test responses are indicative or representative of more enduring personality traits and dispositions that may be evidenced in a wide divergence of stimulus situations. Therefore, a particular response elicited by projective materials is rarely examined in terms of its overt qualities, but it is frequently interpreted within the context of a complex theoretical structure (Abt, 1959). Indeed, Mischel (1972) refers to the system of indirect measurement as an "indirect-sign paradigm". Specifically, he points out that under this system

> It is usually assumed that the person's underlying dispositions are relatively generalized and will manifest themselves pervasively in diverse aspects of his behavior, especially when the situation is unstructured and ambiguous and the examiner's purpose is disguised, as in projective testing. Motives, needs, psychodynamics, complexes, basic attitudes, and other psychic forces are hypothesized as the underlying but not directly observable dispositions that generate numerous diverse manifestations, much as physical disease produces a host of clinical symptoms. (Mischel, 1972, p. 319)

The relationship of information obtained through indirect measurement techniques and the subsequent selection of therapeutic targets and therapeutic techniques is very often unclear. This is in direct contrast to typical medical practice, where frequently a one-to-one relationship is found between diagnosis and treatment. Goldfried and Kent (1972) note that when indirect measurement techniques are used, "The therapeutic approach employed in any given case is usually more a function of the therapist's orientation than it is psychological test findings. . ." (p. 410). Under these circumstances, assessment serves as an academic exercise without a functional purpose. In this connection Goldfried and Pomeranz (1968) contend that ". . . the psychoanalytically oriented therapist would argue that the dynamics which must be dealt with in therapy are universal; little in the way of initial assessment is needed to delineate them" (p. 78).

Although many projective tests were initially developed intuitively, later proponents of indirect techniques have been concerned with basic measurement issues such as reliability and validity (Goldfried and Kent, 1972). Unfortunately, however, indirect measurement techniques have frequently been used in clinical practice despite research data indicating their low predictive validity. For example, Goldfried and Kent (1972) stress the fact that although the interpretation of certain signs on the Bender–Gestalt test (Hutt and Briskin, 1960) was not given empirical support (Goldfried and Ingling, 1964), Hutt (1968), in his revised edition of the Bender–Gestalt manual, apparently discounted these research findings and still recommended the use of questionable interpretations. Chapman and Chapman (1969) also present evidence showing that clinicians using the Rorschach to assess homosexuality used certain signs on an intuitive basis, tended to disregard negative research findings, and rarely used empirically validated indices. It is little wonder, then, that empiricists such as Cronbach

(1956) have been highly critical of the way indirect techniques are being used. He states that "Assessors have been foolhardy to venture predictions of behavior in unanalyzed situations, using tests whose construct interpretations are dubious and personality theory that has more gaps than solid matter" (Cronbach, 1956, p. 174). When important research findings are so blatantly ignored, it becomes clear that scientific standards are no longer being applied to the assessment process but that the intuitive approach has achieved precedence.

Whereas in indirect measurement a particular response is interpreted in terms of a presumed underlying disposition, a response obtained through direct measurement is simply viewed as a sample of a larger population of similar responses elicited under those particular stimulus conditions (Mischel, 1972). Thus, it is hardly surprising that proponents of direct measurement favor the observation of individuals in their natural surroundings whenever possible. When such naturalistic observations are not feasible, analogue situations approximating naturalistic conditions may be developed to study the behavior in question (e.g., the use of a behavioral avoidance test to study the degree of fear of snakes). When neither of these two methods is available or possible, subjects' self-reports are used as criteria. However, it should be underscored that *self-reports are also used as independent criteria* and, at times, may be operating under the control of totally different sets of contingencies than those governing motoric responses (see Begelman and Hersen, 1973; Hersen, 1973b; Skinner, 1953).

Although some of the earlier behavior modifiers eschewed the use of verbal reports and focused primarily on changing rates of motoric behaviors, there is a growing recognition that cognitions, feelings, etc. should also be considered in the behavioral analysis, thus making them appropriate targets for direct modification (Goldfried and Pomeranz, 1968). In addition to the above, direct measurements of physiological response systems (e.g., muscle tension, heart rate, blood pressure, skin conductance, respiration, etc.) are used as criteria. We will describe the interrelationships of the three direct measurement systems in Section 4.6 of this chapter.

A major advantage of direct measurement is that complex interpretations involving an equally complex network of theoretical structures are not required. A further advantage is that predictive validity increases as the discrepancy is diminished between the behavior sampled and the behavior that is predicted. When observations are made under naturalistic conditions, the sampled behaviors and the predicted behaviors are the same. This also applies when the sampled behaviors and the predicted behaviors are elicited in analogue tasks. However, when a behavior sampled in an analogue situation is used to predict a behavior in the natural environment, there would be an expected decrease in predictive validity. Finally, when self-reports are used to predict motoric responses, the gap between the sampled behavior and the predicted behavior is markedly increased. Of course, if the verbal report is highly correlated with the motoric criterion, loss of predictive validity will be minimized.

Under the system of direct measurement, rates of specified behaviors within designated time periods may be used as dependent measures during pre-treatment assessment, actual treatment (ongoing assessment), and post-treatment assessment. Therefore, contrary to the obscure relationship that is characteristic of indirect assessment procedures and treatment, when using direct measurement techniques (as is the case in the behavioral analysis), there is a close relationship between target behaviors selected for modification and the actual process of modification (treatment). Goldfried and Pomeranz (1968) point out that "The selection of any given procedure may be in part a function of the target behavior or situational determinant in need of modification" (p. 82). Similarly, Goldfried and Kent (1972) note that "... the interest in behavioral assessment has been generated by its utility for providing the information essential to the selection and implementation of appropriate behavior modification procedures" (p. 410). Cautela (1968) and Kanfer and Saslow (1969) document in detail the importance of conducting a complete behavioral analysis prior to undertaking modification of a specified behavior or series of behaviors. Cautela (1968) specifically warns the clinician that "The mistaken notion of assessment in Behavior Therapy had led some traditionally trained clinical psychologists, as well as some experimental psychologists, to assume that specific training in behavioral assessment is not a requisite for the practice of Behavior Therapy" (p. 175). Effective behavior modification obviously cannot be implemented without the appropriate selection of target behaviors and a full understanding of all elements contributing to their rates of emission.

Predictive validity

There is an extensive literature on the comparative validity (predictive) of indirect measurement techniques (Mischel, 1968, 1971). Most of these studies involve comparisons of self-report data with data obtained through projective tests and personality inventories. Data obtained via indirect and direct techniques are correlated with independent criteria (e.g., peer ratings, grades, laboratory tasks, etc.). In general, the findings suggest that predictions made on the basis of self-reports are equal or superior to those made on the basis of indirect measurement techniques that are interpreted and scored by "clinical experts." Mischel (1972) indicates that these findings hold true for a wide diversity of content areas (they range from college achievement to successful outcome in psychotherapy).

We will briefly review some of the aforementioned findings. For example,

Carroll (1952) shows that self-reports resulted in higher correlations with an outside criterion than scores obtained on the Guilford–Martion Personnel Inventory. Lindzey and Tejessey (1956) compared self-ratings of aggression and signs of aggression ascertained from TAT stories with criterion determinations of aggression. The results indicated that self-ratings correlated significantly higher with these independent criteria than scores obtained on the TAT. The predictive accuracy of self-reports and projective indices was also evaluated in a study conducted by Wallace and Sechrest (1963). Using peer ratings relating to achievement, hostility, and somatic and religious concerns, it was found that self-ratings correlated ·57 with peer ratings, whereas correlations between projective measures and peer ratings ranged from ·05 to ·14. Similarly, Peterson (1965) demonstrated that two extremely simple self-rating measures could be used as effectively as factor scores derived from complex personality rating schedules.

Comparable results have been reported in more recently conducted investigations (Hase and Goldberg, 1967; Holmes and Tyler, 1968; Scott and Johnson, 1972; Sherwood, 1966). A careful examination of these studies has led Mischel (1972) to conclude that "Given the overall findings regarding the limitations of the indirect-sign paradigm, assessors may want to look more at what people do and say, not as these events serve as indirect signs of underlying dispositions, but rather as direct samples of their behavior" (p. 323). This statement is fully congruent with the theory and principles underlying behavioral diagnosis (Cautela, 1968; Kanfer and Saslow, 1969).

Examiner bias

Not only are there major problems in terms of predictive validity for indirect measurement techniques, but the responses elicited in the test situation are subject to a host of situational and examiner influences. Thus, the specific responses obtained in a given situation rather than reflecting underlying traits and dispositions may simply represent the effects of overt and covert examiner manipulations. The influence of situational and interpersonal variables in the projective test situation was initially reviewed in detail by Masling (1960) some 15 years ago. Since then, additional data have been adduced demonstrating that the psychological examiner is, at best, an imperfect instrument (e.g., Hamilton and Robertson, 1966; Harris and Masling, 1970; Hersen, 1970; Hersen and Greaves, 1971; Marwit, 1969; Marwit and Marcia, 1967; Masling, 1965; Masling and Harris, 1969; Simkins, 1960; Turner and Coleman, 1962). Indeed, he is able to affect both the quantity and quality of his examinees' test responses through his actions.

A few of the above-mentioned investigations will be surveyed to illustrate the points at issue. In a retrospective study of a psychological clinic's test files, Masling and Harris (1969) showed that male examiners had administered significantly more TAT cards that would elicit sexual-romantic themes to female clients than to male clients. This cross-sex discrimination was not made by female examiners. In a subsequent study (again using a retrospective analysis of phychological clinic test files), Harris and Masling (1970) found that male examiners elicited a significantly greater number of Rorschach responses (total N) from female than male clients. These data were corroborated by Hersen (1970) in an independent study in which the files of a state hospital psychology department were examined retrospectively. In both of these studies (Harris and Masling, 1970; Hersen, 1970) this kind of cross-sex discrimination was not evidenced by female examiners.

In attempting to account for how examiner bias might occur in the projective test situation, Hersen and Greaves (1971) used the verbal reinforcement paradigm as an analogue. In this study they demonstrated that the contingent application of verbal reinforcement during Rorschach administrations could be used to increase total number of responses, animal responses, and human responses for selected groups when compared to a no-reinforcement control. These results confirmed Hamilton and Robertson's (1966) earlier findings showing that the examiner's "warmth" could result in "productive" Holtzman Inkblot Technique protocols, and Masling's (1965) investigation in which examiners were influenced on an *a priori* basis to increase subjects' productivity of animal and human responses on the Rorschach.

In reviewing these data, Hersen and Greaves (1971) acknowledged that "mature" examiners operating under clinical conditions were unlikely to reinforce and bias their test results in so blatant a fashion (i.e., when contrasted to data obtained in analogue situations). However, it would appear that the projective test situation is such that it permits many opportunities for the occurrence of extra-test influences. This, then, might lead proponents of direct measurement to point up the superiority of their techniques. But it should be emphasized that direct measurement techniques are also subject to a variety of biases, even though they may be different. For example, when naturalistic observations are obtained, subjects' responses may be influenced by their knowledge that they are being observed (e.g., Moos, 1968; Patterson and Harris, 1968). Goldfried and Kent (1972) argue that "Unless these observations are unobtrusive, the assumption that the behavior remains unaffected by the assessment procedures may turn out to be faulty" (p. 414). Similarly, when self-reports are used as criteria, the issues of faking, impression management, and conformity to experimental demands must be carefully evaluated. Extra-measurement variables interfering with direct assessment procedures are as troublesome for the clinician and researcher as those found in the projective test situation (see Sections 4.3 and 4.4). However, controls are somewhat easier to apply and monitor when using the techniques of direct measurement.

Practical considerations

Aside from the theoretical and empirical considerations relating to the indirect and direct systems of measurement, a question arises as to the suitability of each system when applied to the single case experimental strategy. As the single case experimental strategy has been primarily associated with behavior modification, it is readily understood why direct measurement of behaviors has been favored. When indirect measurement is contrasted with direct measurement, it becomes obvious that the direct techniques are more rapid, less troublesome to administer, easier to score, and considerably less costly. These are important issues as repeated measurements are frequently taken over extended periods of time (weeks to months) in some single case studies.

The practicality of repeating Rorschach, TAT, or MMPI administrations over extended time periods is seriously questioned. Of course, if selected probes are made during designated baseline and experimental phases, the possibility of using indirect techniques is enhanced. However, a distinction then has to be made between *probe measures* (e.g., measurement taken once a week) and *repeated measures* (e.g., measurement taken daily). Certainly, when *continuous measures* are needed (e.g., 24-hour monitoring of activity levels in patients [McFarlain and Hersen, 1974], sleep tracings in nightmare subjects [Hersen, 1972]), indirect techniques are totally useless.

Therefore, based on the foregoing analysis, we must conclude that indirect techniques of measurement have a very limited utility in experimental single case paradigms. Although we have urged "traditional" clinicians to examine the efficacy of their techniques in single case paradigms, it would appear that they would have to select objective criteria to determine improvement and that these criteria would have to be amenable to techniques of direct assessment.

4.3. MOTORIC MEASURES

In their quest for precision in measurement, applied behavioral researchers have used a large variety of motoric behaviors as dependent variables in both single subject and group comparison designs. Following the example set by their non-applied colleagues, applied researchers have tried to automate the measurement process whenever it was feasible. Thus, complicated apparatus consisting of electronic circuitry connected to cumulative recorders are frequently part of the laboratory scene in which behavioral assessments (during baseline and experimental phases) are conducted (see Schwitzgebel, 1968).

For the sake of descriptive purposes, we will refer to this type of measurement procedure as "automated-quantitative." The obvious advantage of automated-quantitative recording devices is that the error element found in human observations is eliminated (Simkins, 1971; Tighe and Elliot, 1968). In that sense, most recording bias is absent during all phases of experimentation. Examples of automated-quantitative systems of measurement for motoric responses include: Mills, Agras, Barlow, and Mills' (1973) "washing pen" for measuring frequency of handwashing rituals in severe obsessive-compulsives, the BRS-Lehigh Valley Modular Human Test System for measuring operant rates of drinking in alcoholics and social drinkers (Miller, Hersen, Eisler, and Hilsman, 1974), and McFarlain and Hersen's (1974) actometer used for measuring activity levels in psychiatric patients.

Other motoric measures do not involve automated recording but are relatively free from human error and bias (referred to as "non-automated-quantitative"). In this kind of measurement situation the task of the observer is simply to record data (from gauges) obtained through the use and application of standardized measurement devices (e.g., tape measures, scales, stopwatches, calorimeters, etc). Under these conditions, the problems in measurement are negligible unless errors in recording are made. Examples of recently employed non-automated-quantitative measures include: weight and caloric intake in anorexia nervosa patients (Agras, Barlow, Chapin, Abel, and Leitenberg, 1974; Elkin, Hersen, Eisler, and Williams, 1973), amounts of alcoholic and non-alcoholic beverages consumed by alcoholics in an analogue "taste-test" situation (Miller and Hersen, 1972; Miller, Hersen, Eisler, and Hemphill, 1973), time spent in phobic situations (Leitenberg, Agras, Butz, and Wincze, 1971), and distance approached toward the phobic stimulus in a behavioral avoidance task for snake phobic subjects (Begelmen and Hersen, 1973).

Despite the advances that have been made with respect to the development and refinement of instrumental recording devices, the bulk of the research presently being conducted in many laboratories requires the use of human observers ("observational-quantitative"). For example, most interactive or social behaviors, particularly under naturalistic conditions, require the use of trained observers who are making judgments (presence or absence of specified behaviors, rates of particular behaviors) in accordance with predetermined behavioral codes. Baer, Wolf, and Risley (1968) initially argued that "The reliable use of human beings to quantify the behavior of other human beings is an area of psychological technology long since well developed, thoroughly relevant, and very often necessary to applied behavior analysis" (p. 93). Unfortunately, Baer, Wolf, and Risley may have been somewhat overly optimistic in their faith in the human recorder, and they may have grossly underestimated the importance of the problems that reliability in measurement pose for the experimental single case strategy (see Johnson and Bolstad, 1973; O'Leary and Kent, 1973; Skindrud, 1973).

The numerous problems involved in human observation and tabulation of motoric responses will be examined in the following subsections. We will be concerned specifically with the observational code, the training of observers, methods of assessing inter-observer reliabilities, and the frequency with which observations are made (Bijou, Peterson, Harris, Allen, and Johnston, 1969).

Observational code

The precision and specificity of the observational code is related to the degree that inter-observer agreement can be achieved. In the initial phases of study (pre-baseline analyses) broad categories of behaviors are examined and some general definitions of these categories may be derived. As the pre-study analysis continues, specific behaviors will be selected as targets, and the definitions will be narrowed so as to avoid overlapping categories. Bijou, Peterson, Harris, Allen, and Johnston (1969) similarly point out that "Initial definitions are likely to be crude and in some cases are refined after evaluations in the actual observational situation. In final form the descriptions of responses in the code should be comprehensive, delineating clearly all responses in each category. It is also desirable to develop codes so that each response class is mutually exclusive–there is no overlapping of definitions" (p. 193).

Some of the most elaborate and carefully worked out codes have been developed by child psychologists (developmental) and child behavior modifiers (e.g., Bijou, Peterson, and Ault, 1968; Bijou, Peterson, Harris, Allen, and Johnston, 1969). In addition to the cited work by Bijou and his colleagues, the reader is referred to the pages of the *Journal of Applied Behavior Analysis* for excellent examples of codes (general and specific) constructed in accordance with the particular problems and populations under consideration. However, let us present two examples to illustrate the question of specificity of definitions. For example, Hart, Allen, Buell, Harris, and Wolf (1964) defined crying behavior in a young child in such a manner that this behavior could be easily differentiated from whining and screaming. In addition to a topographical description of the behavior, elements of time and distance were incorporated to further ensure precision. That is to say, a duration of at least 5 seconds heard within a 50-foot radius was required before the response could be scored as "crying." Turning now to some examples selected from the adult interactive literature,

Eisler, Hersen, and Agras (1973a) defined eye contact and smiling in couples as follows: "... when couples were seated side by side a looking response would in most instances be preceded by a head turn of at least 45° toward the partner with eyes focused somewhere between the top of the head and chin. Smiling responses in most cases required a crease to appear in the subject's cheek, usually but not always with teeth showing" (p. 42). Looking and smiling responses were each scored on an occurrence or nonoccurrence basis within a 10-second time period.

As noted above, some observational codes involve tabulation of responses simply on an occurrence or nonoccurrence basis within a designated time period. When this system of coding is used, the size of the time interval selected for observation should bear close relationship to the natural frequency of the event being tabulated. Thus, for a high-frequency response, short units of measurement should be used to ensure correspondence between tabulated frequencies and actual occurrences of the behavior(s) being observed. On the other hand, when the behavior under study is emitted at a low rate, the time units should be extended in order to better reflect correspondence between actual and recorded frequencies.

In some studies it may be more useful to tabulate total frequencies and durations over extended periods of time (e.g., total frequencies of several behaviors in a daily one-hour period of observation). When this type of code is employed, electromechanical recording devices (a series of pen recorders, with each of the pen recordings representing a separate behavior) may allow the recorder "... to assess more carefully the temporal relationship between stimulus and response events, as well as to record a large number of responses within a given period. On the other hand, paper-and-pencil recording methods are more flexible. They can be used in any setting since they do not require special facilities such as a power supply" (Bijou, Peterson, and Ault, 1968, p. 180). Obviously, the choice between occurrence and nonoccurrence codes and codes involving recordings of total frequencies and durations is determined by the behaviors under study in addition to the conditions under which observations are made (e.g., naturalistic, analogue, videotaped segments, etc.). Similarly, the choice between paper-and-pencil versus electromechanical recording devices is determined by these factors.

Observer training

Observers should be given numerous opportunities to practice before they are required to make recordings in an actual study situation. Usually, they are expected to achieve a level of at least 80 percent agreement during practice trials. However, this criterion may vary from one study to the next, in part, depending on the complexity and quantity of observations that are simultaneously being made. Familiarity with the coding system and the tools used for recording (e.g., data sheets, stopwatches, other timing devices, etc.) will increase the likelihood of obtaining high inter-observer agreement. Careful monitoring by the investigator in addition to the judicious use of feedback during training will also enhance the probability of agreement. If possible, videotaped segments of the behavior under study can be used to standardize and facilitate the observational process (see Bernhardt, Hersen, and Barlow, 1972; Eisler, Hersen, and Agras, 1973a, 1973b; Hersen, Eisler, Miller, Johnson, and Pinkston, 1973; Hersen, Miller, and Eisler, 1973).

An additional method used to ensure reliability of recording is to alternate observation periods with recording periods (O'Leary, O'Leary, and Becker, 1967). Thus, instead of both observing and recording occurrences and nonoccurrences of a particular behavior for consecutive 20-second time periods, observations might be conducted for 20 seconds while the following 10 seconds are used for recording purposes alone. This technique of "non-continuous" observation controls for an observer's possible inattention while recording a particular event.

When observers are placed in naturalistic situations (i.e., where they might be in the same room as the subjects being studied), they will require further instructions with respect to their discouraging subjects from interacting with them (Bijou, Peterson, and Ault, 1968; Bijou, Peterson, Harris, Allen, and Johnston, 1969; Eisler, Hersen, and Agras, 1973a; Johnson and Bolstad, 1973). Indeed, Johnson and Bolstad (1973) note that "Presumably, the more novel and conspicuous the agent of observation, the more distracting are the effects upon the individuals being observed. It would also follow that longer habituation periods would be required for more distracting observational agents in order to achieve stability of data" (p. 45). Not only does the observer affect the behavior of his subject (Moos, 1968) but, in turn, his subject's attempts at interaction (particularly in the case of disruptive children) will interfere with the reliability of the recording process. Of course, one possible solution to this problem is to install audio and videotape equipment in such natural settings and to rate behaviors retrospectively from recordings. Although there are many advantages to this type of recording–namely, that taped segments can be randomized and rated blindly by observers after the experiment has been completed–there are some situations where audio and videotaped recordings in natural environments will not be feasible for a variety of technical reasons. Under these circumstances, extensive training and equally extensive monitoring of the observers should follow.

Calculating reliability coefficients

The careful investigator should be aware that the reliability index not only reflects the degree of inter-observer agreement attained, but also is a function of the type of reliability index used (i.e., how was the reliability coefficient calculated?). Let us consider the following hypothetical example in which two independent observers are recording total number of involuntary movements in a ticquer over a 1-hour period. Observer A records the occurrence of 100 tics while Observer B records the occurrence of 80 tics. One then divides the smaller sum (80) by the larger sum (100) to obtain a percentage of agreement score (80 percent). Although 80 percent agreement might be considered satisfactory in most investigations, when examining large-frequencies over extended periods of time, the possibility exists that the two observers were not recording the same behaviors at precisely the same time. That is, Observer A may have under-recorded in the first one-half hour and over-recorded in the second one-half hour, while Observer B may have followed the reverse pattern. However, the total N recorded by both observers is in relative agreement.

One solution to this problem is to obtain a correlation coefficient based on the number of observations derived from smaller time intervals (e.g., ten 6-minute observation periods). However, when this method is used "... it is possible to obtain high coefficients of correlation when one observer consistently overestimates behavioral rates relative to the other observer. This difference can be rather large, but if it is consistently in one direction, the correlation can be quite high" (Johnson and Bolstad, 1973, p. 14). Indeed, it is possible to obtain a correlation coefficient that approaches 1·00, but where inter-observer agreement is actually quite poor.

Another solution is to use shorter segments of observation in which observer agreements and disagreements on occurrences and nonoccurrences of behavior can be determined. Bijou, Peterson, Harris, Allen, and Johnston (1969) recommend this approach and argue that "As the segments become shorter, one can determine with increasing confidence whether the observers are scoring the same events at the same time" (p. 195). When using such smaller segments, inter-observer agreement is computed by dividing the number of time both observers agreed on the occurrence of the event by the number of times they agreed and disagreed as to its occurrence. Using our ticquer, let us assume that over the first one-half of observation our two observers agreed on the occurrence of 18 tics but disagreed on the occurrence of 7 tics. When the agreement/agreement + disagreement ratio is calculated $(18/(18 + 7) = 18/25)$, the inter-observer agreement reached is 72 percent.

In the above-cited example, inter-observer agreement was calculated on the basis of the number of occurrences of a particular event. Agreement on the number of nonoccurrences was not included in the analysis. However, when a particular behavior is emitted at a low rate, there is the likelihood that the degree of agreement on occurrences may be low while the degree of agreement on nonoccurrences is conversely high. For example, let us assume that there were ten agreements and five disagreements on the occurrence of a particular behavior in a designated period of observation. This leads to a ratio of $10/(10+5) = 10/15 = 67$ percent. Let us also assume that there was inter-observer agreement on the nonoccurrence of the behavior for 30 of the time intervals examined. When recomputed in terms of agreement/(agreement + disagreements) *for both occurrences and nonoccurrences* of the behavior, we now have an entirely new ratio $(40/(40+5) = 40/45 = 89$ percent).

As previously noted, with a low-frequency behavior monitored over an extended period of time, inter-observer agreement for nonoccurrences will be higher than for high-frequency behaviors. A question then arises as to which of the aforementioned computational procedures is to be used. Bijou, Peterson, Harris, Allen, and Johnston (1969) contend that "This problem may be resolved by computing not one but two reliability coefficients, one for occurrence and one for non-occurrence" (p. 198). A further suggestion offered by Bijou and his colleagues is to list as agreements all occurrences that were scored in the adjacent intervals of observation. When this procedure is followed, the inter-observer reliability coefficient is significantly increased.

Regardless of the method that is elected for computing percentage of inter-observer agreement, it is suggested that base-rate chance agreements should also be computed, and that the difference between the obtained agreement and the chance agreement needs to be reported (Johnson and Bolstad, 1973). Methods for computing chance agreement are described by Johnson and Bolstad (1973); the reader is referred to their work for details of the computational procedures.

Monitoring observers

The importance of monitoring the activities of observers during various phases of experimentation in single case research has been examined very carefully in the recent applied behavioral literature (Bijou, Peterson, and Ault, 1968; Bijou, Peterson, Harris, Allen, and Johnston, 1969; Johnson and Bolstad, 1973; McNamara and MacDonough, 1972; O'Leary and Kent, 1973; Reid, 1970; Reid and DeMaster, 1972; Romanczyk, Kent, Diament, and O'Leary, 1973; Skindrud, 1973; Wahler and Leske, 1973). Not only are there initial problems in achieving acceptable levels of agreement between two observers, but the above-cited reports suggest that maintaining acceptable levels of agreement subsequent to training is equally problematic. Moreover, a question has been raised as to whether observers are actually making accurate judgments (i.e., are they recording real events?) in addition to agreeing with one another (Wahler and Leske, 1973). Bijou and his associates (1969) argue that "Even if the observational code employed is adequate, two observers will not necessarily record the occurrence of the same event at the same time unless each has learned his task thoroughly and has been instructed in conducting himself while recording" (p. 193).

Although observers may be quite reliable when they are aware that their recording activities are being monitored, the degree of agreement between any two observers may decrease dramatically if overt monitoring is terminated. This becomes a crucial issue when a study requires an extended period of observation and only spot checks for reliability are possible. For example, O'Leary and Kent (1973) contend that "... an overt reliability assessment performed at regular intervals throughout a study may not reflect the consistency of data generated on a day-to-day basis and that studies reporting reliabilities from ·70 to ·80 may be generating data which is actually reliable at levels from ·30 to ·40" (p. 80). The natural shift from unreliable observation to motoric behaviors has been labeled "observer drift." Let us illustrate this important measurement issue by briefly summarizing the results of three recently published studies.

In the first study Reid (1970) showed that agreement between observers decreased about 25 percent when they were monitored covertly (without their knowledge) as contrasted to their being monitored overtly (with their knowledge). In a subsequent study Reid and DeMaster (1972) replicated these results and, in addition, showed that there was a temporary but marked increase in agreement when overtly conducted spot checks were made during an extended period of covert monitoring. In still another study, Romancyzk, Kent, Diament, and O'Leary (1973) similarly found that during covert assessment procedures agreement of observation decreased significantly. Moreover, underestimates of the targeted behaviors were consistently made by the observers.

Other studies documenting the problem of "observer drift" are reviewed by Johnson and Bolstad (1973) and O'Leary and Kent (1973). Although these studies only reflect problems of obtaining observer agreement with respect to certain types of codes, there is no reason to assume that observer drift is not a more universal phenomenon. Of course, further research is needed to verify this point.

A number of solutions and precautionary measures for counteracting observer drift are possible, but some are obviously more practical (at least from a financial and man-hour point of view) than others. Certainly, the use of spot checks is superior to no checks (Reid and DeMaster, 1972). However, mere interposing of spot checks during various phases of the investigation will not be sufficient to maintain a continuously high level of agreement (as assessed by covert monitoring techniques). A second possibility is to maintain constant monitoring (overt) of observers, but this would undoubtedly be too costly and time-consuming. O'Leary and Kent (1973) have summarized some of the other possibilities as follows:

> In within-subject designs, the critical comparisons involve one experimental condition instituted at one time and another condition instituted subsequently. Assuming that observer drift is a random phenomenon, one might employ a number of independent observer groups across all experimental conditions. For example, if experimental conditions each lasted a week or longer, different groups of observers could be employed on each day of the week. Drift among groups would thus add to the variation of data from each condition but would not distort comparisons of one condition to another. An alternate procedure would involve video taping the behavior of interest during all experimental conditions and showing these recordings to observers in random order. When this is impractical, observation of video tapes of a sample behavior from each experimental condition would provide a measure of the veridicality of behavioral recording obtained *in vivo* across time. (p. 92)

Irrespective of the particular strategy that is used to ensure continuation of reliable observation and recording of data, it is apparent that the issue of reliability (*ipso facto*—the usefulness of the data obtained) has staggering implications for the kinds of conclusions that are to be derived when changes in motoric behaviors are studied in experimental single case designs. Just as the projective tester is an imperfect instrument, the human observer (even if the behaviors to be recorded are precisely described) is an equally imperfect instrument and must be monitored very closely. The occasional reliability check that is obtained during various phases of experimentation is hardly sufficient to ensure continuous reliability of the observers. Unfortunately, in some cases where obtained reliabilities may have been deficient, it is quite possible that erroneous conclusions have been reached. Once again, the importance of obtaining reliable observations of motoric behaviors cannot be emphasized too strongly. This issue undoubtedly deserves the same kind of care and attention that has been given to the selection and definition of target behaviors.

Observer bias

Akin to the issue of "observer drift" is the problem of observer bias in recording motoric behaviors. When observer bias is present, errors in recording are made systematically in a particular direction rather than appearing randomly around the mean. In attempting to identify the sources of observer bias the experimenter must consider both the possibilities that intentional and "unintentional" errors may be committed. That research assistants should ever consider faking or fabricating data is certainly most repugnant to the meticulous investigator. However, distasteful as the idea may appear, this source of bias cannot be summarily discounted (see Azrin, Holz, Ulrich, and Goldiamond, 1961; Rosenthal and Lawson, 1964; Verplanck, 1955). The most flagrant example of data fabrication was discovered by Azrin, Holz, Ulrich, and Goldiamond (1961) while attempting to replicate results originally obtained by Verplanck (1955) in a conditioning experiment. Although undergraduate experimenters were able to replicate Verplanck's (1955) findings, graduate student examiners attempting an additional replication were not successful. Just prior to the graduate students' unsuccessful replication, Azrin and associates (1961) conducted a casual post-experimental inquiry and found that the undergraduate experimenters had falsified their data.

Certainly, careful training with respect to the value of the scientific approach should counteract and discourage an observer's intent to fabricate results consistent with the experimenter's hypothesis. However, even if an observer is scrupulously honest in his recording, a more subtle process (expectancy) may be operating and may contribute to his making unintentional errors (Rosenthal, 1966). Consider the situation where the observer not only records the data but also performs experimental manipulations. Under these conditions, there can be little doubt that a strong possibility for bias is present, particularly as most experimenters communicate (directly or indirectly) their expectations and hypotheses to their research assistants. Although it is hardly recommended that an observer serve in the dual role of experimenter and recorder, this arrangement unfortunately represents the standard procedure practiced in many laboratories. In general, it is preferable that the observer remain unaware of the experimental hypothesis that is being tested. But this is unlikely when observations are required over extended periods of time. Once again, it would appear that the use of audio and videotaped recordings, presented to observers in randomized sequence, represents an ideal method to prevent the intrusion of biasing factors in the observational and recording processes.

Let us now briefly examine some of the research findings on the expectancy phenomenon. Although there is evidence that "expectancy" affects the manner in which the untrained observer performs his functions under laboratory conditions (Fode, 1960, 1965; Rosenthal and Fode, 1963), it is questionable whether well-trained and more sophisticated observers also will evidence such bias (Johnson and Bolstad, 1973). Moreover, the work of Rosenthal and his colleagues has been criticized on methodological grounds, particularly in light of the failure to replicate many of this group's findings (e.g., Barber and Silver, 1968).

Turning to the study of observer bias (expectancy) in the field-experimental situation, Kass and O'Leary (1970) conducted a study in which they experimentally manipulated bias through subtle instructional sets. Videotaped recordings of two child behavior problems were shown to three groups of female undergraduate observers. In the first group the observers were told that rates of disruptive behaviors would increase following mild reprimands from the teacher. In the second group a reverse expectancy was manipulated; these observers were told that rates of disruptive behaviors would decrease following mild reprimands from the teacher. In the third group no expectation was given with respect to the influence of reprimands on disruptive behaviors. The results of this study confirmed the presence of observer bias (i.e., a successful expectancy manipulation) in the two groups where instructional sets were given. However, subsequent attempts to replicate the Kass and O'Leary (1970) findings have proven unsuccessful (Kent, 1972; Skindrud, 1972). In summary, data on the presence or absence of observer bias as a function of expectancy are contradictory. Nevertheless, "Until more information is available on observer bias effects in naturalistic observation, it seems very critical to do everything possible to minimize the potential for their effects. Whenever possible, observers should not have access to information that may give rise to confounding consequences and should be encouraged to reveal the nature and source of any information they do receive" (Johnson and Bolstad, 1973, p. 37). In addition to the use of audio and videotaped recordings of motoric behaviors, O'Leary and Kent's (1973) suggestion with regard to employing teams of observers staggered over different stages of experimentation appears to be a viable solution when research funding is sufficiently large. Less costly procedures will have to be followed in the smaller laboratories.

Reliability and accuracy distinguished

As defined throughout, reliability of observation refers to the extent that two independent observers will agree on the occurrence or nonoccurrence of a specified behavior or set of behaviors. If the percentage of inter-observer agreement exceeds 80 percent, reliability is considered to be sufficiently high and it is presumed that the two observers have agreed upon the occurrence of a *real event.* However, let us consider the possibility that two observers obtain good concurrence on events that never took place. In this instance reliability, although sufficiently high, *does not* necessarily reflect the "accuracy" of their observations. Thus, two observers may be reliable but *inaccurate* in their ratings. Conversely, two observers may be accurate but *unreliable* in their ratings.

An experiment that clearly illustrates this point was recently conducted by Wahler and Leske (1973). In this study, fifteen videotapes of six children engaged in silent reading were prepared. Following a programed script, one of the six children was coached to exhibit distractible behaviors. These behaviors were faded from 75 percent of the time on the first tape to 15 percent of the time on the fifteenth tape. One phase of the study involved presenting these tapes out of sequence (tape 15 contained 15 percent distractible behaviors and followed tape 4, which contained 60 percent distractible behaviors) to two groups of untrained observers. The first group of untrained observers ("subjective") was asked to rate the child's distractibility for each of the 15 tapes on a 1-7 point scale. The second group of untrained observers was also asked to rate each of the tapes on a 1-7 point scale for distractibility. But this group of observers ("objective") was additionally permitted to keep a tally of specific behaviors. This tally was then used to enable raters to arrive at a judgment with respect to overall distractibility (1-7 point scale). Paradoxically, the results of this study showed that "subjective" observers were more *reliable* than their "objective" observer counterparts. By contrast, "objective" observers were more *accurate* than the "subjective" observers. Wahler and Leske (1973) rightly concluded that "One cannot glibly assume that a reliable observer is also an accurate observer" (p. 393).

A careful assessment of the Wahler and Leske (1973) investigation indicates that it is not only concerned with reliability and accuracy of ratings, observer bias, observer training, and monitoring of observers, but it is also concerned with the specificity of the behavioral code that each observer uses. Presumably, the more objective the code (i.e., the precision of definitions used by observers to arrive at overall ratings), the less likely an observer will make an inaccurate summary judgment. However, further research needs to be conducted in this area to clarify these issues, particularly with regard to improving the accuracy level attained by observers. Such research should also involve comparisons of trained and untrained observers in terms of the relationship between reliability and accuracy. *A priori* we would assume that problems regarding accuracy should be minimized for highly trained and carefully monitored observers.

4.4. SELF-REPORT MEASURES

Of the three response systems (motoric, self-report, physiological) that can be monitored in single case strategies, the self-report system is the one that is most subject to conscious distortion on the part of the patient or client. In this connection, it should be noted that in the adult treatment situation the patient's verbal report is most frequently accorded criterional status. Indeed, many clinicians (both behavioral and non-behavioral) rely heavily on patients' descriptions of their behaviors to assess progress in treatment. In his discussion of the self-assessment of fear, Hersen (1973b) argued that "Verbalizations of discomfort from distressed patients cannot be discounted either for clinical, ethical, or moral reasons" (p. 255). There is no doubt that the applied behavioral researcher should be sensitive to the changes and nuances in his patients' verbalizations. However, in an attempt to objectify these verbalizations, he is most likely to have his patients record and report on the rates of defined target behaviors. The applied behavioral researcher must also consider very carefully the possible sources of bias that may interfere with his patients' reliable recording and reporting of targeted behaviors.

In the following subsections we will examine some of these biasing factors. We will specifically outline the problems involved in using self-report measures throughout the course of an experimental single case investigation. Among the issues to be considered are: self-monitoring effects, demand characteristics, impression management, faking, and other more subtle forms of response bias.

Self-monitoring

Adult psychiatric patients, whose treatments are being evaluated in single case strategies, have often been asked to monitor and record specific behaviors on a daily basis, at times extending for weeks. These self-reported data have been tabulated and have served as the major, or one of the major, dependent variables in the study. A large number of self-report measures have been used in single case research, and include such varied behaviors as: total daily number of obsessive thoughts (Gullick and Blanchard, 1973), total daily number of gagging episodes (Epstein and Hersen, 1974), and total daily number of deviant sexual urges (Barlow, Leitenberg, and Agras, 1969).

When patients are requested to monitor their own behaviors, they are administered specific instructions regarding the frequency of and methods for recording data. Moreover, very precise descriptions of the behaviors are presented to enhance the reliability of recording. These descriptions are obtained through protracted discussions held between the patient and the experimental clinician. However, despite these attempts at standardizing procedures for self-monitoring, recent studies (Epstein, Miller, and Webster, 1974; Johnson and White, 1971; McFall, 1970; McFall and Hammen, 1971; Stollak, 1967) suggest that self-observation may be a *reactive* process. That is, by virtue of only monitoring and recording certain behaviors, the rates of these same behaviors will change despite the absence of "therapeutic" intervention. In fact, Johnson and White (1971) consider the possibility of using the reactivity to self-monitoring as a tool for engendering behavioral change.

Let us now examine two studies in which the mere presence of self-observation led to statistically significant rate changes in the behaviors being monitored. In an ingenious experiment, McFall (1970) covertly monitored base rates of smoking for two groups of subjects prior to their being asked to monitor and record their own rates of smoking under two instructional sets–(1) monitoring instances of smoking, (2) monitoring instances of non-smoking. Compared to base rate levels, subjects who monitored instances of smoking actually increased their smoking rates. By contrast, subjects who monitored instances of non-smoking decreased their rates of smoking. McFall (1970) concluded that "... this study has drawn attention to the methodological problems involved in asking *S*s to monitor and record their own behavior. Data collected in this manner are quite likely to show reactivity, or self-monitoring effects; thus, self-monitoring as a data-collection procedure should be employed only with a full awareness of its shortcomings" (p. 142). On the basis of this study, it is quite clear that the direction of reactivity is a function of the instructional set that is given to the subjects.

Johnson and White (1971) corroborated McFall's (1970) findings in an independently conducted investigation. In their study, Johnson and White (1971) asked one group of college subjects to self-observe study behaviors. A second group was asked to self-observe dating behaviors, while a third group served as a control. Grades obtained on weekly quizzes was the dependent measure. The results of this study indicated that in the latter portion of the academic term the study group achieved significantly higher grades than control subjects. Dating group subjects obtained lower grades than study group subjects but higher grades than the controls. However, neither of these differences was significant. Johnson and White (1971) concluded that their "... study helps confirm the suspicion that self-observation procedures may have important reactive effects on the observed behaviors, but perhaps more important, the results suggest that reactive effects may be used to produce desired behavior change" (p. 496).

When we consider the fact that subjects in these two investigations were not motivated to change their behaviors (Johnson and White, 1971; McFall, 1970), the implications are all the more considerable for motivated subjects who are making repeated self-observations as part of their treatment regime. Working under the assumption that self-observation is truly a reactive process, how should the single case researcher proceed when he is interested in using self-reports of discrete events as dependent measures? McFall's (1970) answer to this question is that under certain circumstances (i.e., when base rates are compared with treatment rates) it may not be important to have an uncontaminated baseline measure. Rather, the primary consideration should be that baseline data achieve stability "... and that such base rates be sensitive to subsequent interventions" (McFall, 1970, p. 141). However, if during the baseline assessment rates of behavior either increase or decrease systematically, then the investigator must examine the source of reactivity. Obviously, baseline measurement should be continued until a stable pattern emerges. If after extended measurement a particular trend continues, the investigator might consider manipulating the instructional set to ascertain its controlling effects over reactivity. Even when baseline stability is obtained, data taken from the treatment phase will represent a possible interaction between the treatment process and reactivity to self-monitoring.

One possible control is to obtain covert reliability checks on the self-monitoring process. But this kind of check will have only limited application, as many of the behaviors that might undergo self-observation are classified as "private events" (e.g., obsessive thoughts, depressive feelings, sexual urges). Moreover, covert reliability checks are difficult to engineer, even for discrete events that are open to public observation. With respect to private events, unless the subject's immediate reaction to each event can be defined and reliably observed, the use of covert reliability checks is not possible.

As previously mentioned, the experimenter must be extremely careful about the kind of instructional set that he administers to his subject, especially as the instructional set may affect the direction of reactivity. Particular attention should be given to maintaining identical instructional sets across the different phases of experimentation.

In summary, reactivity to self-observation cannot be discounted and certainly needs to be acknowledged when reporting the results of single subject research. Conclusions and generalizations made on the basis of self-report data are to be limited to the unique conditions under which data were collected (e.g., frequency of data collection, times selected for self-observation, and instructional sets administered).

Demand characteristics

That demand characteristics can operate as a major artifact in behavioral research has been fairly well-documented in a number of separate investigations (see Orne, 1962, 1969, 1970; Orne and Evans, 1965; Orne and Holland, 1968). In these paradigms it has been shown that experimental subjects frequently respond in accordance with the experimenter's hypotheses and expectations as soon as they become aware of their existence. Recent studies by Horton, Larson, and Maser (1972) and Johnson and Lobitz (1972) indicate that even data obtained under naturalistic conditions may represent confounding, possibly as a result of demand characteristics and other kinds of response sets.

In single case strategies, where repeated measures are taken and where baseline phases are alternated with treatment phases, the possibilities for "demand" sets are numerous. While physiological responses are less likely to be affected significantly by demand characteristics of the experiment (e.g., Barlow, Agras, Leitenberg, Callahan, and Moore, 1972), motoric measures (under voluntary control) are more subject to such artifacts. Moreover, self-report measures, whether they involve self-observations of specified behaviors or written responses to standardized inventories (e.g., fear or depression scales), are obviously the most vulnerable to the effects of experimental demand.

Let us consider a typical A-B-A-B strategy in which self-report measures are used as a major dependent variable. In baseline, little in the way of therapeutic activity is carried out and, of consequence, the patient's expectation for improvement is minimized. By contrast, when treatment is instituted in the next phase considerable therapeutic attention is given to the patient. Under these circumstances, the patient is more likely to assume that changes in his condition will occur. Thus, when a marked change in the patient's verbal report is seen following initiation of treatment, a question arises as to how much of the change represents "true" improvement and how much is purely a function of the patient's expectation of improvement and sheer compliance with "therapeutic" demand. Of course, if initial change in verbal reports is mainly due to demand characteristics of the therapeutic situation, when such treatment is continued a plateau will be attained followed by a possible reversion, but the final treatment goal is unlikely to be achieved.

Even when the final portion of the design is implemented (second A-B sequence) and the expected rate changes in self-reports follow, these changes might still be a function of renewed conformity to experimental demand. This is most likely to occur when the contrast between baseline and treatment operations is extensive. Therefore, one method to control and/or counteract the artifact of experimental demand is to lessen the discrepancy between baseline and treatment phases by replacing placebo treatments for initial baseline assessment. Placebo treatments might also be instituted following baseline measurement and prior to actual treatment, with changes between baseline and placebo and placebo and treatment being evaluated. Another strategy is to avoid sole reliance on verbal reports by monitoring physiological and motoric response systems as well (see Section 4.6 of this chapter regarding inter-correlation of these three response systems). A final precautionary measure is to introduce a negative expectation (via instructional sets) during active treatment phases to ensure that resulting data reflect changes due to the treatment variable *per se* rather than to artifacts in measurement (Barlow, Agras, Leitenberg, Callahan, and Moore, 1972).

To recapitulate, when using verbal reports in experimental single case research, the investigator must view his results with a healthy skepticism and must also consider alternative explanations for the "apparent" success of his treatment manipulations. Only when these other factors have been carefully ruled out can the investigator safely attribute changes to the introduction of his treatment procedure.

Impression management and faking

The patient's ability to engage in impression management is a frequently omitted consideration when single case strategies are carried out with hospitalized adult psychiatric patients. Contrary to the prevailing view that such patients were helpless, the victims of environmental control, frightened, inert, and incapacitated individuals, Braginsky and his colleagues (Braginsky and Braginsky, 1967; Braginsky, Grosse, and Ring, 1966) showed that, depending on their perception of a given situation, both acute and chronic psychiatric patients were able to present themselves as either "healthy" or "sick" in structured interviews and on standardized self-report inventories. Thus, in these studies it was found that patients made conscious efforts to manipulate the impressions that they made on the experimenter. In one study (Braginsky and Braginsky, 1967), when the open ward status for chronic schizophrenics was questioned, these patients presented themselves as "healthy" in a standardized interview situation. By contrast, when their residency status in the hospital was questioned, a similar group of chronic schizophrenics presented themselves as "sick" in the identical standardized interview situation. Braginsky and Braginsky (1967) interpreted their results, ". . . as supporting assumptions of patient effectiveness in implementing goals" (p. 543).

As in the case of demand characteristics, when impression management is considered with respect to the three response systems (physiological, motoric, and self-report), the self-report system again appears to be the one most vulnerable to bias. For example, biases in verbal reporting including lying and faking have been carefully evaluated by the constructors of standardized self-report inventories such as the Minnesota Multiphasic Inventory (MMPI). The validity scales of the MMPI (L, F, K) were specifically designed to counteract this kind of response bias (Dahlstrom and Welsh, 1960). The F-K ratio ("fake bad") and its converse, the K-F ratio ("fake good"), were developed to facilitate the *correct* interpretation of resulting MMPI protocols.

Specifically, in terms of experimental single case research with psychiatric inpatients, it is possible that their self-reports, even under highly standardized experimental conditons, contain aspects of impression management. After all, frequently the patient is not informed that he is being treated experimentally. Moreover, even when he is aware of the experimental nature of his treatment, participation in the experiment represents only a small fraction of the total time he spends on the inpatient unit. Let us also consider the possibility of the patient who specifically exaggerated his symptomatology to gain admission to the inpatient unit in order to avoid unpleasant circumstances in his natural environment. However, after a brief stay (e.g., 2 weeks) the patient now becomes dissatisfied with the hospital milieu. Let us also assume that baseline assessment commenced shortly after admission and that active treatment coincided with the patient's dissatisfaction with hospital life. If this same patient then assumes that a substantial change in his self-report will facilitate his being discharged (at the 2-week point), resulting data will obviously represent confounding between possible change due to treatment and change attributable to the patient's attempt to convince staff to discharge him. We might also note at this point that impression management is not necessarily limited to inpatient settings, but that its effects are more dramatic and more readily identified in these environments.

Given the aforementioned example of possible confounding, the single case researcher working in inpatient settings should not only examine graphed changes in designated target behaviors, but he should also consider other sources of information (e.g., observations made by other professionals in these settings) before deriving conclusions from his data. Again, we would like to underscore that data reported in single case studies usually represent only a very small portion of the patient's 24-hour day, unless his behavior is monitored continuously.

Additional forms of response bias

Personality scale constructors have identified a number of more subtle forms of response bias that may be of some interest to the single case researcher who relies on responses to standardized inventories as a major dependent variable. "Yeasaying" and "naysaying" (Couch and Keniston, 1960; Jackson and Messick, 1958) and social desirability responding (Crowne and Marlowe, 1964) are two forms of bias that attracted considerable attention in the personality literature in the late 1950s and early 1960s. In the former the subject tends to respond in a uni-direction to true-false items irrespective of specific content. In the latter the subject tends to respond to items in terms of what he perceives as the most socially acceptable position. In that sense social desirability and responsivity to experimental demand are akin, and the strategies for counteracting demand characteristics would appear to apply equally well to the social desirability response set.

Crowne and Marlowe (1964) and Jackson and Messick (1958) have interpreted these response sets in terms of stylistic variables associated with particular kinds of personality patterns. However, these formulations have been challenged by Rorer (1965). A thorough examination of the two conflicting points of view is well beyond the scope of this discussion. But, suffice it to say, these kinds of reponse biases confound the measurement process. And, when examined in light of repeated measurement procedures, their possible presence cannot be ignored.

As previously mentioned, the single case researcher might avoid primary reliance on self-report measures when feasible. But when self-report measures are used as criterional, the results should be bolstered with data obtained via motoric and physiological measures. Some of the specific pitfalls of self-report measures could also be avoided. First, "yeasaying" and "naysaying" on true-false inventories can be counteracted by either presenting items in a forced-choice format or by presenting items in scaled format (e.g., 1-4 point scale for each item). Secondly, items can be keyed in both directions (one-half positively worded; one-half negatively worded) as in Zung's (1965) 20-item Self-Rating Depression Scale. Finally, alternate forms of the scale might be administered in successive probe days to prevent familiarity with the test items (Eisler and Hersen, 1973).

4.5. PHYSIOLOGICAL MEASURES

The exponentially mounting data in the area of biofeedback (e.g., Blanchard and Young, 1973; Katkin and Murray, 1968; Shapiro, Barber, DiCara, Kamiya, Miller, and Stoyva, 1973) definitely indicate that autonomic variables are subject to modification via instructional, feedback, and reinforcement procedures. However, in these paradigms considerable training is required before changes approach clinical significance. Moreover, the results of training, to date, are rather ephemeral as indicated by long-term follow-ups. In light of these factors and in spite of their direct modifiability, physiological responses seem to be less subject to conscious distortion on the part of the patient or client when used as dependent variables in experimental single case research. Only a highly sophisticated and equally highly trained subject (in terms of biofeedback for the particular response modality under consideration) would have the capacity to willfully control his physiological responses.

A large variety of physiological measurement techniques have recently been used in single subject experimental strategies. They include such measures as: penile circumference (Abel, Levis, and Clancy, 1970; Barlow and Agras, 1973; Harbert, Barlow, Hersen, and Austin, 1974), EMG (Agras and Marshall, 1965; Budzynski, Stoyva, and Adler, 1970; Epstein, Hersen, and Hemphill, 1974), anal sphincter pressure (Kohlenberg, 1973), heart rate (Boulougouris and Bassiakos, 1973; Leitenberg, Agras, Butz, and Wincze, 1971; Scott, Peters, Gillespie, Blanchard, Edmunson, and Young, 1973), pulse rate (Boulougouris and Bassiakos, 1973), and GSR (Boulougouris and Bassiakos, 1973).

When physiological response systems are monitored, the typical problems encountered with motoric and self-report measures are somewhat obviated. Certainly the question of observer reliability is not at issue as recording of physiological responses is usually automatic. Also, as previously noted, subjects' willful distortion of physiological data, as might occur when self-report measures are used (e.g., response bias, conformity to perceived experimental demands, faking, and impression management), are only remote possibilities. However, a whole host of new problems appear when physiological measurements are taken. In the following subsections we will briefly examine some of the difficulties that are associated with the monitoring of physiological responses in single case research. We will discuss such issues as mechanical problems, adaptation phases, resetting baselines, decreasing responsivity with repeated presentations of stimuli, experimenter and contextual variables, and the question of stimulus-response specificity in the absence of confirmatory verbal reports. We will also discuss the advantage of using the single case strategy when individual differences in autonomic reactivity are considered (White, 1956, pp. 445-451–the "autonomic constitution").

Mechanical problems

The sensitive equipment that is used to monitor physiological responses is frequently plagued by mechanical failure. Although in any experimental endeavor that involves the use of laboratory equipment there is the expectation that mechanical problems will be minimized, the importance of maintaining physiological recorders in good working condition is heightened when single case research is being conducted. Missing data as a function of mechanical failures, although troublesome in group comparison studies, can be remedied by systematically eliminating subjects from the analysis (randomly from other groups) or by filling in hypothetical data for individual subjects based on group means and trends. However, the application of these procedures to single subject research is either unfeasible or inappropriate. In short, the clinical researcher is left with empty data points on his graph. When examined from the clinical context, if the recording of physiological data is directly connected with the treatment process as in biofeedback paradigms (see Epstein, Hersen, and Hemphill, 1974; Scott, Peters, Gillespie, Blanchard, Edmunson, and Young, 1973), the accurate functioning of the equipment is mandatory.

To recapitulate, laboratory assistants should be thoroughly familiar with the equipment, its proper maintenance, its calibration, in addition to the particular requirements posed by the experimental design. They would also be well advised to first practice all procedures with non-clinical subjects before actually monitoring physiological reactivity during experimental treatment.

Adaptation phases and resetting baselines

In addition to achieving stability over trials and days during the initial baseline assessment phase (see Section 3.3 in Chapter 3), the experimenter taking physiological measurements must allow sufficient time for adaptation during each trial prior to baseline and treatment assessments. Being attached with electrodes and other lead wires to physiological monitoring equipment, in itself, will result in response reactivity in most subjects. Consider also the most extreme condition where Freund's penile plethysmograph is employed in sexual research with male sexual deviates (Freund, Sedlacek, and Knob, 1965). Initially when applied this device may actually serve to stimulate the penis. Obviously, when this kind of measuring instrument is used, a lengthy adaptation period will be required before an accurate estimate of baseline responding can be obtained. A second feature peculiar to the measures is the necessity of resetting baselines after each period of arousal. Several investigators have noted that penile circumference will not return to the initial baseline, but stabilizes at a level 5 to 10 percent higher (e.g., Barlow, Becker, Leitenberg, and Agras, 1970). Further responding must be calculated relative to this new baseline. This underlines the importance of becoming familiar with the idiosyncracies of a given physiological measure before undertaking clinical research.

Adaptation phases have varied in length in different experiments in accordance with the type of measure used and the particular dictates of the design. For example, Scott, Peters, Gillespie, Blanchard, Edmunson, and Young (1973) were concerned with heart rate measurements and allowed 20 minutes of a 40-minute session for purposes of adaptation. Epstein, Hersen, and Hemphill (1974) set aside a 10-minute adaptation phase for their headache subject in each 20-minute EMG evaluation session. In addition to the time element, some experimenters set an *a priori* variability limit before the adaptation phase is considered to be complete. Scott and his colleagues (1973) specifically defined stability of heart rate ". . . as less than 15 percent variability (that is, the HR value for one trial was within ± 7·5 percent of the mean HR for the three trials) in *S*'s HR across three consecutive trials" (p. 181).

There are many factors that affect stability of measurement for physiological responses, and these have been carefully reviewed by Montague and Coles (1966) with respect to the galvanic skin response (GSR). Aside from experimenter and other contextual variables, there are some specific environmental conditions such as temperature (Venables, 1955), humidity (Venables, 1955), and time of day (Farmer and Chambers, 1925) that have been shown to affect GSR independently of the experimental manipulations. As these variables are frequently under only limited experimental control (especially temperature and humidity), they may, in part, contribute to some of the variability observed in baseline.

Indeed, the baseline will often have to be reset from trial to trial, and effects of therapeutic variables should therefore be evaluated in terms of proportional changes rather than in terms of absolute changes (Lacey, 1956). This type of simple statistical operation will then control for the day-to-day variability patterns that are essentially beyond current experimental control. Similarly, Zuckerman (1971) argues that "Covariance techniques. . . can be used to remove the influence of the baseline measure from the response measure" (pp. 325-326).

The effects of repeated measurement

Although we have cited the advantages of using repeated measurement techniques in applied behavioral research, there are some distinct disadvantages associated with their use with physiological responses (particularly GSR). Montagu and Coles (1966), in their excellent discussion of the galvanic skin response, point out that "It has long been known that the GSR tends to diminish in size with repetition of the stimulus. The diminution may be observed when the stimulus is repeated at short intervals within the session. . . or at longer intervals from day to day. . ." (p. 276). This phenomenon has alternately been labeled habituation and adaptation. For instance, Solyom and Beck (1967) found that repeated presentation of sexually arousing stimuli to sexual deviates resulted in decreased GSR amplitude over trials. Similar kinds of findings were reported by Woodmansee (1966) in his examination of the pupillary response system. Zuckerman (1971) argues that "The reaction of the first trial may be as much a function of novelty or surprise as the nature of the stimulus itself" (p. 325). On the other hand, GSR has, at times, been quite reliable, holding up over many months of repeated measures (Barlow, Leitenberg, and Agras, 1969).

Despite the fact that all physiological response systems do not show equal decrements in the face of repeated presentations, the careful researcher must ensure that when treatment is instituted decremental responding results primarily as a function of the variable introduced rather than mere habituation to a standard stimulus. Of course, different trends in the data reflecting introduction and removal of treatment variables should ensure that the dependent measure is reliable. However, the possible interaction of the treatment's effectiveness with naturally occurring decremental responding cannot be overlooked.

Especially when using GSR as the dependent measure, the experimenter might consider developing a number of standard stimuli that on pre-test elicit a relatively equal amount of physiological responsiveness. These stimuli, then, might be alternated throughout both baseline and treatment phases, thereby decreasing the possibility of the subject's habituation to them.

Experimenter and contextual variables

The research studies conducted by Rosenthal and his colleagues (Rosenthal, 1966; Rosenthal and Fode, 1963; Rosenthal and Lawson, 1964) on experimenter bias and other contextual variables have sensitized most investigators to the interacting effects of their behavior and subjects' responses. These interactive aspects must also be considered when physiological measures are used, especially in single subject research where the experimental clinician often performs therapeutic functions in addition to monitoring and recording of responses. In reviewing the literature on physiological measures of sexual arousal, Zuckerman (1971) ". . . found that even physiological responses, GSR, pupil size, and urinary acid phosphatase secretion, may be influenced by set induced instructions, or the characteristics and behavior of the experimenter" (p. 326).

Let us now examine a group comparison study that illustrates the above point. Chapman, Chapman, and Brelje (1969) compared the effects of using two different experimenters (Experimenter 1–an aloof business-like graduate student; Experimenter 2–a casual outgoing undergraduate) on pupillary dilation of undergraduate males to slides of nude and partially clad women, men, and control scenes. Subjects who were shown slides by the first experimenter evidenced an equal amount of dilation to both male and female slides. By contrast, subjects who were shown slides by the second experimenter (casual undergraduate) evidenced significantly more dilation to pictures of women than of men. Chapman, Chapman, and Brelje (1969) conclude that "If the *E*s had been working in different laboratories, they would have reported contradictory findings. The result is consistent with the hypothesis that differing modes of interaction with *S* produce different pupillary responses" (p. 399). Although the specific cues that influenced subjects' responses were not experimentally isolated in this study, it is quite clear that the experimenter's attitudes and actions do influence subjects even at the physiological level. These kinds of findings have been corroborated in other contexts (Barclay, 1969; Martin, 1964; Zuckerman, Persky, and Link, 1969).

It becomes apparent, then, that the investigator must ensure that the actions of the individual responsible for obtaining the data should not inhibit the specific physiological response system under study. This is of particular importance when one monitors physiological response systems that are under partial voluntary control (e.g., penile erection) (see Laws and Rubin, 1969).

Stimulus-response specificity

When an experimenter monitors a specified physiological response system (e.g., GSR) in reaction to presentation of standard stimuli, he may assume that responding over baseline levels is evidence of a discrete emotional reaction (e.g., positive interest). For example, undergraduate males' increased pupillary response to slides of nude females might be taken as an indication of a positive interest in these stimuli. However, if the same magnitude of pupillary response is subsequently elicited by slides of attractive muscular males, the experimenter's conclusion about his data would undergo radical modification. Continuing our analysis of this example, reactivity to attractive male slides in a normal male heterosexual sample might simply result as a function of the novelty of the stimulus rather than an expression of "deviant" sexual interest (Zuckerman, 1971).

Consider a further example in which a blood and knife phobic and a sadist both react equally (in some physiological response system) to a scene involving torture, blood, and knives. While both individuals evidence physiological arousal upon presentation of this scene (e.g., increased heart rate), the specific emotion experienced by each is entirely different (fear in the phobic and sexual excitement in the sadist).

It becomes obvious that when physiological response systems such as GSR, blood pressure, heart rate, pupillary response, etc., are used as indices of emotional arousal, the particular emotion experienced by the subject cannot be *assumed* by the experimenter in the absence of a confirmatory verbal report by his subject. Zuckerman's thorough (1971) review of physiological measures of sexual arousal indicates that individual responsivity to the same stimuli may be associated with either positive or negative affect. Examined from the vantage point of single case experimental research, the concurrent monitoring of different response systems (motoric, self-report, and physiological) is once again recommended. Secondly, instead of individual differences in physiological responsivity being canceled out in the group statistical design, these same differences in responding can become subjects for an experimental analysis in the single case strategy.

Individual differences in autonomic reactivity

Lacey and his co-workers (Lacey, 1950, 1956; Lacey and Van Lehn, 1952) demonstrated that very early in life individual profiles of autonomic reactivity were evidenced by human subjects in response to stress. That is to say, regardless of the stressor, the individual reacts with a particular physiological response system. In short, it would appear that the choice of physiological system in response to stress is constitutionally determined (White, 1956). Some subjects are GSR responders, some are heart rate responders, others respond with elevated blood pressure, and still others might respond via increased rates of respiration. Zuckerman (1971) notes that "Most measures from different peripheral autonomic systems are minimally or inconsistently correlated across subjects" (p. 325).

These inconsistent correlations present difficulties when group comparison designs are used to evaluate treatment variables. When a specific physiological response system is used as a dependent measure in evaluating success of a treatment (e.g., flooding in phobics), it is quite possible that despite the treatment's apparent efficacy (as measured motorically and subjectively), the results of the analysis with respect to the physiological measure will not attain statistical significance. As different subjects respond with specific systems (at the physiological level), these differences will be "averaged out" in the statistical analysis. Therefore, it is recommended that several physiological systems be monitored concurrently to ascertain which is the most sensitive indicator of change. On the other hand, the problems involved in "averaging out" differences due to specificity of physiological responding can be avoided if single case strategies are used to evaluate therapeutic efficacy (see Chapter 2). Indeed, once the most sensitive response systems are identified, direct or systematic replication series can be conducted over subjects. In the systematic replications different physiological response systems for different subjects can be monitored to evaluate efficacy of treatment in accordance with preferred physiological reactivity.

4.6. RELATIONSHIPS AMONG MOTORIC, SELF-REPORT, AND PHYSIOLOGICAL MEASURES

Our examination of the motoric, self-report, and physiological measurement systems shows us that none of the systems is devoid of major problems and that all three are subject to a variety of intrusive variables that serve to diminish the reliability and validity of the measurement process. Faced with these factors, the applied behavioral researcher must decide which system or which combination of systems will be used to evaluate outcome of treatment. Essentially the question that must be asked is: what is the criterion for improvement? The answer to this question is simplified for the clinical researcher who is primarily concerned with the treatment of young children. These children obviously do not seek treatment at their own initiative but are presented at clinics and hospitals by their parents or authorities. Direct observation of motoric behaviors in laboratory settings and at home in addition to parental reports of their observations of specified behaviors frequently may serve as the criteria for improvement. Less frequently, physiological measures may be taken. Self-reports are rarely used.

In the adult paradigm the choices are more complex. Here the patient or client actively seeks treatment (with the exception of those who are legally committed), and in addition to motoric and physiological target measures, the clinical researcher is most concerned with his patient's perceptions of improvement (i.e., the self-report). We will not belabor the enormous problems associated with the use of patients' self-reports (see Hersen, 1973b). However, we would like to examine how the three types of measurement vary with respect to the proportion of improvement reflected in each for different treatment paradigms and different measurement situations. We will first examine some representative examples derived from the analogue and clinical research literatures. Then we would like to offer some commentary and conclusions with regard to this most challenging issue.

Research findings

The relationships among the three response systems have been carefully evaluated by researchers studying the fear of small animals in college populations (e.g., snakes, dogs, rats, spiders, etc.) and to a lesser extent by investigators examining changes during the behavioral treatment of clinically incapacitating phobias and compulsions (e.g., Leitenberg, Agras, Butz, and Wincze, 1971; Mills, Agras, Barlow, and Mills, 1973). Hersen's (1973b) review of the self-assessment of fear indicates that the correlations obtained between subjects' self-reports of fear and motoric and physiological indices of fear are not uniformly high. For example, Geer (1966) found that undergraduate subjects reporting a high degree of fear of spiders evidenced significantly greater and longer GSRs in response to viewing slides of spiders than low fear-of-spider subjects. On the other hand, Cooke (1966) reported a relatively low correlation ($r = \cdot 35$) between female undergraduate subjects' self-reported fear of rats and overt behavior in a laboratory approach task. Lang (1968) found a low negative correlation of $-\cdot 26$ between self-reported fear of snakes and behavior in a laboratory approach task. Using schizophrenic patients as subjects, Begelman and Hersen (1973) found that high fear ratings of snakes and later behavior in a laboratory approach task correlated only ·06. After completion of the approach task, these subjects were asked to rate their fear a second time. This fear rating correlated ·40 with the approach task. Normal subjects, however, evidenced consistently high correlations, both pre- ($r = \cdot 73$) and post-approach task ($r = \cdot 72$).

Although the extent of the relationship between measures in the aforementioned studies varies considerably, these subjects are not clinically phobic. Moreover, these measures are not monitored repeatedly during a treatment paradigm. Finally, descriptions of the feared stimuli are not consistent with what the subject is expected to do in the behavioral approach situation. For example, the subject might be asked to rate his overall fear of nonpoisonous snakes on a 1-6 scale, but he has absolutely no idea of the task he will be required to perform in the laboratory situation (i.e., approaching, looking at, touching, and holding the feared stimulus). It is quite possible that if the subject were asked to rate a full and specific description of the laboratory task subsequently required of him, greater concordance between the verbal rating and his behavioral performance might be obtained. Research along these lines is warranted.

Leitenberg, Agras, Butz, and Wincze (1971) examined the relationship between heart rate and motoric indices of fear during the behavioral treatment of nine phobic subjects. In five of these subjects self-reports of anxiety were also taken. The results of this study yielded a number of different relationships among the three response systems. In two subjects phobic behavior and heart rate diminished concurrently. In three subjects phobic behavior decreased while heart rate remained unchanged. In one subject heart rate decreased following motoric change, while in a second subject heart rate remained unaffected following motoric change. In four of the five cases where phobic behavior declined self-report ratings of anxiety decreased regardless of the direction of heart rate (no change, increase, or decrease). Leitenberg, Agras, Butz, and Wincze (1971) concluded that "Physiologically defined anxiety need not always be inhibited first in order to obtain desired behavioral change during treatment of phobia. In fact, anxiety reduction may sometimes be a consequence rather than a cause of behavioral change" (p. 59). However, these conclusions might have to be tempered somewhat when one considers the fact that heart rate was the only physiological response system monitored. As noted in Section 4.5 of this chapter, physiological arousal manifests itself through different systems for different subjects. Thus, the absence of concordance between physiological and motoric measures in this study might result from individual differences in autonomic reactivity. In a second study where marked discrepancies between physiological and self-report measures were noted (Barlow, Agras, Leitenberg, Callahan, and Moore, 1972), four homosexuals were told that a specific treatment–covert sensitization–would make them worse. Three of the four subjects improved based on decreases in homosexual arousal measured by penile circumference change, but it was simultaneously indicated that they were getting worse on self-report measures.

The discrepancies between response systems have also been noted when unassertive subjects are treated through modeling, instructions, and behavior rehearsal (assertive training). In these analogue studies changes in targeted motoric responses are frequently seen but self-reports on assertive questionnaires do not change significantly on a pre-post basis (e.g., Hersen, Eisler, and Miller, 1973; Hersen, Eisler, Miller, Johnson, and Pinkston, 1973). Of course, a problem in analogue studies is that experimental treatment is too brief to enable attitudinal change to appear. Elsewhere, this phenomenon has been labeled "attitudinal lag" (Hersen, 1973b).

Turning now to a different study, Mills, Agras, Barlow, and Mills (1973) examined the effects of response prevention as a treatment for obsessive-compulsive patients. Frequency of compulsive rituals actually performed and self-reports of urges to perform rituals were monitored and obtained daily during baseline, instructions, placebo, and response prevention phases. A visual inspection of the data for four subjects indicates overall concordance between motoric and self-report measures, although self-report measures lagged well behind motoric indices.

Williams, Barlow, and Agras (1972) developed motoric measures of severe depression that involved nursing assistants making repeated daily ratings of the presence or absence of the following three behaviors: (1) talking, (2) smiling, and (3) motoric activity. In addition to obtaining behavioral ratings for ten severe depressives, self-reports of depression (Beck Depressive Inventory) and physicians' ratings of depression (Hamilton Rating Scale) were made. All three measures were taken during the course of inpatient pharmacological treatment. The correlations among the three measures were as follows: Beck and Hamilton, $r = \cdot 82$; Hamilton and behavioral, $r = \cdot 71$; Beck and behavioral, $r = \cdot 67$. The results also indicated that in some cases the behavioral ratings predicted post-hospital adjustment more accurately than either the Hamilton Rating Scale or the Beck Depressive Inventory.

Finally, in a recent single case study involving covert sensitization treatment of incestuous behavior, good concordance between self-reports of incestuous urges and a physiological measure of deviant sexual interest (mean penile circumference change in reaction to photographs and audiotapes of the incestuous object and situation) was noted (Harbert, Barlow, Hersen, and Austin, 1974).

Commentary

In discussing the criterion problem with respect to outcome research in psychotherapy, Paul (1967) contends that "While multiple measures of outcome are necessary, the dependent variable in any outcome evaluation must be... change in the distressing behaviors which brought the client to treatment" (p. 112). Therefore, even though many behavior modifiers have eschewed the use of the verbal report as criterional, this obviously cannot be done in the clinical situation (Hersen, 1973b). However, in addition to the patient's description of the problem, the clinician (clinical researcher) will make his detailed analysis of the case. Targeted behaviors will be selected and these will then be subjected to modification. Very frequently the verbal report will not be modified directly, but motoric measures will be chosen as targets for treatment (e.g., in the phobic, avoidance behavior will be treated with expectation that change may follow in verbal and physiological response systems). If one considers the possibility that the three response systems are being controlled by different sets of contingencies (Begelman and Hersen, 1973; Hersen, 1973b; Lang, 1968), then it is unlikely that change in all three systems will follow a similar pattern. In short, there would be no reason to assume a high degree of correlation among the three response systems.

Lang (1968) suggests ". . . that we should apply specific techniques to the different behavioral systems that we are trying to change–verbal, overt-motor, and somatic–and that therapy be a self-conscious, multidimensional process" (p. 92). Similarly, in his discussion of the modification of fear, Hersen (1973b) has called attention to the fact that each of the three response systems must undergo some modification if permanent results are to be achieved. Under these circumstances, greater concordance among response measures is likely to be attained.

4.7. SPECIFIC TECHNIQUES

Applied behavioral researchers have been inventive and resourceful in developing measurement techniques for use in their clinical research endeavors. A wide variety of motoric, self-report, and physiological measurement systems have been employed both in single subject research and in group comparison designs. In the following subsections we would like to briefly describe some representative examples of measurement techniques that either have been used in single case designs or are considered to be suitable for such analyses. Of course, these measures can also be applied in group comparison studies. We will present specific techniques designed for studying neurotic, psychotic, interpersonal, alcoholic, sexual, psychosomatic, and child behavior disorders.

Neurotic disorders

Compulsions. Mills, Agras, Barlow, and Mills (1973) constructed an automatic measuring device to record frequency of hand-washing episodes in hospitalized compulsive handwashers. A "washing pen," consisting of a 6 x 8-foot wooden board surrounded by a gate, enclosed the wash basin in the patient's room. Each time the patient approached the basin for purposes of hand-washing, eight switches wired in sequence underneath the board activated a cumulative recorder placed in an adjoining room. Thus, cumulative frequency of hand-washing was obtained on a 24-hour basis. Nurses' reports of hand-washing correlated ·96 with the obtained cumulative frequency.

Obsessions. When certain private events such as obsessions are evaluated, the investigator is compelled to rely on the patient's self-reports. Gullick and Blanchard (1973) examined the effects of several psychotherapeutic techniques in a single case analysis using reports of number of obsessive thoughts per day as the dependent measure. However, recordings of these kinds of data can be made more precise by providing patients with counters or tally sheets.

Phobias. Some rather straightforward motoric indices have been used in the assessment of phobic behavior. For example, with claustrophobics and knife phobics (Agras, Leitenberg, Barlow, and Thomson, 1969; Leitenberg, Agras, Butz, and Wincze, 1971; Leitenberg, Agras, Thomson, and Wright, 1968), the measurement situation simply involved monitoring amount of time spent in the phobic situation under standardized conditions. With agoraphobics (Agras, Leitenberg, and Barlow, 1968), distance walked away from the hospital was added to the time dimension.

Physiological measurements have also been taken concurrently with motoric and self-report indices of fear. Specifically, Leitenberg, Agras, Butz, and Wincze (1971) monitored their phobic patients' heart rate on a trial-by-trial basis using a Cambridge Verascribe Mark II portable EKG machine.

In the clinical situation assessment has primarily been limited to self-reports of anxiety and discomfort. In evaluating suitability of the hieracrchy in systematic desensitization and overall progress of his patients with respect to specific fears, Wolpe (1969) has developed a 100-scale point measure labeled SUDS (subjective units of disturbance). On this measure, 0 represents absolute calm and 100 the worst possible anxiety experienced. This measure is quite similar to the 10-point fear thermometer initially employed by Walk (1956) in evaluating fear of airborne trainees practicing parachute jumps from a 34-foot tower. On both measures, the patient or subject is asked to make a subjective estimate of experienced anxiety at stated intervals.

Hysterical Spasmodic Torticollis. Bernhardt, Hersen, and Barlow (1972) were able to quantify percentage of torticollis movements emitted during 10-minute videotaped sessions. The patient was videotaped while seated in a profile arrangement, and his behavior was observed on the monitor in a room adjacent to the television studio. A piece of clear plastic containing superimposed black lines (spaced one-quarter of one-half inch apart) was placed over the television monitor. When a horizontal line intersected both the auditory meatus and the patient's nostril, or when the patient's nostril was below the horizontal line, this was considered to represent the normal position. If, on the other hand, the patient's nostril were above a horizontal line intersecting the auditory meatus, this position was operationally defined as an instance of torticollis movement. Inter-observer agreements for this behavioral measure ranged from 71·84 percent to 94·18 percent (mean = 85·41 percent).

In addition to measuring motoric improvements, Agras and Marshall (1965) assessed the effects of hypnosis and negative practice on a torticollis patient by monitoring electromyographic activity from the right sternocleidomastoid muscle, which was extremely hypertrophied. These measures were taken before, during, and after treatment. However, this assessment technique could be readily adapted on a repeated measurement basis in experimental single-case analyses for patients suffering from spasmodic torticollis and other forms of tics.

Anorexia Nervosa. In recent experimental studies (Agras, Barlow, Chapin, Abel, and Leitenberg, 1974; Elkin, Hersen, Eisler, and Williams, 1973; Leitenberg, Agras, and Thomson, 1968) standardized methods for monitoring caloric intake and weight have been used for assessing the effects of treatment in anorexia nervosa patients. For example, Elkin, Hersen, Eisler, and Williams (1973) presented their anorexia nervosa patient with three daily meals consisting of 1000 calories each. The patient dined in his room alone for a 20-minute period, but no comments were offered by staff with respect to amounts of food consumed. Caloric intake was then computed by the dietary service by subtracting caloric value remaining on the patient's tray from the original 1000 calories presented. This was accomplished for each meal without the patient's knowledge. Similarly, weight was monitored daily at a stated time under specific conditions (patient in shorts with his back turned to the scale) by the same research assistant.

Astasia-Abasia. Standardized measurement situations for assessing motoric progress in astasia-abasia patients have also appeared in the recent literature (e.g., Agras, Leitenberg, Barlow and Thomson, 1969; Hersen, Gullick, Matherne, and Harbert, 1972; Turner and Hersen 1975). Hersen, Gullick, Matherne, and Harbert (1972) assessed the effects of instructions and social reinforcement on mean number of steps taken and mean distance walked for an astasia-abasia patient. A young female research assistant visited the patient in his room thrice daily at stated intervals. She first spent 10 minutes chatting about topics unrelated to his disorder, and then requested him to walk as far as he could within the confines of his room (back and forth). Number of steps taken and distance walked were recorded by this assistant for each session during all phases of experimentation. Similar procedures have been used by Agras, Leitenberg, Barlow, and Thomson (1969) and by Turner and Hersen (1975).

Psychotic disorders

Delusions. Liberman, Teigen, Patterson, and Baker (1973) were concerned with the effects of social reinforcement contingencies on paranoid and grandiose delusions of four schizophrenics. Each patient was interviewed four times daily for 10 minutes by a member of the nursing staff. A list of suitable topics for discussion had been prepared to enable nurses to prompt conversation with their patients when needed. While engaged in conversation with the patient, the nurse monitored surreptitiously (via stopwatch) amount of time spent talking in rational conversation. Whenever delusional talk was noted, the watch was stopped. The nurse was instructed to maintain the flow of conversation by acknowledging the patient's comments every 15 seconds with head nods, "mmm-hmmm," "sounds good," etc. If the patient remained silent for 30 seconds or more, the nurse was instructed to prompt further conversation by selecting another topic from the list. Thus, percentage of delusional talk was obtained for each session. Inter-rater agreement for this measure ranged from ·75 to ·95 (mean = ·82).

Wincze, Leitenberg, and Agras (1972) also developed a measure for assessing delusional behavior in schizophrenic patients. For each of ten patients, a pool of 105 questions, designed to elicit delusional responses, was compiled. In each of two daily test sessions the patient was asked to respond to fifteen questions, drawn at random from the larger pool of 105. The patient was allowed a 20-second time limit in response to each question. When the response was monosyllabic, the patient was asked to elaborate upon his answer. Twenty-five percent of these interviews were tape recorded in order to obtain independent reliability checks. Responses were rated as either delusional or non-delusional. Percentage of agreement was greater than 90 percent for eight patients. Agreement for the remaining two was 85 percent and 90 percent respectively.

Schizophrenic Withdrawal. Liberman, Davis, Moon, and Moore (1973) used a very simple measurement technique for evaluating effects of a placebo and stelazine on social participation in a withdrawn schizophrenic patient. This patient was requested eighteen times daily to partake in short casual chats with various members of the nursing staff. Number of refusals to participate was the dependent measure. This measure reflected changes in administration of the placebo and drug. In addition, willingness to partake in chats appeared to be correlated with rate of rational speech.

Severe Depression. As noted in Section 4.6, Williams, Barlow, and Agras (1972) developed some simple motoric measures of severe depression that predicted post-hospital adjustment with greater accuracy than the Hamilton Rating Scale and the Beck Depressive Inventory. Specifically, nursing assistants were asked to make sixteen behavioral ratings per day (averaging one per half-hour) for different inpatients between the hours of 8:00 a.m. and 4:00 p.m. Surreptitious ratings of patients' activities (talking, smiling, and motor activity) were scored on a presence or absence basis. Scores could range from 0-48 (one point for presence of each activity per rating multiplied by the number of ratings). Thus, a high total score (e.g., 45) indicated low depression, whereas a low total score (e.g., 10) indicated high depression. Inter-rater agreement averaged 96 percent for these three behavioral categories. We might note that this measure is also sensitive to changes in neurotic depression for patients who are treated in token economic programs (see Eisler and Hersen, 1973; Hersen, Eisler, Alford, and Agras, 1973).

Zung (1965) devised a 20-item self-rating scale for severe depression (SDS) that incorporates several features typical of the symptom complex of depression (i.e., affective, physiological, and psychological components). Each item is presented in a 1-4 point scale. An index for the total SDS is obtained by dividing the sum of the raw scores by the maximum score of 80. Thus, the index can range from ·00 to a maximum of 1·00. One important aspect of this scale is that one-half of the items are worded symptomatically positive, whereas the other half are worded symptomatically negative. This presumes to control, in part, for possible response bias. Pre-post indices for patients diagnosed and treated for depression were ·74 and ·39. The mean index for the control group was ·33, while patients originally classified as depressives but discharged with different diagnoses obtained a mean index of ·53.

Interpersonal disorders

Marital Discord. Eisler, Hersen, and Agras (1973a, 1973b), Eisler, Miller, Hersen, and Alford (1974), and Hersen, Miller, and Eisler (1973) have videotaped married couples while they were discussing their particular difficulties. Specific verbal and nonverbal target behaviors were rated retrospectively from tapes of *ad libitum* interactions between husbands and wives. These target behaviors have served as dependent measures both in experimental single case analyses and in pre-post designs. Typically, the husband and wife are instructed to discuss their marital difficulties for a stated period of time (e.g., 24 minutes) following a short "warm-up" session to familiarize them with the equipment and the videotape studio. During the actual recording, the therapist and the research assistants observe the interaction on the television monitor from an adjoining room. Absence of the therapist and/or the research assistants is considered to be a facilitative factor in addition to approximating the naturalistic situation more closely.

In one of a series of cases (Eisler, Miller, Hersen, and Alford, 1974) the following behaviors were examined separately for the husband and wife: eye contact, speech duration, positive reactions, negative reactions, questions, and smiles. Inter-observer agreement for frequency of occurrence measures ranged from 94 percent to 100 percent; inter-observer correlations for eye contact and speech duration were ·96 and ·93. In a second case, which involved an alcoholic and his wife, four target behaviors were rated: speech duration, eye contact, references to drinking, and requests for new behavior. Inter-observer agreements for references to alcohol and requests for new behavior were 100 percent; inter-observer correlations for eye contact and speech duration were ·99. In both of the aforementioned cases pre-post videotapes of 24 minutes' duration were obtained, with data presented in three blocks of 8 minutes for each tape.

Unassertiveness. Eisler, Miller, and Hersen (1973) developed an analogue measure of assertive behavior (Behavioral Assertiveness Test) in which a number of verbal and nonverbal components of assertiveness are identified. The Behavioral Assertiveness Test consists of 14 standard interpersonal situations that require an assertive response on the part of the patient or subject. The subject is videotaped while seated next to a female role model who prompts the response. Narrations of the interpersonal encounters are presented over an intercom from an adjoining room. One of the situations is described here for illustration. *Narrator*: "You're in a restaurant with some friends. You order a very rare steak. The waitress brings a steak to the table which is so well done it looks burned." *Role Model Waitress*: "I hope you enjoy your dinner, sir." Responses to these situations are then rated retrospectively on the following variables: duration of looking, smiles, duration of reply, latency of response, loudness of speech, fluency of speech, compliance content, requests for new behavior, affect, and overall assertiveness. Inter-observer agreements and correlations for all measures are typically in the 90s. In the initial study subjects judged to be high in overall assertiveness were differentiated from their low assertive counterparts on five of these component measures. In addition, the Wolpe–Lazarus (1966) Assertiveness Questionnaire significantly differentiated subjects judged to be high and low on overall assertiveness on the basis of their responses to the 14 situations.

The Behavioral Assertiveness Test has been used as a dependent measurement system in experimental single case analyses, pre-post designs, and group comparison designs (see Eisler, Hersen, and Miller, 1973; Eisler, Hersen, and Miller, 1974; Eisler, Miller, Hersen, and Alford, 1974; Hersen, Eisler, and Miller, 1974; Hersen, Eisler, Miller, Johnson, and Pinkston, 1973).

Alcoholism

Analogue Measures. Miller and Hersen (1972) developed a "taste test" measurement system, adapted from Schacter's (1971) work with the obese, to assess alcohol consumption in alcoholics. The inpatient alcoholic is requested to participate in an experiment conducted by the psychology service. When he arrives for the experiment, he is seated at a table on which are placed six opaque glasses, each containing 100 cc of beverage. Three of the beverages are non-alcoholic (coke, gingerale, water), while three consist of 10 cc bourbon, vodka, or scotch mixed with 90 cc of water. The subject is given semantic-differential type rating sheets and is administered the following instructions: "This is a taste experiment. We want you to judge each beverage on the taste dimensions (sweet, sour, etc.) listed on the sheets. Taste as little or as much as you want in making your judgments. The important thing is that your ratings be as accurate as possible." The subject is allowed a total of 20 minutes for tasting and making ratings. After he leaves the experimental room, the precise amount consumed for each glass is computed. Thus, both proportional and absolute amounts of alcoholic and nonalcoholic beverages are available. The Miller and Hersen (1972) "taste test" is very similar to the "tasting test" independently devised by Marlatt, Demming, and Reid (1973).

The "taste test" measure has been used as a dependent variable in both clinical and analogue research (see Miller and Hersen, 1972; Miller, Hersen, Eisler, and Elkin, 1974; Miller, Hersen, Eisler, and Hemphill, 1973). Although calculations of beverage consumption are obtained without the subject's presumed knowledge, it is quite possible that under repeated measurement conditions he becomes aware of the "true" nature of the task. However, in a recent study (Miller, Hersen, Eisler, and Elkin, 1974) consumption of alcohol on this measure differentiated ($p < \cdot 06$) successfully and unsuccessfully treated alcoholics.

Operant analyses of drinking have also been used in clinical and analogue projects (e.g., Mello and Mendelson, 1965; Miller, Hersen, Eisler, and Elkin, 1974; Miller, Hersen, Eisler, Epstein, and Wooten, 1974; Miller, Hersen, Eisler, and Hilsman, 1974; Nathan and O'Brien, 1971). In these analyses specific amounts of alcohol (e.g., FR 50 ratio) may be used as reinforcers for designated behavioral responses. For example, Miller, Hersen, Eisler, and Hilsman (1974) used the BRS-Lehigh Valley Modular Human Test System in an analogue study of social stress with alcoholics and social drinkers. The apparatus consists of a metal paneled console which is placed on a small table. An insert on the console contains a 1½-ounce shot glass. Suspended over the glass is a polyethelyne bottle containing a mixture of 30 percent bourbon and 70 percent water. A Lindsley-type manipulandum extends from the console adjacent to the shot glass. This arrangement is connected to an electric switch on a separate relay rack in an adjoining room. In this study the rack was programed such that 50 presses of the manipulandum resulted in a 5 cc squirt of the alcohol mixture into the glass (FR 50 schedule). Patients were allowed 10 minutes to respond during each test session. Number of presses were automatically cumulated by a counter on the rack. As in the case of the "taste test," responses on the operant task differentiated ($p < \cdot 05$) successfully and unsuccessfully treated alcoholics (Miller, Hersen, Eisler, and Elkin, 1974).

In Vivo Measures. In vivo measures of blood/alcohol concentration taken at randomly scheduled intervals in the natural environment (home and work) have proven to be useful (e.g., Miller, Hersen, Eisler, and Watts, 1974). The test simply involves the subject being asked to inflate a balloon-like device (Breathalyzer or Sobermeter; Stevenson Corporation, Red Bank, New Jersey, and Luckey Laboratories in California). The device can then be used for rapid screening purposes (although false-negatives are frequently identified) or the air collection can be analyzed via gas chromatography. This biochemical procedure results in an accurate estimate of blood/alcohol concentration–correlates ·80 to ·91 with actual blood/alcohol analyses (Roberts and Fletcher, 1969). In the Miller, Hersen, Eisler, and Watts (1974) study with a "skid-row" alcoholic, blood/alcohol levels were obtained within 1 hour of collection. Moreover, relatively few logistical problems were encountered during collection of data during this experimental single case analysis. We might note in passing that these Breathalyzer devices are the same kind that the authorities typically use in their screening of intoxicated motorists.

Sexual deviation

In recent studies two basic measurement techniques (Card Sort and Mechanical Strain Gauge for Recording Penile Circumference Change) have been used to assess the effects of behavioral treatment for a variety of sexual disorders including homosexuality, pedophilia, and incest (see Barlow and Agras, 1973; Barlow, Agras, Leitenberg, Callahan, and Moore, 1972; Barlow, Leitenberg, and Agras, 1969; Harbert, Barlow, Hersen, and Austin, 1974; Herman, Barlow, and Agras, 1971; Herman, Barlow, and Agras, 1974). Rather than documenting the use of these measures with all of the specific sexual disorders, we will simply describe the two basic techniques as they can be easily adapted for measuring any kind of male sexual deviation.

Card Sort Technique. A hierarchy of sexually deviant arousing scenes (low to high) is constructed and then typed on individual index cards. In the Barlow, Leitenberg, and Agras (1969) study two subjects were asked to place each of the cards (32 and 45, respectively) for their particular hierarchies in one of five envelopes labeled 0-4. The specific instructions were as follows: "The numbers on the envelopes represent amount of sexual arousal, 0 equals no arousal, 1 equals a little arousal, 2 a fair amount, 3 much, and 4 very much arousal. I would like you to read the description of each scene and place the card in the envelope which comes closest to how arousing the scene is to you at this moment" (p. 598). This procedure was carried out during probe days, and a total score on the Card Sort was computed.

Harbert, Barlow, Hersen, and Austin (1974) modified this procedure somewhat with their subject (incestuous disorder) and included five deviant scenes with the daughter and five scenes of normal father-daughter interactions. Thus, measurement of both deviant and non-deviant interests were obtained. As can be seen from the above descriptions, this technique is quite flexible and can be varied in accordance with the requisities of the particular case or experimental design.

Strain Gauge. Barlow, Becker, Leitenberg, and Agras (1970) developed a mechanical strain gauge for recording penile circumference changes in relation to different stimuli (e.g., reactivity to heterosexual and homosexual slides in a male homosexual). The ring portion of the strain gauge, which is made of corrosion-resistant surgical spring material, is placed on the mid-shaft of the penis with the gauge resting on the dorsal side. This apparatus is adapted for use with a Grass pre-amplifier, Model 7P1A.

The measurement situation involves the subject being seated in a comfortable chair with the ring placed on the penis. In pre-experimental trials maximum erection diameter is determined. Subsequent responses to normal and deviant sexual stimuli are scored on the basis of percentage of full erection. Response latency to stimuli may vary from 60 to 120 seconds. Inter-trial intervals are determined by the amount of time needed to return to baseline levels of responding. This may require from 30 to 300 seconds. At times, initial baseline levels are not retrieved, and new baseline levels are established. Barlow, Becker, Leitenberg, and Agras point out "that small but reliable circumference increases monitored by the ring, in the vicinity of 3 percent maximum, are often not subjectively reported by the subject" (p. 76). In short, this measurement system is most sensitive to the smallest changes in penile diameter. As penile erection is under partial voluntary control, this measure is considered to represent both motoric and physiological components.

Psychosomatic disorders

In some of the psychosomatic disorders mere frequency counts or duration of specific episodes have served as dependent measures under study. In others, more complicated measurement procedures involving physiological recordings have been required. In this subsection we will provide examples of both types of measurement strategies.

Asthma. Neisworth and Moore (1972) treated asthmatic responding in a 7-year-old boy as an operant and examined the contingent effects of parental attention on duration of episodes. Because of the difficulties of obtaining a reliable frequency count of the number of asthmatic episodes occurring throughout the day, a duration measure was recorded nightly after the child was put to bed by his parents. Thus, length of each asthmatic episode, consisting of coughing and wheezing, was monitored under varying contingency conditions.

Tension Headache. Epstein, Hersen, and Hemphill (1974) used electromyographic recordings from the frontalis muscle of a headache sufferer as the index of tension. A Grass 7P-3 hi-gain pre-amplifier with a bandpass width of 10-100 Hz was used to amplify EMG in this biofeedback study. The EMG signal was integrated with a 7P-3 integrator, providing mean microvolt level for each ·05 second interval. Measurement sessions consisted of 10 minutes of adaptation and 10 minutes of recording while the subject sat relaxed on a comfortable chair. A criterion level of ten microvolts was chosen as more than 50 percent of the 1-second intervals during baseline contained integrated responses greater than ten microvolts. The dependent measure was mean seconds/minute above criterion. Self-reports of levels of headache intensity were also obtained during baseline and feedback phases. The results indicated that subjective appraisals of tension coincided with frontalis muscle tension recorded physiologically during these phases.

Child Behavior Disorders

As the measurement of target behaviors in most child behavior disorders simply involves frequency counts by observers, it would be of little heuristic value to present a list of specific disorders (i.e., target behaviors). However, we would like to describe the particular coding technique initially developed by Allen, Hart, Buell, Harris, and Wolf (1964), which is the prototype for others used by applied behavioral researchers in the last decade.

Allen and his colleagues (1964) were specifically concerned with measuring the effects of social reinforcement on isolate behavior of a nursery school child. The child's interactions with adults and other children were observed daily by two independent research assistants in the nursery school setting. Proximity to and interactions with children and adults were recorded on a presence or absence basis for 10-second intervals. Proximity was operationally defined as being within 3 feet of another adult or child. Interaction involved either talking to, smiling at, looking toward, or touching or helping another child or adult.

Data recording involved placing slashes (/) for proximity or x's (x) for interaction in two lines consisting of 30 boxes each. The first line of 30 boxes related to adults, while the second line of 30 boxes related to children. With 60 boxes (two lines of 30 each), 5 minutes of recording were possible for these target behaviors. Each tally sheet contained 12 sets of two lines of 30 boxes each, thus permitting 60 minutes of recording proximity or interaction with adults and children for 10-second intervals on a presence or absence basis. For example, if within a 10-second interval the child were interacting with an adult but also within 3 feet of another child, the tally sheet would reflect this observation–i.e., an (x) in the adult line and a (/) in the child line for the same 10-second box.

The above-described schemata can be easily adapted for recording a large variety of behavioral disorders in children. That is, additional target behaviors may be recorded, or length of observation intervals may be shortened (e.g., 5 seconds) or increased (e.g., 20 seconds) in accordance with the dictates of the specific case or experimental design. This type of recording system has been used in single case experimental studies in school settings, laboratory situations, and natural (home) environments. It is inexpensive, practical, and relatively easy to teach to research personnel.

References

Abel, G. G., Levis, D. J., and Clancy, J. Aversion therapy applied to taped sequences of deviant behavior in exhibitionism and other sexual deviations: A preliminary report. *Journal of Behavior Therapy and Experimental Psychiatry*, 1970, **1**, 59-66.

Abt, L. E. A theory of projective psychology. In L. E. Abt and L. Bellak (Eds.). *Projective psychology: Clinical approaches to the total personality*. Pp. 33-66. New York: Grove Press, 1959.

Agras, W. S., Barlow, D. H., Chapin, H. N., Abel, G. G., and Leitenberg, H. Behavior modification of anorexia nervosa. *Archives of General Psychiatry*, 1974, **30**, 279-286.

Agras, W. S., Leitenberg, H., and Barlow, D. H. Social reinforcement in the modification of agoraphobia. *Archives of General Psychiatry*, 1968, **19**, 423-427.

Agras, W. S., Leitenberg, H., Barlow, D. H., and Thomson, L. E. Instructions and reinforcement in the modification of neurotic behavior. *American Journal of Psychiatry*, 1969, **125**, 1435-1439.

Agras, W. S., and Marshall, D. The application of negative practice to spasmodic torticollis. *American Journal of Psychiatry*, 1965, **122**, 579-582.

Allen, K. E., Hart, B., Buell, J. S., Harris, F. R., and Wolf, M. M. Effects of social reinforcement on isolate behavior of a nursery school child. *Child Development*, 1964, **35**, 511-518.

Azrin, N. H., Holz, W., Ulrich, R., and Goldiamond, I. The control of the content of conversation through reinforcement. *Journal of the Experimental Analysis of Behavior*, 1961, **4**, 25-30.

Baer, D. M., Wolf, M. M., and Risley, T. R. Some current dimensions of applied behavior analysis. *Journal of Applied Behavior Analysis*, 1968, **1**, 91-97.

Barber, T. X., and Silver, M. J. Fact, fiction, and the experimental bias effect. *Psychological Bulletin Monograph*, 1968, **70**, No. 6, Part II, 1-29.

Barclay, A. M. The effect of hostility on physiological and fantasy responses. *Journal of Personality*, 1969, **37**, 651-667.

Barlow, D. H., and Agras, W. S. Fading to increase heterosexual responsiveness in homosexuals. *Journal of Applied Behavior Analysis*, 1973, **6**, 353-366.

Barlow, D. H., Agras, W. S., Leitenberg, H., Callahan, E. J., and Moore, R. C. The contribution of therapeutic instruction to covert sensitization. *Behaviour Research and Therapy*, 1972, **10**, 411-415.

Barlow, D. H., Becker, R., Leitenberg, H., and Agras, W. S. A mechanical strain gauge for recording penile circumference change. *Journal of Applied Behavior Analysis*, 1970, **3**, 73-76.

Barlow, D. H., and Hersen, M. Single-case experimental designs: Uses in applied clinical research. *Archives of General Psychiatry*, 1973, **29**, 319-325.

Barlow, D. H., Leitenberg, H., and Agras, W. S. Experimental control of sexual deviation through manipulation of the noxious scene in covert sensitization. *Journal of Abnormal Psychology*, 1969, **74**, 596-601.

Begelman, D. A., and Hersen, M. An experimental analysis of the verbal-motor discrepancy in schizophrenia. *Journal of Clinical Psychology*, 1973, **31**, 175-179.

Bernhardt, A. J., Hersen, M., and Barlow, D. H. Measurement and modification of spasmodic torticollis: An experimetal analysis. *Behavior Therapy*, 1972, **3**, 294-297.

Bijou, S. W., Peterson, R. F., and Ault, M. H. A method to integrate descriptive and experimental field studies at the level of data and empirical concepts. *Journal of Applied Behavior Analysis*, 1968, **1**, 175-191.

Bijou, S. W., Peterson, R. F., Harris, F. R., Allen, K. E., and Johnston, M. S. Methodology for experimental studies of young children in natural settings. *Psychological Record*, 1969, **19**, 177-210.

Blanchard, E. B., and Young, L. B. Self-control of cardiac functioning: A promise as yet unfulfilled. *Psychological Bulletin*, 1973, **79**, 145-163.

Boulougouris, J. C., and Bassiakos, L. Prolonged flooding in cases with obsessive-compulsive neurosis. *Behaviour Research and Therapy*, 1973, **11**, 227-231.

Braginsky, B. M., and Braginsky, D. D. Schizophrenic patients in the psychiatric interview: An experimental study of their effectiveness at manipulation. *Journal of Consulting Psychology*, 1967, **31**, 543-547.

Braginsky, B. M., Grosse, M., and Ring, K. Controlling outcomes through impression-management: An experimental study of the manipulative tactics of mental patients. *Journal of Consulting Psychology*, 1966, **30**, 295-300.

Browning, R. M., and Stover, D. O. *Behavior modification in child treatment: An experimental and clinical approach.* Chicago and New York: Aldine-Atherton, 1971.

Budzynski, T., Stoyva, J., and Adler, C. Feedback-induced muscle relaxation: Application to tension headache. *Journal of Behavior Therapy and Experimental Psychiatry*, 1970, **1**, 205-211.

Carroll, J. B. Ratings on traits measured by a factored personality inventory. *Journal of Abnormal and Social Psychology*, 1952, **47**, 626-632.

Cautela, J. R. Behavior therapy and the need for behavioral assessment. *Psychotherapy: Theory, Research and Practice*, 1968, **5**, 175-179.

Chapman, L. J., and Chapman, J. P. Illusory correlations as an obstacle to the use of valid psychodiagnostic signs. *Journal of Abnormal Psychology*, 1969, **74**, 271-280.

Chapman, L. J., Chapman, J. P., and Brelje, T. Influence of the experimenter on pupillary dilation to sexually provocative pictures. *Journal of Abnormal Psychology*, 1969, **74**, 396-400.

Cooke, G. The efficacy of two desensitization procedures: An analogue study. *Behaviour Research and Therapy*, 1966, **4**, 17-24.

Couch, A., and Keniston, K. Yeasayers and naysayers: Agreeing response set as a personality variable. *Journal of Abnormal and Social Psychology*, 1960, **60**, 161-174.

Cronbach, L. J. Assessment of individual differences. *Annual Review of Psychology*, 1956, **7**, 173-196.

Crowne, D. P., and Marlowe, D. *The approval motive.* New York: Wiley, 1964.

Dahlstrom, W. F., and Welsh, G. S. *An MMPI handbook.* Minneapolis: University of Minnesota Press, 1960.

Eisler, R. M., and Hersen, M. The A-B design: Effects of token economy on behavioral and subjective measures in neurotic depression. Paper read at American Psychological Association, Montreal, August 29, 1973.

Eisler, R. M., Hersen, M., and Agras, W. S. Videotape: A method for the controlled observation of non-verbal interpersonal behavior. *Behavior Therapy*, 1973a, **4**, 420-425.

Eisler, R. M., Hersen, M., and Agras, W. S. Effects of videotape and instructional feedback on non-verbal marital interaction: An analog study. *Behavior Therapy*, 1973b, **4**, 551-558.

Eisler, R. M., Hersen, M., and Miller, P. M. Effects of modeling on components of assertive behavior. *Journal of Behavior Therapy and Experimental Psychiatry*, 1973, **4**, 1-6.

Eisler, R. M., Hersen, M., and Miller, P. M. Shaping components of assertiveness with instructions and feedback. *American Journal of Psychiatry*, 1974, **131**, 1344-1347.

Eisler, R. M., Miller, P. M., and Hersen, M. Components of assertive behavior. *Journal of Clinical Psychology*, 1973, **29**, 295-299.

Eisler, R. M., Miller, P. M., Hersen, M., and Alford, H. Effects of assertive training on marital interaction. *Archives of General Psychiatry*, 1974, **30**, 643-649.

Elkin, T. E., Hersen, M., Eisler, R. M., and Williams, J. G. Modification of caloric intake in anorexia nervosa: An experimental analysis. *Psychological Reports*, 1973, **32**, 75-78.

Epstein, L. H., and Hersen, M. Behavioral control of hysterical gagging. *Journal of Clinical Psychology*, 1974, **30**, 102-104.

Epstein, L. H., Hersen, M., and Hemphill, D. P. Music feedback as a treatment for tension headache: An experimental case study. *Journal of Behavior Therapy and Experimental Psychiatry*, 1974, **5**, 59-63.

Epstein, L. H., Miller, P. M., and Webster, J. S. The effects of reinforcing concurrent behavior on self-monitoring. Unpublished manuscript, 1974.

Farmer, E., and Chambers, E. G. Concerning the use of psychogalvanic reflex in psychological experiments. *British Journal of Psychology*, 1925, **15**, 237-254.

Fode, K. L. The effect of non-visual and non-verbal interaction on experimenter bias. Unpublished master's thesis, University of North Dakota, 1960.

Fode, K. L. The effect of experimenters' and subjects' anxiety and social desirability on experimenter outcome bias. Unpublished doctoral dissertation, University of North Dakota, 1965.

Freund, K., Sedlacek. F., and Knob, K. A simple transducer for mechanical plethysmography of the male genital. *Journal of the Experimental Analysis of Behavior*, 1965, **8**, 169-170.

Geer, J. H. Fear and autonomic arousal. *Journal of Abnormal Psychology*, 1966, **71**, 253-255.

Goldfried, M. R., and Ingling, J. H. The connotative and symbolic meaning of the Bender-Gestalt. *Journal of Projective Techniques and Personality Assessment*, 1964, **28**, 185-191.

Goldfried, M. R., and Kent, R. N. Traditional versus behavioral personality assessment. *Psychological Bulletin*, 1972, **77**, 409-420.

Goldfried, M. R., and Pomeranz, D. M. Role of assessment in behavior modification. *Psychological Reports*, 1968, **23**, 75-87.

Greenspoon, J., and Gersten, C. D. A new look at psychological testing: Psychological testing from the standpoint of a behaviorist. *American Psychologist*, 1967, **22**, 848-853.

Gullick, E. L., and Blanchard, E. B. The use of psychotherapy and behavior therapy in the treatment of an obsessional disorder: An experimental case study. *Journal of Nervous and Mental Disease*, 1973, **156**, 427-431.

Hamilton, R. G., and Robertson, M. H. Examiner influence on the Holtzman Inkblot Technique. *Journal of Projective Techniques and Personality Assessment*, 1966, **30**, 553-558.

Harbert, T. L., Barlow, D. H., Hersen, M., and Austin, J. B. Measurement and modification of incestuous behavior. A case study. *Psychological Reports*, 1974, **34**, 79-86.

Harris, S., and Masling, J. Examiner sex, subject sex, and Rorschach productivity. *Journal of Consulting and Clinical Psychology*, 1970, **34**, 60-63.

Hart, B. M., Allen, K. E., Buell, J. S., Harris, F. R., and Wolf, M. M. Effects of social reinforcement on operant crying. *Journal of Experimental Child Psychology*, 1964, **1**, 145-153.

Hase, H. D., and Goldberg, L. R. Comparative validity of different strategies of constructing personality inventory scales. *Psychological Bulletin*, 1967, **67**, 231-248.

Herman, S. H., Barlow, D. H., and Agras, W. S. Exposure to heterosexual stimuli: An effective variable in treating homosexuality? Paper read at American Psychological Association, Washington, D. C., September 1971.

Herman, S. H., Barlow, D. H., and Agras, W. S. An experimental analysis of classical conditioning as a method of increasing heterosexual arousal in homosexuals. *Behavior Therapy*, 1974, **5**, 33-47.

Hersen, M. Sexual aspects of Rorschach administration. *Journal of Projective Techniques and Personality Assessment*, 1970, **34**, 104-105.

Hersen, M. Nightmare behavior: A review. *Psychological Bulletin*, 1972, **78**, 37-48.

Hersen, M. Different design options, Paper read at Association for Advancement of Behavior Therapy, Miami, December 7, 1973a.

Hersen, M. Self-assessment of fear. *Behavior Therapy*, 1973b, **4**, 241-257.

Hersen, M., Eisler, R. M., Alford, G. S., and Agras, W. S. Effects of token economy on neurotic depression: An experimental analysis. *Behavior Therapy*, 1973, **4**, 392-397.

Hersen, M., Eisler, R. M., and Miller, P. M. Development of assertive responses: Clinical, measurement, and research considerations. *Behaviour Research and Therapy*, 1973. **11**, 505-522.

Hersen, M., Eisler, R. M., and Miller, P. M. An experimental analysis of generalization in assertive training. *Behaviour Research and Therapy*, 1974, **12**, 295-310.

Hersen, M., Eisler, R. M., Miller, P. M., Johnson, M. B., and Pinkston, S. G. Effects of practice, instructions, and modeling on components of assertive behavior. *Behaviour Research and Therapy*, 1973, **11**, 443-451.

Hersen, M., and Greaves, S. T. Rorschach productivity as related to verbal reinforcement. *Journal of Personality Assessment*, 1971, **35**, 436-441.

Hersen, M., Gullick, E. L., Matherne, P. M., and Harbert, T. L. Instructions and reinforcement in the modification of a conversion reaction. *Psychological Reports*, 1972, **31**, 719-722.

Hersen, M., Miller, P. M., and Eisler, R. M. Interactions between alcoholics and their wives: A descriptive analysis of verbal and nonverbal behavior. *Quarterly Journal of Studies on Alcohol*, 1973, **34**, 516-520.

Holmes, D. S., and Tyler, J. D. Direct versus projective measurement of achievement motivation. *Journal of Consulting and Clinical Psychology*, 1968, **32**, 712-717.

Horton, G. O., Larson, J. L., and Maser, A. L. The generalized reduction of student teacher disapproval behavior. Unpublished manuscript, University of Oregon, Eugene, 1972.

Hutt, M. L. *The Hutt adaptation of the Bender-Gestalt test: Revised.* New York: Grune and Stratton, 1968.

Hutt, M. L., and Briskin, G. J. *The clinical use of the revised Bender-Gestalt test.* New York: Grune and Stratton, 1960.

Jackson, D. N., and Messick, S. Content and style in personality assessment. *Psychological Bulletin*, 1958, **55**, 243-252.

Johnson, S. M., and Bolstad, O. D. Methodological issues in naturalistic observation: Some problems and solutions for field research. In L. A. Hamerlynck, L. C. Handy, and E. J. Marsh (Eds.), *Behavior change: Methodology, concepts, and practice*. Pp. 7-68, Champaign, Ill.: Research Press, 1973.

Johnson, S. M., and Lobitz, G. Demand characteristics in naturalistic observation. Unpublished manuscript, University of Oregon, Eugene, 1972.

Johnson, S. M., and White, G. Self-observation as an agent of behavioral change. *Behavior Therapy*, 1971, **2**, 488-497.

Kanfer, F.H., and Saslow, G. Behavioral diagnosis. In C. M. Franks (Ed.), *Behavior therapy: Appraisal and Status.* Pp. 417-444. New York: McGraw-Hill, 1969.

Kass, R. E., and O'Leary, K. D. The effects of observer bias in field-experimental settings. Paper read at a symposium entitled "Behavior Analyses in Education," University of Kansas, Lawrence, April 1970.

Katkin, E. S., and Murray, E. N. Instrumental conditioning of autonomically mediated behavior: Theoretical and methodological issues. *Psychological Bulletin*, 1968, **70**, 52-68.

Kazdin, A. E. Methodological and assessment considerations in evaluating reinforcement programs in applied settings. *Journal of Applied Behavior Analysis*, 1973, **6**, 517-531.

Kent, R. The human observer: An imperfect cumulative recorder. Paper read at Fourth Banff Conference on Behavior Modification, Banff, Alberta, Canada, 1972.

Kohlenberg, R. J. Operant conditioning of human anal sphincter pressure. *Journal of Applied Behavior Analysis*, 1973, **6**, 201-208.

Lacey, J. I. Individual differences in somatic response patterns. *Journal of Comparative and Physiological Psychology*, 1950, **113**, 338-350.

Lacey, J. I. The evaluation of autonomic responses: Toward a general solution. *Annals of the New York Academy of Sciences*, 1956, **67**, 123-164.

Lacey, J. I. Psychophysiological approaches to the evaluation of psychotherapeutic process and outcome. In F. Rubinstein and M. B. Parloff (Eds.) *Research in psychotherapy*. Pp. 160-208. Washington, D. C.: American Psychological Association, 1958.

Lacey, J. I., and Van Lehn, R. Differential emphasis in somatic response to stress. *Psychosomatic Medicine*, 1952, **4**, 71-81.

Lang, P. J. Fear reduction and fear behavior: Problems in treating a construct. In J. M. Shlien (Ed.), *Research in psychotherapy*. Vol. 3, pp. 90-101. Washington, D.C.: American Psychological Association, 1966.

Laws, D. R., and Rubin, H. B. Instructional control of an autonomic sexual response. *Journal of Applied Behavior Analysis*, 1969, **2**, 93-99.

Leitenberg, H. The use of single-case methodology in psychotherapy research. *Journal of Abnormal Psychology*, 1973, **82**, 87-101.

Leitenberg, H., Agras, W. S., Butz, R., and Wincze, J. Relationship between heart rate and behavioral change during the treatment of phobics. *Journal of Abnormal Psychology*, 1971, **78**, 59-68.

Leitenberg, H., Agras, W. S., and Thomson, L. E. A sequential analysis of the effect of selective positive reinforcement in modifying anorexia nervosa. *Behaviour Research and Therapy*, 1968, **6**, 211-218.

Leitenberg, H., Agras, W. S., Thomson, L., and Wright, D. E. Feedback in behavior modification: An experimental analysis in two phobic cases. *Journal of Applied Behavior Analysis*, 1968, **1**, 131-137.

Liberman, R. P., Davis, J., Moon, W., and Moore, J. Research design for analyzing drug-environment-behavior interactions. *Journal of Nervous and Mental Disease*, 1973, **156**, 432-439.

Liberman, R. P., Teigen, J., Patterson, R., and Baker, V. Reducing delusional speech in chronic paranoid schizophrenics. *Journal of Applied Behavior Analysis*, 1973, **6**, 57-64.

Lindzey, G., and Tejessey, C. Thematic Apperception Test: Indices of aggression in relation to measures of overt and covert behavior. *American Journal of Orthopsychiatry*, 1956, **26**, 567-576.

Marlatt, G. A., Demming, B., and Reid, J. V. Loss of control drinking of alcoholics: An experimental analogue. *Journal of Abnormal Psychology*, 1973, **81**, 233-241.

Martin, B. Expression and inhibition of sex motive arousal in college males. *Journal of Abnormal and Social Psychology*, 1964, **68**, 307-312.

Marwit, S. J. Communication of tester bias by means of modeling. *Journal of Projective Techniques and Personality Assessment*, 1969, **33**, 345-352.

Marwit, S. J., and Marcia, J. Tester bias and response to projective instruments. *Journal of Consulting Psychology*, 1967, **37**, 253-258.

Masling, J. The influence of situational and interpersonal variables in projective testing. *Psychological Bulletin*, 1960, **57**, 65-85.

Masling, J. Differential indoctrination of examiners and Rorschach response. *Journal of Consulting Psychology*, 1965, **29**, 198-201.

Masling, J., and Harris, S. Sexual aspects of TAT administration. *Journal of Consulting and Clinical Psychology*, 1969, **33**, 166-169.

McFall, R. M. Effects of self-monitoring on normal smoking behavior. *Journal of Consulting and Clinical Psychology*, 1970, **35**, 135-142.

McFall, R. M., and Hammen, C. L. Motivation structure and self-monitoring: The role of nonspecific factors in smoking reduction. *Journal of Consulting and Clinical Psychology*, 1971, **37**, 80-86.

McFarlain, R. A., and Hersen, M. Continuous measurement of activity level in psychiatric patients. *Journal of Clinical Psychology*, 1974, **30**, 37-39.

McNamara, J. R., and MacDonough, T. S. Some methodological considerations in the design and implementation of behavior therapy research. *Behavior Therapy*, 1972, **3**, 361-378.

Mello, N. K., and Mendelson, J. G. Operant analysis of drinking patterns of chronic alcoholics. *Nature*, 1965, **206**, 43-46.

Miller, P. M. Behavioral assessment in alcoholism research and treatment: Current techniques. *International Journal of the Addictions*, 1973, **8**, 831-839.

Miller, P. M., and Hersen, M. Quantitative changes in alcohol consumption as a function of electrical aversive conditioning. *Journal of Clinical Psychology*, 1972, **28**, 590-593.

Miller, P. M., Hersen, M., Eisler, R. M., and Elkin, T. E. A retrospective analysis of alcohol consumption on laboratory tasks as related to therapeutic outcome. *Behaviour Research and Therapy*, 1974, **12**, 73-76.

Miller, P. M., Hersen, M., Eisler, R. M., Epstein, L. H., and Wooten, L. S. Relationship of alcohol cues to the drinking behavior of alcoholics and social drinkers: An analogue study. *Psychological Record*, 1974, **24**, 61-66.

Miller, P. M., Hersen, M., Eisler, R. M., and Hemphill, D. P. Electrical aversion therapy with alcoholics: An analogue study. *Behaviour Research and Therapy*, 1973, **11**, 491-498.

Miller, P. M., Hersen, M., Eisler, R. M., and Hilsman, G. Effects of social stress on operant drinking of alcoholics and social drinkers. *Behaviour Research and Therapy*, 1974, **12**, 65-72.

Miller, P. M., Hersen, M., Eisler, R. M., and Watts, J. G. Contingent reinforcement of lowered blood/alcohol levels in an outpatient chronic alcoholic. *Behaviour Research and Therapy*, 1974, **12**, 261-263.

Mills, H. L., Agras, W. S., Barlow, D. H., and Mills, J. R. Compulsive rituals treated by response prevention: An experimental analysis. *Archives of General Psychiatry*, 1973, **28**, 524-529.

Mischel, W. *Personality and assessment*. New York: Wiley, 1968.

Mischel, W. *Introduction to personality*. New York: Holt, Rinehart and Winston, 1971.

Mischel, W. Direct versus indirect personality assessment: Evidence and implications. *Journal of Consulting and Clinical Psychology*, 1972, **38**, 319-324.

Montague, J. D., and Coles, E. M. Mechanism and measurement of the galvanic skin response. *Psychological Bulletin*, 1966, **65**, 261-279.

Moos, R. H. Behavioral effects of being observed: Reactions to a wireless radio transmitter. *Journal of Consulting and Clinical Psychology*, 1968, **32**, 383-388.

Nathan, P. E., and O'Brien, J. S. An experimental analysis of the behavior of alcoholics and nonalcoholics during prolonged experimental drinking: A necessary precursor to behavior therapy? *Behavior Therapy*, 1971, **2**, 455-476.

Neisworth, J. T., and Moore, F. Operant treatment of asthmatic responding with the parent as therapist. *Behavior Therapy*, 1972, **3**, 95-99.

O'Leary, K. D., and Kent, R. Behavior modification for social action: Research tactics and problems. In L. A. Hamerlynck, L. C. Handy, and E. J. March (Eds.), *Behavior change: Methodology, concepts, and practice*. Pp. 69-96. Champaign, Ill.: Research Press, 1973.

O'Leary, K. D., O'Leary, S. G., and Becker, W. C. Modification of a deviant sibling interaction pattern in the home. *Behaviour Research and Therapy*, 1967, **5**, 113-120.

Orne, M. T. On the social psychology of the psychological experiment: With particular reference to demand characteristics and their implications. *American Psychologist*, 1962, **17**, 776-783.

Orne, M. T. Demand characteristics and the concept of quasi-controls. In R. Rosenthal and R. Rosnow (Eds.), *Artifacts in behavioral research.* Pp. 144-177. New York: Academic Press, 1969.

Orne, M. T. From the subject's point of view, when is behavior private and when is it public: Problems of inference. *Journal of Consulting and Clinical Psychology*, 1970, **35**, 143-147.

Orne, M. T., and Evans, F. J. Social control in the psychological experiment: Antisocial behavior and hypnosis. *Journal of Personality and Social Psychology*, 1965, **1**, 189-200.

Orne, M. T., and Holland, C. H. On the ecological validity of laboratory deception. *International Journal of Psychiatry*, 1968, **6**, 282-293.

Patterson, G. R., and Harris, A. Some methodological considerations for observation procedures. Paper read at American Psychological Association, San Francisco, September 1968.

Paul, G. L. Strategy of outcome research in psychotherapy. *Journal of Consulting Psychology*, 1967, **31**, 104-118.

Peterson, D. R. Scope and generality of verbally defined personality factors. *Psychological Review*, 1965, **72**, 48-59.

Reid, J. B. Reliability assessment of observation data: A possible methodological problem. *Child Development*, 1970, **41**, 1143-1150.

Reid, J. B., and DeMaster, B. The efficacy of the spot-check procedure in maintaining the reliability of data collected by observers in quasi-natural settings: Two pilot studies. *Oregon Research Bulletin*, 1972, **12**, No. 8.

Risley, T. R., and Wolf, M. M. Strategies for analyzing behavioral change over time. In J. Nesselroade and H. Reese (Eds.), *Life-span developmental psychology: Methodological issues.* Pp. 175-183. New York: Academic Press, 1972.

Roberts, D. L., and Fletcher, D. A. Comparative study of blood alcohol testing devices. *Rocky Mountain Medical Journal*, 1969, **66**, 37-39.

Romanczyk, R. G., Kent, R. N., Diament, C., and O'Leary, K. D. Measuring the reliability of observational data: A reactive progress. *Journal of Applied Behavior Analysis*, 1973, **6**, 175-184.

Rorer, L. R. The great response style myth. *Psychological Bulletin*, 1965, **63**, 129-156.

Rosenthal, R. *Experimenter effects in behavioral research.* New York: Appleton-Century-Crofts, 1966.

Rosenthal, R., and Fode, K. L. The effect of experimenter bias on the performance of the albino rat. *Behavior Science*, 1963, **8**, 183-189.

Rosenthal, R., and Lawson, R. A longitudinal study of the effects of experimenter bias on the operant learning of laboratory rats. *Journal of Psychiatric Research*, 1964, **2**, 61-72.

Schacter, S. Some extraordinary facts about obese humans and rats. *American Psychologist*, 1971, **26**, 129-144.

Schwitzgebel, R. L. Survey of electromechanical devices for behavior modification. *Psychological Bulletin*, 1968, **70**, 444-459.

Scott, R. W., Peters, D., Gillespie, W. J., Blanchard, E. B., Edmunson, E. D., and Young, L. D. The use of shaping and reinforcement in the operant acceleration and deceleration of heart rate. *Behaviour Research and Therapy*, 1973, **11**, 179-185.

Scott, W. A., and Johnson, R. C. Comparative validities of direct and indirect personality tests. *Journal of Consulting and Clinical Psychology*, 1972, **38**, 301-318.

Shapiro, D., Barber, T. X., DiCara, L. V., Kamiya, J., Miller, N. E., and Stoyva, J. (Eds.) *Biofeedback and self-control 1972.* Chicago: Aldine, 1973.

Sherwood, J. J. Self-report and projective measures of achievement and affiliation. *Journal of Consulting Psychology*, 1966, **30**, 329-337.

Simkins, L. Examiner reinforcement and situational variables in a projective testing situation. *Journal of Consulting Psychology*, 1960, **24**, 541-547.

Simkins, L. The reliability of self-recorded behaviors. *Behavior Therapy*, 1971, **2**, 83-87.

Skindrud, K. An evaluation of observer bias in experimental field studies of social interaction. Unpublished doctoral dissertation, University of Oregon, Eugene, 1972.

Skindrud, K. Field evaluation of observer bias under overt and covert monitoring. In L. A. Hamerlynck, L. C. Handy, and E. J. March (Eds.), *Behavior change: Methodology, concepts, and practice.* Pp. 97-118. Champaign, Ill.: Research Press, 1973.

Skinner, B. F. *Science and human behavior.* New York: Macmillan, 1953.

Solyom, L., and Beck, P. R. GSR assessment of aberrant sexual behavior. *International Journal of Neuropsychiatry*, 1967, **3**, 52-59.

Stollak, G. E. Weight loss obtained under different experimental procedures. *Psychotherapy: Theory, Research and Practice*, 1967, **4**, 61-64.

Thoresen, C. E. The intensive design: An intimate approach to counseling research. Paper read at American Educational Research Association, Chicago, April 1972.

Tighe, T. J., and Elliott, R. A technique for controlling behavior in natural life settings. *Journal of Applied Behavior Analysis*, 1968, **1**, 263-266.

Turner, G. C., and Coleman, J. Examiner influence on Thematic Apperception responses. *Journal of Projective Techniques*, 1962, **26**, 478-486.

Turner, S. M., and Hersen, M. Instructions and reinforcement in modification of case of astasia-abasia. *Psychological Reports*, 1975, **36**, 606-612.

Venables, P. H. The relationship between PGR scores and temperature and humidity. *Quarterly Journal of Experimental Psychology*, 1955, **7**, 12-18.

Verplanck, N. S. The control and the content in conversation: Reinforcement of statements of opinion. *Journal of Abnormal and Social Psychology*, 1955, **31**, 668-676.

Wahler, R. G., and Leske, G. Accurate and inaccurate observer summary reports: Reinforcement theory interpretation and investigation. *Journal of Nervous and Mental Disease*, 1973, **156**, 386-394.

Walk, R. D. Self ratings of fear in a fear-invoking situation. *Journal of Abnormal and Social Psychology*, 1956, **22**, 171-178.

Wallace, J., and Sechrest, L. Frequency hypothesis and content analysis of projective techniques. *Journal of Consulting Psychology*, 1963, **27**, 387-393.

White, R. W. *The abnormal personality.* New York: Ronald Press, 1956.

Williams, J. G., Barlow, D. H., and Agras, W. S. Behavioral measurement of severe depression. *Archives of General Psychiatry*, 1972, **27**, 330-333.

Wincze, J. P., Leitenberg, H., and Agras, W. S. The effects of token reinforcement and feedback on the delusional verbal behavior of chronic paranoid schizophrenics. *Journal of Applied Behavior Analysis*, 1972, **5**, 247-262.

Wolf, M. M., and Risley, T. R. Reinforcement: Applied research. In R. Glaser (Ed.), *The nature of reinforcement.* Pp. 310-325. New York: Academic Press, 1971.

Wolpe, J. *The practice of behavior therapy.* New York: Pergamon Press, 1969.

Wolpe, J., and Lazarus, A. A. *Behavior therapy techniques.* New York: Pergamon Press, 1966.

Woodmansee, J. J. Methodological problems in pupillographic experiments. *Proceedings of the 74th Annual Convention of the American Psychological Association*, 1966, **1**, 133-134.

Zuckerman, M. Physiological measures of sexual arousal in the human. *Psychological Bulletin*, 1971, **75**, 297-329.

Zuckerman, M., Persky, H., and Link, K. E. The influence of set and diurnal factors on autonomic responses to sensory deprivation. *Psychophysiology*, 1969, **5**, 612-624.

Zung, W. W. K. A self-rating depression scale. *Archives of General Psychiatry*, 1965, **12**, 63-70.

CHAPTER 5

Basic A-B-A Designs

5.1. INTRODUCTION

In this chapter we will examine the prototype of experimental single case research–the A-B-A design–and its many variants. The primary objective is to inform and familiarize the reader as to the advantages and limitations of each design strategy while illustrating from the clinical, child, and behavior modification literatures. The development of the A-B-A design will be traced, beginning with its roots in the clinical case study and in the application of "quasi-experimental designs" (Campbell and Stanley, 1966). Procedural issues discussed at length in Chapter 3 will also be evaluated here for each of the specific design options as they apply. Both "ideal" and "problematic" examples, selected from the applied research area, will be used for illustrative purposes.

Limitations of the case study approach

For many years, descriptions of uncontrolled case histories have predominated in the psychoanalytic, psychotherapeutic, and psychiatric literatures (see Chapter 1). Despite the development of applied behavioral methodology (presumably based on sound theoretical underpinnings) in the late 1950s and early to mid-1960s, the case study approach was still the primary method for demonstrating the efficacy of innovative treatment techniques (e.g., Ashem, 1963; Lazarus, 1963; Ullmann and Krasner, 1965; Wolpe, 1958).

Although there can be no doubt that the case history method yields interesting (albeit uncontrolled) data, that it is a rich source for clinical speculation, and that ingenious technical developments derive from its application, the multitude of uncontrolled factors present in each study do not permit sound cause-and-effect conclusions. Even when the case study method is applied at its best (e.g., Lazarus, 1973), the absence of experimental control and the lack of precise measures for target behaviors under evaluation remain mitigating factors. Of course, proponents of the case study method (e.g., Lazarus and Davison, 1971) are well aware of its inherent limitations as an evaluative tool, but they show

how it can be used to advantage to generate hypotheses that later may be subjected to more rigorous experimental scrutiny. Some of these advantages are listed below. The case study method can be used to: (1) foster clinical innovation, (2) cast doubt on theoretic assumptions, (3) permit study of rare phenomena (e.g., Gilles de la Tourette's Syndrome), (4) develop new technical skills, (5) buttress theoretical views, (6) result in refinement of techniques, and (7) provide clinical data to be used as a departure point for subsequent controlled investigations.

With respect to the latter point, Lazarus and Davison (1971) refer to the use of "objectified single case studies." Included are the A-B-A experimental designs that allow for an analysis of the controlling effects of variables, thus permitting scientifically valid conclusions. However, in the more typical case study approach a subjective description of treatment interventions and resulting behavioral changes is made by the therapist. Most frequently, several techniques are administered simultaneously, therefore precluding an analysis of the relative merits of each procedure. Moreover, evidence for improvement is usually based on the therapist's "global" clinical impressions. Not only is there the strong possibility of bias in these evaluations, but controls for the treatment's "placebo" value are unavailable. Finally, the effects of time (maturational factors) are confounded with application of the treatment(s), and the specific contribution of each of the factors is obviously not distinguished.

A very modest improvement over the uncontrolled case study method elsewhere (Browning and Stover, 1971) has been labeled the "B Design". In this "Design," baseline measurement is omitted but the investigator monitors one or a number of target measures throughout the course of treatment. One might also categorize this procedure as the simplest of the time series analyses (see Glass, Willson, and Gottman, 1973). Although this strategy obviously yields a more objective appraisal of the patient's progress, the confounds that typify the case study method apply equally here. In that sense the "B Design" is essentially an uncontrolled case study with objective measures taken repeatedly.

5.2. A-B DESIGN

The A-B design, although the simplest of the experimental strategies, corrects for some of the deficiencies of the case study method and those of the "B Design." In this design the target behavior is clearly specified and repeated measurement is taken throughout the A and B phases of experimentation. As in all single case experimental research, the A phase involves a series of baseline observations of the natural frequency of the target behavior(s) under study. In the B phase the treatment variable is introduced and changes in the dependent measure are noted. Thus, with *some major reservations*, changes in the dependent variable are attributed to the effects of treatment (Barlow and Hersen, 1973; Campbell, 1969; Campbell and Stanley, 1966; Eisler and Hersen, 1973; Gottman, 1973; Gottman, McFall, and Barnett, 1969; Hersen, 1973; Risley and Wolf, 1972; Wolf and Risley, 1971).

Let us now examine some of the important reservations. In their evaluation of the A-B strategy, Wolf and Risley (1971) argue that "The analysis provided no information about what the natural course of the behavior would have been had we not intervened with our treatment condition" (pp. 314–315). That is to say, it is very possible that changes in the B phase might have occurred regardless of the introduction of treatment or that changes in B might have resulted as a function of correlation with some fortuitous (but uncontrolled) event. When considered in this light, the A-B strategy does not permit a full experimental analysis of the controlling effects of the treatment inasmuch as its correlative properties are quite apparent. Indeed, Campbell and Stanley (1966) refer to this strategy as a "quasi-experimental design."

Risley and Wolf (1972) present an interesting discussion of the limitations of the A-B design with respect to predicting or "forecasting" the B phase on the basis of data obtained in A. Two hypothetical examples of the A-B design are depicted, with both showing a mean increase in the amount of behavior in B over A. However, in the first example, a steady and stable trend in baseline is followed by an abrupt increase in B, which is then maintained. In the second case, the upward trend in A is continued in B. Therefore, despite the equivalence of means and variances in the two cases, the importance of the trend in evaluating the data is underscored. Some tentative conclusions can be reached on the basis of the first example, but in the second example the continued linear trend in B permits no conclusions as to the controlling effects of the B treatment variable.

In further analyzing the difficulties inherent in the A-B strategy, Risley and Wolf (1972) contend that "The weakness in this design is that the data in the experimental condition is compared with a forecast from the prior baseline data. The accuracy of an assessment of the role of the experimental procedure in producing the change rests upon the accuracy of that forecast. A strong statement of causality therefore requires that the forecast be supported. This support is accomplished by elaborating the A-B design" (p. 5). Such elaboration is found in the A-B-A design to be discussed and illustrated in Section 5.3 of this chapter.

Despite these aforementioned limitations, it is shown how in some settings (where control group analysis or repeated introduction and withdrawals of treatment variables are not feasible) the A-B design can be of some utility (Campbell and Stanley, 1966). For example, the use of the A-B strategy in the private practice setting has previously been recommended in Section 3.2 of Chapter 3.

Campbell (1969) presents a comprehensive analysis of the use of the A-B strategy in field experiments where more traditional forms of experimentation are not at all possible (e.g., the effects of modifying traffic laws on the documented frequency of accidents). However, when using the "quasi-experimental" design, Campbell cautions the investigator as to the numerous threats to internal validity (history, maturation, instability, testing, instrumentation, regression artifacts, selection, experimental mortality, and selection-maturation interaction) and external validity (interaction effects of testing, interaction of selection and experimental treatment, reactive effects of experimental arrangements, multiple-treatment interference, irrelevant responsiveness of measures, and irrelevant replicability of treatments) that may be encountered. The interested reader is referred to Campbell's (1969) excellent article for a full discussion of the issues involved in large-scale retrospective or prospective field studies.

In summary, it should be apparent that the use of a "quasi-experimental design" such as the A-B strategy results in rather weak conclusions. This design is subject to the influence of a host of confounding variables and is best applied as a last resort measure when circumstances do not allow for more extensive experimentation. Examples of such cases will now be illustrated.

A-B with single target measure and extended follow-up

Epstein and Hersen (1974) used an A-B design with an extended follow-up procedure to assess the effects of reinforcement on frequency of gagging in a 26-year-old psychiatric inpatient. The patient's symptomatology had persisted for approximately 2 years despite repeated attempts at medical intervention. During baseline (A phase), the patient was instructed to record time and frequency of each gagging episode on an index card, collected by the experimenter the following morning at ward rounds. Treatment (B phase) consisted of presenting the patient with $2.00 in canteen books (exchangeable at the hospital store for goods) for a decrease (N-1) from the previous daily frequency. In addition, zero rates of gagging were similarly reinforced. In order to facilitate maintenance of gains after treatment, no instructions were given as to how the patient might control his gagging. Thus, emphasis was placed on self-management of the disorder. At the conclusion of his hospital stay, the patient was requested to continue recording data at home for a period of 12 weeks. In this case, treatment conditions were not withdrawn during the patient's hospitalization because of clinical considerations.

Results of this study are plotted in Fig. 5-1. Baseline frequency of gagging fluctuated between eight and seventeen episodes per day but stabilized to some extent in the last 4 days. Institution of reinforcement procedures in the B phase resulted in a decline to zero within 6 days. However, on Day 15, frequency of gagging rose again to seven daily episodes. At this point, the criterion for obtaining reinforcement was reset to that originally planned for Day 13. Renewed improvement was then noted between Days 15-18, and treatment was continued through Day 24. Thus, the B phase is twice as long as baseline, but it was extended for very obvious clinical considerations.

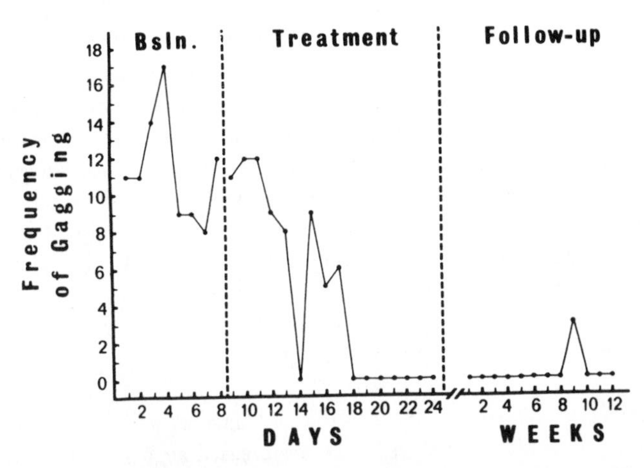

Fig. 5-1. Frequency of gagging during baseline, treatment, and follow-up. (Fig. 1, p. 103, from: Epstein, L. H., and Hersen, M. Behavioral control of hysterical gagging. *Journal of clinical Psychology*, 1974, **30**, 102-104. Reproduced by permission.)

The 12-week follow-up period reveals a zero level of gagging with the exception of Week 9, when three gagging episodes were recorded. Follow-up data were corroborated by the patient's wife, thus precluding the possibility that treatment only affected the patient's verbal report rather than diminution of actual symptomatology.

Although treatment appeared to be the effective ingredient of change in this study, particularly in light of the longevity of the patient's disorder, it is conceivable that some unidentified variable coincided with the application of reinforcement procedures and actually accounted for observed changes. However, the A-B design does not permit a definitive answer to this question. It might also be noted that the specific use of this design (baseline, treatment, and follow-up) could readily have been carried out in an outpatient facility (clinic or private practice setting) with a minimum of difficulty and with no deleterious effects to the patient.

A-B with multiple target measures

In our next example we will examine the use of an A-B design in which a number of target behaviors were monitored simultaneously (Eisler and Hersen, 1973). The effects of token economy on points earned, behavioral ratings of depression (Williams, Barlow, and Agras, 1972), and self-ratings of depression (Beck Depressive Inventory–Beck, Ward, Mendelsohn, Mock, and Erbaugh, 1961) were assessed in a 61-year-old reactively depressed male patient. In this study the treatment variable was not withdrawn due to time limitations. During baseline (A), the patient was able to earn points for a variety of specified target behaviors (designated under general rubrics of work, personal hygiene, and responsibility), but these points earned were exchangeable for ward privileges and material goods in the hospital canteen. During each phase, the patient filled out a Beck Depressive Inventory (three alternate forms were used to prevent possible response bias) at daily morning "Banking Hours," at which time points previously earned on the token economy were tabulated. In addition, behavioral ratings (talking, smiling, motor activity) of depression (high ratings indicate low depression) were obtained surreptitiously on the average of one per hour between the hours of 8:00 a.m. and 10:00 p.m. during non-work-related activities.

The results of this study appear in Fig. 5-2. Inspection of these data indicates that number of points earned in baseline increased slightly but then stabilized. Baseline ratings of depression show stability, with evidence of greater daytime activity. Beck scores ranged from 19-28. Insitution of token economy on Day 5 resulted in a marked linear increase in points earned, a substantial increase in day and evening behavioral ratings of depression, and a linear decrease in self-reported Beck Inventory scores.

Thus, it appears that token economy effected improvement in this patient's depression as based on both objective and subjective indices. However, as was previously pointed out, this design does not permit a direct analysis of the controlling effects of the therapeutic variable introduced (token economy), as does our example of an A-B-A design seen in Fig. 5-5 (Hersen, Eisler, Alford, and Agras, 1973). Nonetheless, the use of an A-B design in this case proved to be useful for two reasons. First, from a clinical standpoint, it was possible to obtain *some* objective estimate of the treatment's success during the patient's abbreviated hospital stay. Second, the results of this study prompted the further investigation of the effects of token economic procedures in three additional reactively depressed subjects (Hersen, Eisler, Alford, and Agras, 1973). In that investigation more sophisticated experimental strategies confirmed the controlling effects of token economy in neurotic depression.

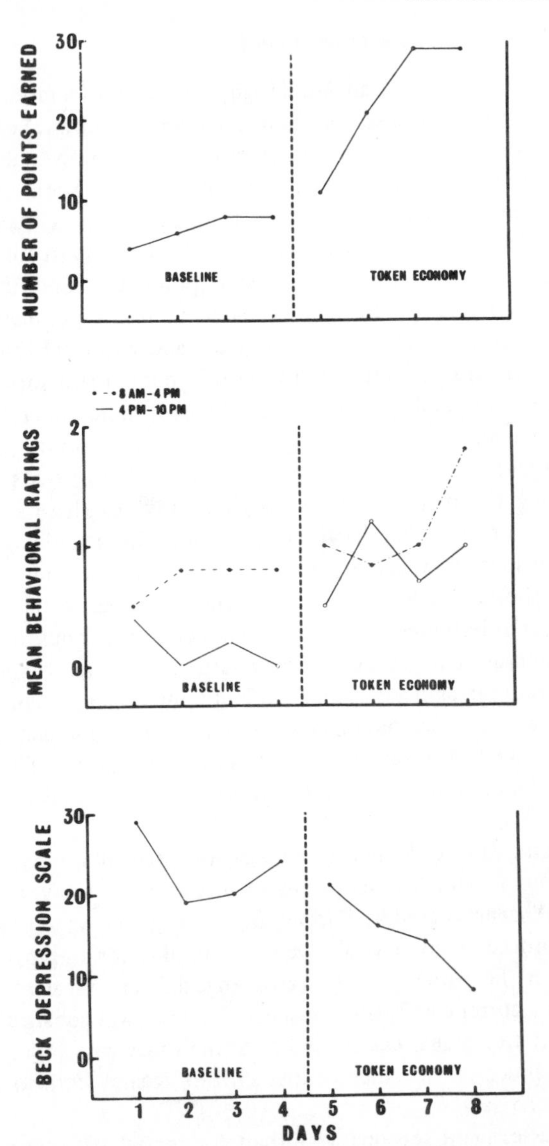

Fig. 5-2. Number of points earned, mean behavioral ratings, and Beck Depression Scale scores during baseline and token economy in a reactively depressed patient. (Fig. 1, from: Eisler, R.M., and Hersen, M. The A-B design: Effects of token economy on behavioral and subjective measures in neurotic depression. Paper read at American Psychological Association, Montreal, August 29, 1973.)

A-B with follow-up and booster treatment

In our next illustration of an A-B design, clinical considerations necessitated a short baseline period and also contraindicated the withdrawal of treatment procedures (Harbert, Barlow, Hersen, and Austin, 1974). However, during the course of extended follow-up assessment, the patient's condition deteriorated and required the reinstatement of treatment in booster sessions. Renewed improvement immediately followed, thus lending additional support for the treatment's efficacy. When examined from a design standpoint, the conditions of the more complete A-B-A-B strategy are approximated in this experimental case study.

More specifically, Harbert, Barlow, Hersen, and Austin (1974) examined the effects of covert sensitization therapy on self-report (card sort technique) and physiological (mean penile circumference changes) indices in a 52-year-old male inpatient who complained of a long history of incestuous episodes with his adolescent daughter. The card sort technique consisted of ten scenes (typed on cards) depicting the patient and his daughter. Five of these scenes were concerned with normal father-daughter relations; the remaining five involved descriptions of incestuous activity between father and daughter. The patient was asked to rate the ten scenes, presented in random sequence, on a 0-4 basis, with 0 representing no desire and 4 representing much desire. Thus, measures of both deviant and non-deviant aspects of the relationship were obtained throughout all phases of study. In addition, penile circumference changes scored as a percentage of full erection were obtained in response to audiotaped descriptions of incestuous activity and in reaction to slides of the daughter. Three days of self-report data and 4 days of physiological measurements were taken during baseline (A phase).

Covert sensitization treatment (B phase) consisted of approximately 3 weeks of daily sessions in which descriptions of incestuous activity were paired with the "nauseous" scene as used by Barlow, Leitenberg, and Agras (1969). However, as "nausea" proved to be a weak aversive stimulus for this patient, a "guilt" scene–in which the patient is discovered engaging in sexual activity with the daughter by his current wife and a respected priest–was substituted during the second week of treatment. The flexibility of the single case approach is exemplified here inasmuch as a "therapeutic shift of gears" follows from a close monitoring of the data.

Follow-up assessment sessions were then conducted after termination of the patient's hospitalization at 2-week, 1-, 2-, 3-, and 6-month intervals. After each follow-up session, brief booster covert sensitization was administered.

The results of this study appear in Figs. 5-3 and 5-4. Inspection of Fig. 5-3 indicates that mean penile circumference changes to audiotapes in baseline ranged from 18 percent to 35 percent (mean = 22·8 percent). Penile circumference changes to slides ranged from 18 percent to 75 percent (mean = 43·5 percent). Examination of Fig. 5-4 shows that non-deviant scores remained at a maximum of 20 for all three baseline probes; deviant scores achieved a level of 17 throughout.

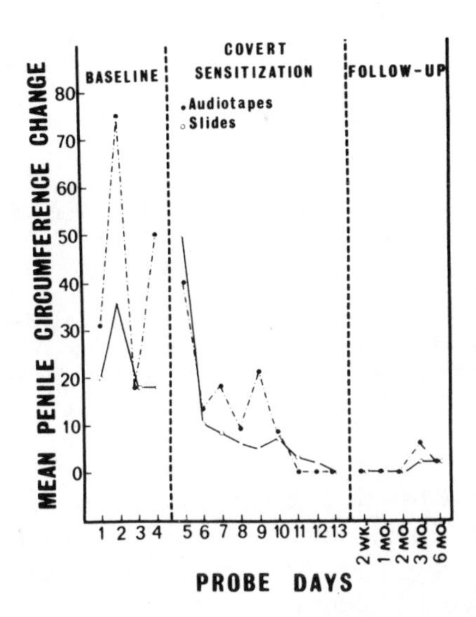

Fig. 5-3. Mean penile circumference change to audiotapes and slides during baseline, covert sensitization, and follow-up. (Fig. 1, p. 83, from: Harbert, T.L., Barlow, D.H., Hersen, M., and Austin, J.B. Measurement and modification of incestuous behavior: A case study. *Psychological Reports*, 1974, **34,** 79-86. Reproduced by permission.)

Introduction of standard covert sensitization followed by use of the "guilt" imagery resulted in decreased penile responding to audiotapes and slides (see Fig. 5-3), and a substantial decrease in the patient's self-reports of deviant interests in his daughter (see Fig. 5-4). Non-deviant interests, however, remained at a high level.

Follow-up data in Fig. 5-3 reveal that penile circumference changes remained at zero during the first three probes but increased slightly at the 3-month assessment. Similarly, Fig. 5-4 data show a considerable increase in deviant interests at the 3-month follow-up. This coincides with the patient's reports of marital disharmony. In addition, non-deviant interests diminished during follow-up (at that point the patient was angry at his daughter for rejecting his positive efforts at being a father).

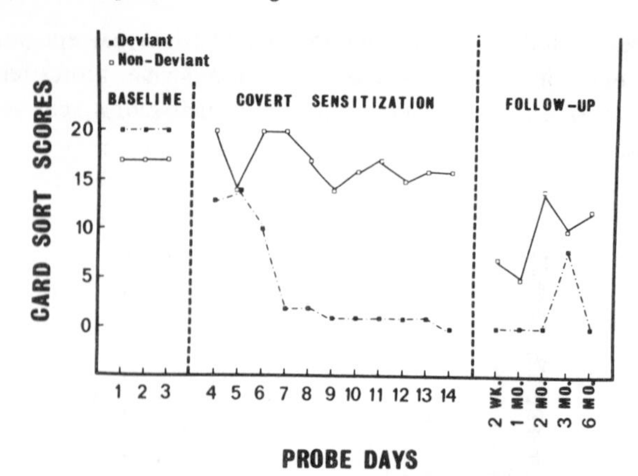

Fig. 5-4. Card sort scores on probe days during baseline, covert sensitization, and follow-up. (Fig. 2, p. 84, from: Harbert, T.L., Barlow, D.H., Hersen, M., and Austin, J.B. Measurement and modification of incestuous behavior: A case study. *Psychological Reports*, 1974, **34**, 79-86. Reproduced by permission.)

As there appeared to be some deterioration at the 3-month follow-up, an additional course of outpatient covert sensitization therapy was carried out in three weekly sessions. The final assessment period at 6 months appears to reflect the effects of additional treatment in that (1) penile responding was negligible, and (2) deviant interests had returned to a zero level.

5.3. A-B-A DESIGN

The A-B-A design is the simplest of the experimental analysis strategies in which the treatment variable is introduced and then withdrawn. Whereas the A-B design only permits tentative conclusions as to a treatment's influence, the A-B-A design allows for an analysis of the controlling effects of its introduction and subsequent removal. If after baseline measurement (A) the application of a treatment (B) leads to improvement and conversely results in deterioration after it is withdrawn (A), one can conclude with a high degree of certainty that the treatment variable is the agent responsible for observed changes in the target behavior. Unless the natural history of the behavior under study were to follow identical fluctuations in trends, it is *most improbable* that observed changes are due to any influence (e.g., some correlated or uncontrolled variable) other than the treatment variable that is systematically changed. Also, replication of the A-B-A design in different subjects strengthens conclusions as to power and controlling forces of the treatment (see Chapter 9).

Although the A-B-A strategy is acceptable from an experimental standpoint, it has one major undesirable feature when considered from the clinical context. Unfortunately for the patient or subject, this paradigm ends on the A or baseline phase of study, therefore denying him the full benefits of experimental treatment. Along these lines, Barlow and Hersen (1973) have argued that "On an ethical and moral basis it certainly behooves the experimenter-clinician to continue some form of treatment to its ultimate conclusion subsequent to completion of the research aspects of the case. A further design, known as the A-B-A-B design, meets this criticism as study ends on the B or treatment phase" (p. 321). However, despite this limitation, the A-B-A design is a useful research tool when time factors (e.g., premature discharge of a patient) or clinical aspects of a case (e.g., necessity of changing the level of medication in addition to reintroducing a treatment variable after the second A phase) interfere with the correct application of the more comprehensive A-B-A-B strategy.

A second problem with the A-B-A strategy concerns the issues of multiple treatment interference and sequential confounding (Bandura, 1969; Campbell and Stanley, 1966). The problem of sequential confounding in an A-B-A design and its variants also limits generalization to the clinic somewhat. As Bandura (1969) and Kazdin (1973b) note, the effectiveness of a therapeutic variable in the final phase of an A-B-A design can only be interpreted in the context of the previous phases. Change occurring in this last phase may not be comparable to changes that would have occurred if the treatment had been introduced initially. For instance, in an A-B-BC-B design, when A is baseline and B and C are two therapeutic variables, the effects of the BC phase may be more or less powerful than if they had been introduced initially. This point has been demonstrated in studies by O'Leary and his associates (O'Leary and Becker, 1967; O'Leary, Becker, Evans, and Saudargas, 1969), who noted that the simultaneous introduction of two variables produced greater change than the sequential introduction of the same two variables.

Similarly, the second introduction of variable A in a withdrawal A-B-A design may affect behavior differently than the first introduction. (Generally our experience is that behavior improves more rapidly with a second introduction of the therapeutic variable.) In any case, the reintroduction of therapeutic phases is a feature of A-B-A designs that differs from the typical applied clinical situation when the variable is introduced only once. Thus, appropriate cautions must be exercised in generalizing results from phases occurring late in an experiment to the clinical situation.

In dealing with this problem, the clinical researcher should keep in mind that the purpose of subsequent phases in an A-B-A design is to confirm the effects of the independent variable (internal validity) rather than to generalize to the clinical situation. The results that are most generalizable, of course, are data from the first introduction of the treatment. When two or more variables are introduced in sequence, the purpose again is to test the separate effects of each variable. Subsequently, order effects and effects of combining the variable can be tested in systematic replication series, as was the case with the O'Leary, Becker, Evans, and Saudergas (1969) study.

Two examples of the A-B-A design, one selected from the clinical literature and one from the child development area, will be used for illustration. Attention will be focused on some of the procedural issues outlined in Chapter 3.

A-B-A from clinical literature

In pursuing their study of the effects of token economy on neurotic depression, Hersen and his colleagues (Hersen, Eisler, Alford, and Agras, 1973) used A-B-A strategies with three reactively depressed subjects. The results for one of these subjects (52-year-old, white, married farmer who became depressed after

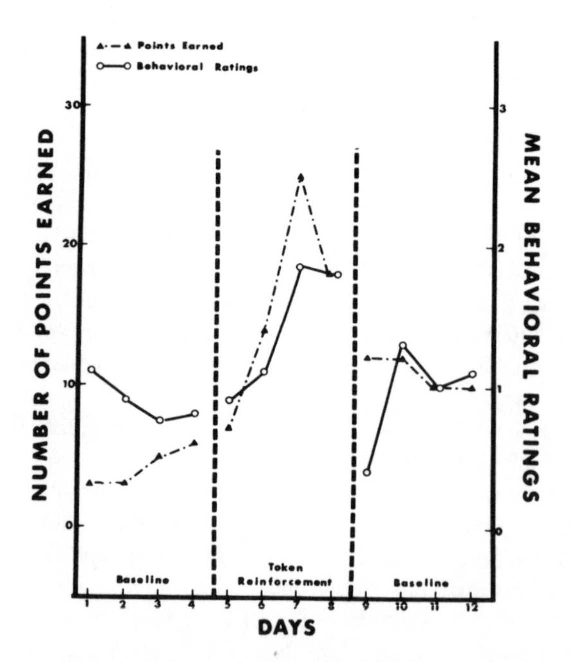

Fig. 5-5. Number of points earned and mean behavioral ratings for Subject 1. (Fig. 1, p. 394, from: Hersen, M., Eisler, R.M., Alford, G.S., and Agras, W.S. Effects of token economy on neurotic depression: An experimental analysis. *Behavior Therapy*, 1973, **4**, 392-397. Reproduced by permission.)

the sale of his farm) appear in Fig. 5-5. As in the Eisler and Hersen (1973) study, previously described in detail in Section 5.2 of this chapter, points earned in baseline (A) had no exchange value, but during the token reinforcement phase (B) they were exchangeable for privileges and material goods. Unlike the Eisler and Hersen (1973) study, however, token reinforcement procedures were withdrawn and a return to baseline conditions (A) took place during Days 9-12. The effects of introducing and removing token economy were examined on two target behaviors—points earned and behavioral ratings (higher ratings indicate lowered depression).

A careful examination of baseline data reveals a slightly decreased trend in behavioral ratings, thus indicating some very minor deterioration in the patient's condition. As was noted in Section 3.3 of Chapter 3, the "deteriorating" baseline is considered to be an acceptable trend. However, there appeared to be a concomitant but slight increase in points earned during baseline. It will be recalled that an improved trend in baseline is not the most desirable trend. However, as the slope of the curve was not extensive and in light of the primary focus on behavioral ratings (depression), we proceeded with our change in conditions on Day 5. Had there been unlimited time, baseline conditions would have been maintained until number of points earned daily stabilized to a greater extent.

We might note parenthetically at this point that all of the "ideal" conditions (procedural rules) outlined in our discussion in Chapter 3 are rarely approximated when conducting single case experimental research. Our experience shows that procedural variations from the "ideal" are required, as data simply *do not* conform to theoretical expectation. Moreover, experimental finesse is sometimes sacrificed at the expense of time and clinical considerations.

Continued examination of Fig. 5-5 indicates that instigation of token economic procedures on Day 5 resulted in a marked linear increase in both points earned and behavioral ratings. The abrupt change in slope of the curves, particularly in points earned, strongly suggests the influence of the token economy variable, despite the slightly upward trend initially seen in baseline. Removal of token economy on Day 9 led to an initially large drop in behavioral ratings, which then stabilized at a somewhat higher level. Points earned also declined but maintained stability throughout the second 4-day baseline period. The obtained decrease in target behaviors in the second baseline phase confirms the controlling effects of token economy over neurotic depression in this paradigm. We might also point out here that an equal number of data points appear in each phase, thus facilitating interpretation of the trends.

These results were replicated in two additional reactively depressed subjects (Hersen, Eisler, Alford, and Agras, 1973), lending further credence to the notion that token economy exerts a controlling influence over the behavior of neurotically depressed individuals.

A-B-A from child literature

Walker and Buckley (1968) used an A-B-A design in their functional analysis of the effects of an individualized educational program for a $9\frac{1}{2}$-year-old boy whose extreme distractibility in a classroom situation interfered with task-oriented performance. During baseline assessment (A), percentage of attending behavior was recorded in 10-minute observation sessions while the subject was engaged in working on programed learning materials. Following baseline measurement, a reinforcement contingency (B) was instituted whereby the subject earned points (exchangeable for a model of his choice) for maintaining his attention (operationally defined for him) to the learning task. During this phase, a progressively increasing time criterion for attending behaviors over sessions was

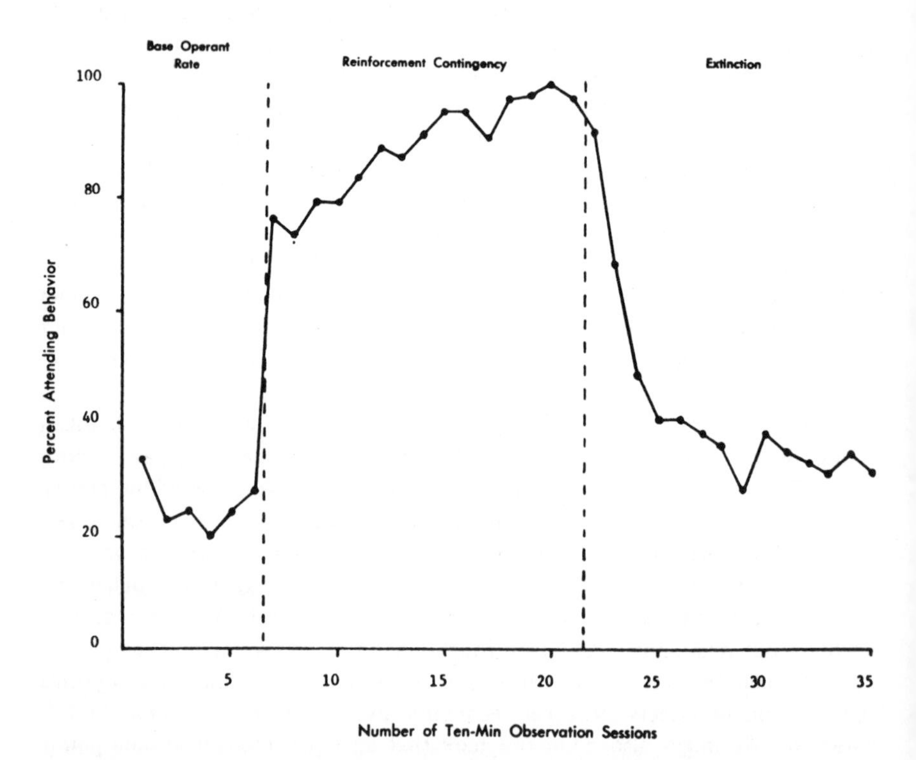

Fig. 5-6. Percentage of attending behavior in successive time samples during the individual conditioning program. (Fig. 2, p. 247, from: Walker, H.M., and Buckley, N.K. The use of positive reinforcement in conditioning attending behavior. *Journal of Applied Behavior Analysis,* 1968, 1, 245-250. Copyright by Society for the Experimental Analysis of Behavior, Inc. Reproduced by permission.)

required (30 to 600 seconds of attending per point). The extinction phase (A) involved a return to original baseline conditions.

Examination of baseline data shows a slightly decreasing trend followed by a slightly increasing trend, but within stable limits (mean = 33 percent). Insitution of reinforcement procedures led to an immediate improvement, which then increased to its asymptote in accordance with the progressively more difficult criterion. Removal of the reinforcement contingency in extinction resulted in a decreased percentage of attending behaviors to approximately baseline levels. After completion of experimental study, the subject was returned to his classroom where a variable-interval reinforcement program was used to increase and maintain attending behaviors in that setting.

With respect to experimental design issues, we might point out that Walker and Buckley (1968) used a short baseline period (6 data points) followed by longer B (15 data points) and A phases (14 data points). However, in view of the fact that an immediate and large increase in attention was obtained during reinforcement, the possible confound of time when using disparate lengths of phases (see Section 3.6, Chapter 3) *does not* apply here. Moreover, the shape of the curve in extinction (A) and the relatively equal lengths of the B and A phases further dispel doubts that the reader might have as to the confound of time.

Secondly, with respect to the decreasing-increasing baseline obtained in the first A phase, although it might be preferable to extend measurement until full stability is achieved (see Section 3.3, Chapter 3), the range of variability is very constricted here, thus delimiting the importance of the trends.

5.4. A-B-A-B DESIGN

The A-B-A-B strategy, referred to as an "equivalent time-samples design" by Campbell and Stanley (1966), controls for the deficiencies present in the A-B-A design. Specifically, the A-B-A-B design ends on a treatment phase (B), which then can be extended beyond the experimental requirements of study for clinical reasons (e.g., Miller, 1973). In addition, this design strategy provides for *two* occasions (B to A and then A to B) for demonstrating the positive effects of the treatment variable. This, then, strengthens the conclusions that can be derived as to its controlling effects over target behaviors under observation (Barlow and Hersen, 1973).

In the succeeding subsections we will provide four examples of the use of the A-B-A-B strategy. In the first we will present an example from the child literature which illustrates the "ideal" in procedural considerations. In the second we will examine the problems encountered in interpretation when improvement fortuitously occurred during the second baseline period. In the third we will illustrate the use of the A-B-A-B design when concurrent behaviors are monitored in addition to targeted behaviors of interest. Finally, in the fourth we will examine the advantages and disadvantages of using the A-B-A-B strategy without the experimenter's knowledge of results throughout the different phases of study.

A-B-A-B from child literature

An excellent example of the A-B-A-B design strategy appears in a study conducted by Hall, Fox, Willard, Goldsmith, Emerson, Owen, Davis, and Porcia (1971). In this study the effects of contingent teacher attention were examined in a 10-year-old retarded boy whose "talking-out" behaviors during special education classes proved to be disruptive as other children then emulated his actions. Baseline observations of "talk-outs" were recorded by the teacher (reliability checks indicated 84 percent to 100 percent agreement) during five daily 15-minute sessions. During these first five sessions, the teacher responded naturally to "talk-outs" by paying attention to them. However, in the next five sessions, the teacher was instructed to ignore "talk-outs" but to provide

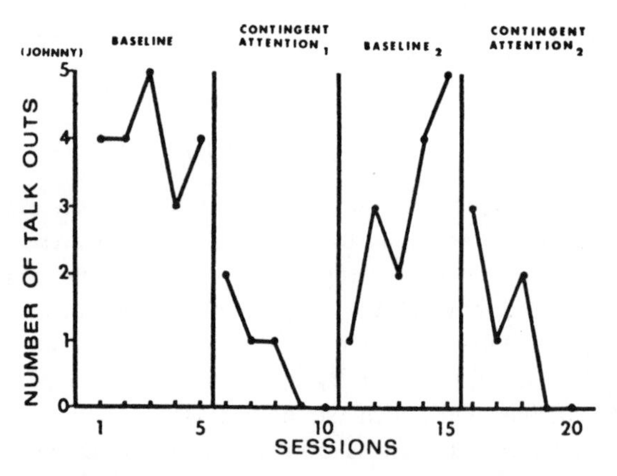

Fig. 5-7. A record of talking-out behavior of an educable mentally retarded student. Baseline$_1$–before experimental conditions. Contingent Teacher Attention$_1$–systematic ignoring of talking-out and increased teacher attention to appropriate behavior. Baseline$_2$–reinstatement of teacher attention to talking-out behavior. Contingent Teacher Attention$_2$–return to systematic ignoring of talking-out and increased attention to appropriate behavior. (Fig. 2, p. 143, from: Hall, R.V., Fox, R., Willard, D., Goldsmith, L., Emerson, M., Owen, M., Davis, T., and Porcia, E. The teacher as observer and experimenter in the modification of disputing and talking-out behaviors. *Journal of Applied Behavior Analysis*, 1971, **4**, 141-149. Copyright by Society for the Experimental Analysis of Behavior, Inc. Reproduced by permission.)

increased attention to the child's productive behaviors. The third series of five sessions involved a return to baseline conditions while the last series of five sessions consisted of reinstatement of contingent attention.

The results of this study are plotted in Fig. 5-7. The presence of equal phases in this study facilitates the analysis of results. Baseline data are stable and range from three to five "talk-outs", with three of the five points at a level of four "talk-outs" per session. Institution of contingent attention resulted in a marked decrease that achieved a zero level in Sessions 9 and 10. Removal of contingent attention led to a linear increase of "talk-outs" to a high of five. However, reinstatement of contingent attention once again brought "talk-outs" under experimental control. Thus, application and withdrawal of contingent attention clearly demonstrates its controlling effects on "talk-out" behaviors. This is twice-documented as seen in the decreasing and increasing data trends in the second set of A and B phases.

A-B-A-B with unexpected improvement in baseline

In our next example we will illustrate the difficulties that arose in interpretation when unexpected improvement took place during the latter half of the second series of baseline (A) measurements. Epstein, Hersen, and Hemphill (1974) used an A-B-A-B design in their assessment of the effects of feedback on frontalis muscle activity in a patient who had suffered from chronic headaches for a 16-year period. EMG recordings were taken for 10 minutes following 10 minutes of adaptation during each of the six baseline (A) sessions. EMG data were obtained while the patient relaxed in a reclining chair in the experimental laboratory. During the six feedback (B) sessions, the patient's favorite music (prerecorded on tape) was automatically turned on whenever EMG activity decreased below a pre-set criterion level. Responses above that level conversely turned off recordings of music. Instructions to the patient during this phase were to "keep the music on." In the next six sessions baseline (A) conditions were reinstated, while the last six sessions involved a return to feedback (B). Throughout all phases of study, the patient was asked to keep a record of the intensity of headache activity.

Examination of Fig. 5-8 indicates that EMG activity during baseline ranged from 28 to 50 seconds (mean = 39·18) per minute that contained integrated responses above the criterion microvolt level. Institution of feedback procedures resulted in decreased activity (mean = 23·18). Removal of feedback in the second baseline initially resulted in increased activity in Sessions 13-15. However, an *unexplained but decreased trend* was noted in the last half of that phase. This downward trend, to some extent, detracts from the interpretation that music feedback was the responsible agent of change during the first B phase. In addition, the importance of maintaining equal lengths of phases is highlighted here. Had baseline measurement been concluded on Day 15, an unequivocal interpretation (though probably erroneous) would have been made. However, despite the downward trend in baseline, mean data for this phase (30·25) were higher than the previous feedback phase (23·18).

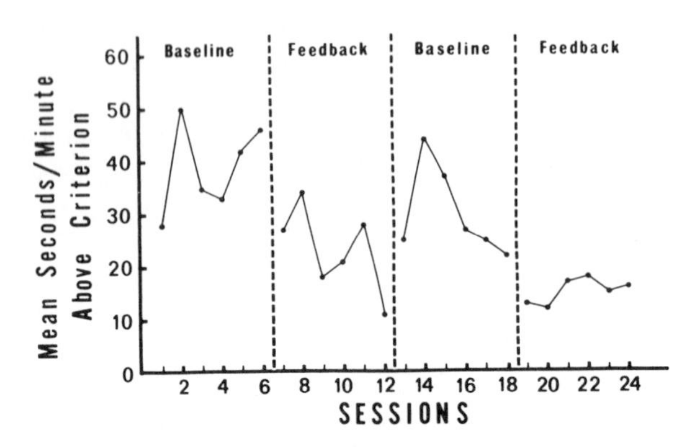

Fig. 5-8. Mean seconds per minute that contained integrated responses above criterion microvolt level during baseline and feedback phases. (Fig. 1, p. 61, from: Epstein, L.H., Hersen, M., and Hemphill, D.P. Music feedback as a treatment for tension headache: An experimental case study. *Journal of Behavior Therapy and Experimental Psychiatry*, 1974, 5, 59-63. Reproduced by permission.)

In the final phase, feedback resulted in a further decline that was generally maintained at low levels (mean = 14·98). Unfortunately, it is not fully clear whether this further decrease might have occurred naturally without the benefits of renewed introduction of feedback. Therefore, despite the presence of statistically significant differences between baseline and feedback phases and confirmation of EMG differences by self-reports of decreased headache intensity during feedback, the downward trend in the second baseline prevents a definitive interpretation of the controlling effects of the feedback procedure.

When the aforementioned data patterns results, it is recommended, where possible, that variables possibly leading to improvement in baseline be examined through additional experimental analyses. However, time limitations and pressing clinical needs of the patient or subject under study usually preclude such additional study. Therefore, the next best strategy involves a replication of the procedure with the same subject—as was done on an outpatient basis by Epstein, Hersen, and Hemphill (1974)—or with additional subjects bearing the same kind of diagnosis (see Chapter 9).

A-B-A-B with monitoring of concurrent behaviors

When using the withdrawal strategy, such as the A-B-A-B design, most experimenters have been concerned with the effects of their treatment variable on one behavior—the targeted behavior. However, in recent reports (Kazdin, 1973a; Kazdin, 1973b; Lovaas and Simmons, 1969; Risley, 1968; Sajwaj, Twardosz, and Burke, 1972; Twardosz and Sajwaj, 1972) the importance of monitoring concurrent (non-targeted) behaviors is documented. This is of particular importance when side effects of treatment are possibly negative (see Sajwaj, Twardosz, and Burke, 1972). Kazdin (1973b) has listed some of the potential advantages in monitoring the multiple effects of treatment in operant paradigms. "One initial advantage is that such assessment would permit the possibility of determining response generalization. If certain response frequencies are increased or decreased, it would be expected that other related operants would be influenced. It would be a desirable addition to determine generalization of beneficial response changes by looking at behavior related to the target response. In addition, changes in the frequency of responses might also correlate with topographical alterations" (p. 527). We might note here that the examination of collateral effects of treatment should not be restricted to operant paradigms when using experimental single case designs.

In our following example the investigators (Twardosz and Sajwaj, 1972) used an A-B-A-B design to evaluate the efficacy of their program to increase sitting in a 4-year-old hyperactive, retarded boy, who was enrolled in an experimental preschool class. In addition to assessment of the target behavior of interest (sitting), the effects of treatment procedures on a variety of concurrent behaviors (posturing, walking, use of toys, proximity of children) were monitored. Observations of this child were made during a free-play period (one-half hour) in which class members were at liberty to choose their playmates and toys. During baseline (A), the teacher gave the child instructions (as she did to all others in class) but did not prompt him to sit or praise him when he did. Institution of the sitting program (B) involved prompting the child (placing him in a chair with toys before him on the table), praising him for remaining seated and for evidencing other positive behaviors, and awarding him tokens (exchangeable for candy) for in-seat behavior. In the third phase (A) the sitting program was withdrawn and a return to baseline conditions took place. Finally, in phase four (B) the sitting program was reinstated.

The results of this study appear in Fig. 5-9. Examination of the top part of the graph shows that the sitting program, with the exception of the last day in the first treatment phase, effected improvement over baseline conditions on both occasions. Continued examination of the figure reveals that posturing decreased during the sitting program, but walking remained at a consistent rate

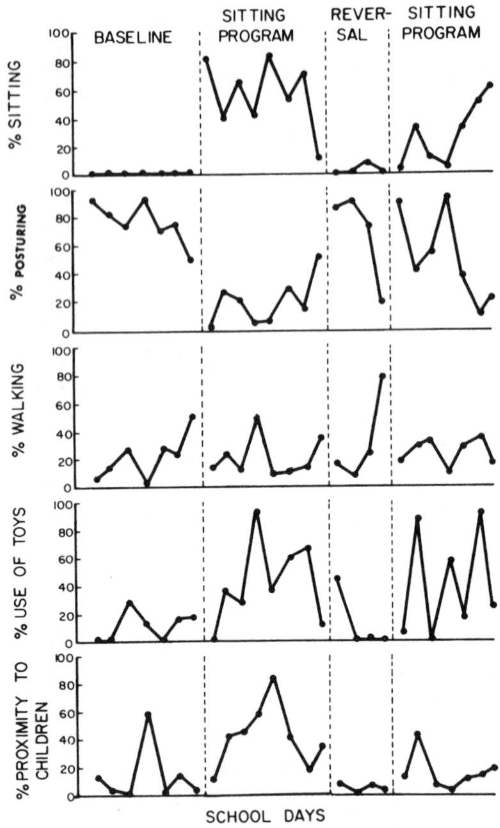

Fig. 5-9. Percentages of Tim's sitting, posturing, walking, use of toys, and proximity to children during freeplay as a function of the teacher ignoring him when he did not obey a command to sit down. (Fig. 1, p. 75, from: Twardosz, S., and Sajwaj, T. Multiple effects of a procedure to increase sitting in a hyperactive retarded boy. *Journal of Applied Behavior Analysis*, 1972, **5**, 73-78. Copyright by Society for the Experimental Analysis of Behavior, Inc. Reproduced by permission.)

throughout all phases of study. Similarly, use of toys and proximity to children increased during administrations of the sitting program. In discussing their results, Twardosz and Sajwaj (1972) state that "This study . . . points out the desirability of measuring several child behaviors, although a modification procedure might focus on only one. In this way the preschool teacher can assess the efficacy of her program based upon changes in other behaviors as well as the behavior of immediate concern" (p. 77). However, in the event that non-targeted behaviors remain unmodified or that deterioration occurs in others, additional behavioral techniques can then be applied (Sajwaj, Twardosz, and Burke, 1972). Under these circumstances it might be preferable to use a multiple baseline strategy (Barlow and Hersen, 1973) in which attention to each behavior can be programed in advance (see Chapter 7).

A-B-A-B with no feedback to experimenter

A major advantage of the single case strategy (cited in Section 3.2 of Chapter 3) is that the experimenter is in a position to alter therapeutic approaches in accordance with the dictates of the case. Such flexibility is possible as repeated monitoring of target behaviors is taking place. Thus, changes from one phase to the next are accomplished with the experimenter's full knowledge of prior results. Moreover, specific techniques are then applied with the expectation that they will be efficacious. Although these factors are of benefit to the experimental clinician, they present certain difficulties from a purely experimental standpoint. Indeed, critics of the single case approach have concerned themselves with the possibilities of bias in evaluation and in actual application and withdrawal of specified techniques. One method of preventing such "bias" is to determine lengths of baseline and experimental phases on an *a priori* basis, while keeping the experimenter uninformed as to trends in the data during their collection. A problem with this approach, however, is that decisions regarding choice of baselines and those concerned with appropriate timing of institution and removal of therapeutic variables are left to chance.

The above-discussed strategy was carried out in an A-B-A-B design in which target measures were rated from videotape recordings for all phases on a post-experimental basis. Hersen, Miller, and Eisler (1973) examined the effects of varying conversation topics (non-alcohol and alcohol-related) on duration of looking and duration of speech in four chronic alcoholics and their wives in *ad libitum* interactions videotaped in a television studio. Following 3 minutes of "warm-up" interaction, each couple was instructed to converse for 6 minutes (A phase) about any subject *unrelated* to the husband's drinking problem. Instructions were repeated at 2-minute intervals over a two-way intercom from an adjoining room to ensure maintenance of the topic of conversation. In the next 6 minutes (B phase) the couple was instructed to converse *only* about the husband's drinking problem (instructions were repeated at 2-minute intervals). The last 12 minutes of interaction consisted of identical replications of the A and B phases.

Mean data for the four couples are presented in Fig. 5-10. Speech duration data show no trends across experimental phases either for husbands or wives. Similarly, duration of looking for husbands across phases does not vary greatly.

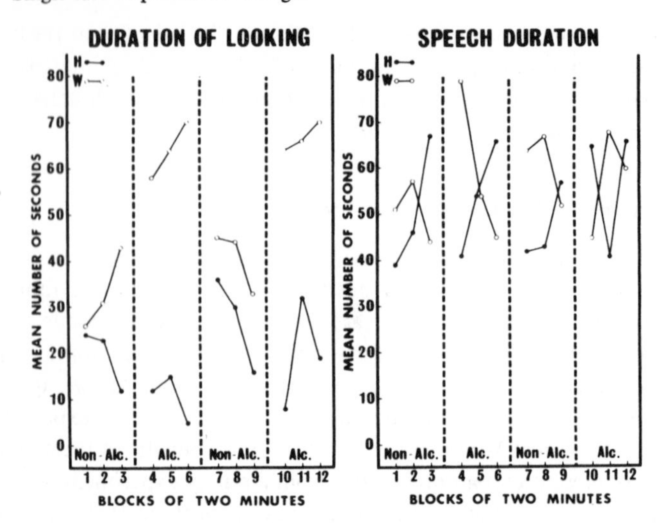

Fig. 5-10. Looking and speech duration in non-alcohol- and alcohol-related interactions of alcoholics and their wives. Plotted in blocks of 2 minutes. Closed circles–husbands; open circles–wives. (Fig. 1, p. 518, from: Hersen, M., Miller, P.M., and Eisler, R.M. Interactions between alcoholics and their wives: A descriptive analysis of verbal and non-verbal behavior. *Quarterly Journal of Studies on Alcohol*, 1973, **34,** 516-520. Copyright by Journal of Studies on Alcohol, Inc. New Brunswick, N.J. 08903. Reproduced by permission.)

However, duration of looking for wives was significantly greater during alcohol- than non-alcohol-related segments of interaction. In the first non-alcohol phase, looking duration ranged from 26 to 43 seconds, with an upward trend in evidence. In the first alcohol phase (B), duration of looking ranged from 57 to 70 seconds, with a continuation of the upward linear trend. Reintroduction of the non-alcohol phase (A) resulted in a decrease of looking (38 to 45 seconds). In the final alcohol segment (B), looking once again increased, ranging from 62 to 70 seconds.

An analysis of these data do not allow for conclusions with respect to the initial A and B phases inasmuch as the upward trend in A continued into B. However, the decreasing trend in the second A phase succeeded by the increasing trend in the second B phase suggests that topic of conversation had a controlling influence on the wives' rates of looking. We might note here that if the experimenters were in position to monitor their results throughout all experimental phases, the initial segment probably would have been extended until the wives' looking duration achieved stability in the form of a plateau. Then, the second phase would have been introduced.

5.5. B-A-B DESIGN

The B-A-B design has frequently been used by investigators evaluating effectiveness of their treatment procedures (Agras, Leitenberg, and Barlow, 1968; Ayllon and Azrin, 1965; Leitenberg, Agras, Thomson, and Wright, 1968; Mann and Moss, 1973; Rickard and Saunders, 1971). In this experimental strategy the first phase (B) usually involves the application of a treatment, in the second phase (A) the treatment is withdrawn, and in the final phase (B) it is reinstated. Some investigators (e.g., Agras, Leitenberg, and Barlow, 1968) have introduced an abbreviated baseline session prior to the major B-A-B phases. The B-A-B design is superior to the A-B-A design, described earlier in Section 5.3, in that the treatment variable is in effect in the terminal phase of experimentation. However, absence of an initial baseline measurement session precludes an analysis of the effects of treatment over the natural frequency of occurrence of the targeted behaviors under study (i.e., baseline). Therefore, as previously pointed out by Barlow and Hersen (1973), the use of the more complete A-B-A-B design is preferred for assessment of singular therapeutic variables.

We will illustrate the use of the B-A-B strategy with one example selected from the operant literature and a second drawn from the Rogerian framework. In the first, an entire group of subjects underwent introduction, removal, and reintroduction of a treatment procedure in sequence (Ayllon and Azrin, 1965). In the second, a variant of the B-A-B design was imployed by proponents of client-centered therapy (Truax and Carkhuff, 1965) in an attempt to experimentally manipulate levels of therapeutic conditions.

B-A-B with group data

Ayllon and Azrin (1965) used the B-A-B strategy on a group basis in their evaluation of the effects of token economy on the work performance of 44 "back ward" schizophrenic subjects. During the first 20 days (B phase) of the experiment, subjects were awarded tokens (exchangeable for a large variety of "back-up" reinforcers) for engaging in hospital ward work activities. In the next 20 days (A phase) subjects were given tokens on a non-contingent basis, regardless of their work performance. Each subject received tokens daily, based on the mean daily rate obtained in the initial B phase. In the last 20 days (second B phase) the contingency system was reinstated. We might note at this point that this design could alternately be labeled B-C-B as the middle phase is not a "true" measure of the natural frequency of occurrence of the target measure (see Section 5.6).

Work performance data (total hours per day) for the three experimental phases appear in Fig. 5-11. During the first B phase, total hours per day worked by the entire group averaged about 45 hours. Removal of the contingency in A resulted in a marked linear decrease to a level of 1 hour per day on Day 36. Reinstitution of the token reinforcement program in B led to an immediate increase in hours worked to a level approximating the first B phase. Thus, Ayllon and Azrin (1965) presented the first experimental demonstration of the controlling effects of token economy over work performance in state hospital psychiatric patients.

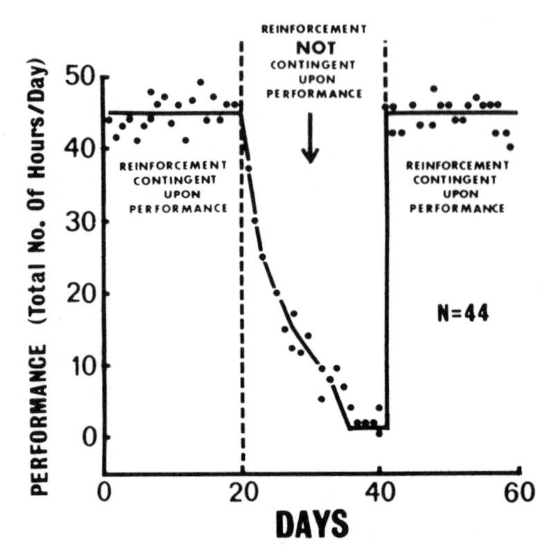

Fig. 5-11. The total number of hours of the on-ward performance by a group of 44 patients, Exp. III (Fig. 4, p. 393, redrawn from: Ayllon, T., and Azrin, N.H. The measurement and reinforcement of behavior of psychotics. *Journal of the Experimental Analysis of Behavior*, 1965, 8, 357-383. Copyright by Society for the Experimental Analysis of Behavior, Inc. Reproduced by permission.)

It should be pointed out here that when experimental single case strategies, such as the B-A-B design, are used on a group basis, it behooves the experimenter to show that a majority of those subjects exposed to and then withdrawn from treatment provide supporting evidence for its controlling effects. Individual data presented for selected subjects can be quite useful, particularly if data trends differ. Otherwise, difficulties inherent in the traditional group comparison approach (e.g., averaging out of effects, effects due to a small minority while the majority remains unaffected by treatment) will be carried over to the experimental analysis procedure. In this regard, Ayllon and Azrin (1965) showed that 36 of their 44 subjects decreased their performance from contingent to non-contingent reinforcement. Conversely, 36 of 44 subjects increased their performance from non-contingent to contingent reinforcement. Eight subjects were totally unaffected by contingencies and maintained a zero level of performance in all phases.

B-A-B from Rogerian framework

Although the withdrawal design has been used in physiological research for years, as well as being associated with the operant paradigm, the experimental strategies that are applied there can easily be employed in the investigation of non-operant (both behavioral and "traditional") treatment procedures. In this connection, Truax and Carkhuff (1965) systematically examined the effects of high and low "therapeutic conditions" on the responses of three psychiatric patients during the course of initial 1-hour interviews. Each of the interviews consisted of three 20-minute phases. In the first phase (B) the therapist was instructed to evidence high levels of "accurate empathy" and "unconditional positive warmth" in his interactions with the patient. In the following A phase the therapist experimentally lowered these conditions, while in the final phase (B) they were reinstated at a high level.

Each of the three interviews was audiotaped. From these audiotapes, five 3-minute segments for each phase were obtained and rerecorded on separate spools. These were then presented to raters (naive as to which phase the tape originated) in random order. Ratings made on the basis of the Accurate Empathy Scale and the Unconditional Positive Regard Scale confirmed (graphically and statistically) that the therapist followed directions as indicated by the dictates of the experimental design (B-A-B).

The effects of high and low therapeutic conditions were then assessed in terms of depth of the patient's intrapersonal exploration. Once again, three-minute segments from the A and B phases were presented to "naive" raters in randomized order. These new ratings were made on the basis of the Truax Depth of Interpersonal Exploration Scale (reliability of raters per segment = ·78). Data with respect to depth of intrapersonal exploration are plotted in Fig. 5-12. Visual inspection of these data indicate that depth of intrapersonal exploration, despite considerable overlapping in adjacent phases, was *somewhat* lowered during the middle phase (A) for each of the three patients. This was confirmed statistically with F and t tests.

Although these data are far from perfect (i.e., overlap between phases), the study does illustrate that the controlling effects of *non-behavioral therapeutic variables can be investigated systematically using the experimental analysis of behavior model.* Those of non-behavioral persuasion might be encouraged to assess the effects of their technical operations more frequently in this fashion.

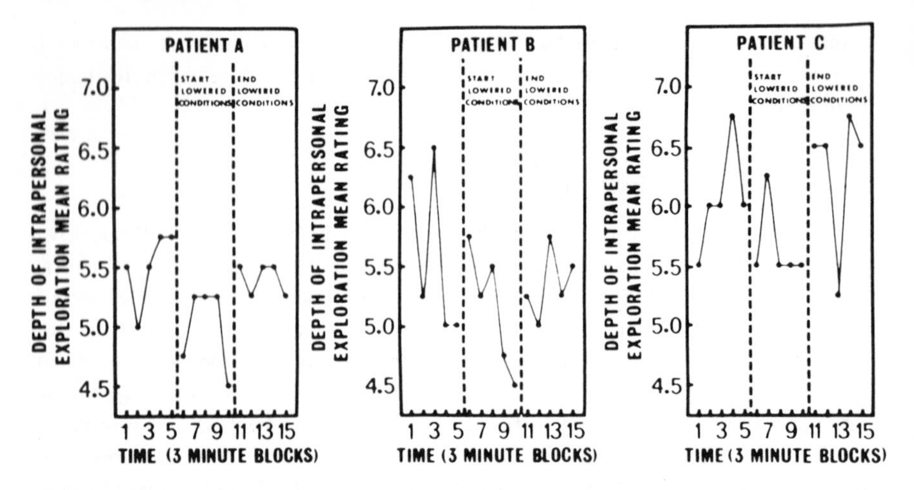

Fig. 5-12. Depth of intrapersonal exploration. (Fig. 4, p. 122, redrawn from: Truax, C.B., and Carkhuff, R.R. Experimental manipulation of therapeutic conditions. *Journal of Consulting Psychology*, 1965, **29**, 119-124. Copyright 1965 by the American Psychological Association. Reproduced by permission.)

5.6. A-B-C-B DESIGN

The A-B-C-B design, a variant of the A-B-A-B design, has been used to evaluate the effects of reinforcement procedures. Whereas in the A-B-A-B strategy, baseline and treatment (e.g., contingent reinforcement) are alternated in sequence, in the A-B-C-B strategy only the first two phases of experimentation consist of baseline and contingent reinforcement. In the third phase (C), instead of returning to baseline observation, reinforcement is administered in proportions equal to the preceding B phase but on a totally *non-contingent* basis. This phase controls for the added attention ("attention-placebo") that a subject receives for being in a treatment condition and is analogous to the A_1 phase (placebo) used in drug evaluations (see Chapter 6, Section 6.4). In the final phase, contingent reinforcement procedures are reinstated. Thus, the last three phases of study are identical to those used by Ayllon and Azrin (1965) in the example described in Section 5.5 of this chapter (however, there the study is labeled as B-A-B).

In the A-B-C-B design the A and C phases are not comparable inasmuch as experimental procedures differ. Therefore, the main experimental analysis is derived from the B-C-B portion of study. However, baseline observations are of some value as the effects of B over A are suggested (here we have the limitations of the A-B analysis). We will illustrate the use of the A-B-C-B design with one example concerned with the control of drinking in a chronic alcoholic.

A-B-C-B with a biochemical target measure

Miller, Hersen, Eisler, and Watts (1974) examined the effects of monetary reinforcement in a 48-year-old "skid row" alcoholic. During all phases of study, a research assistant obtained "breathalyzer" samples, analyzed biochemically shortly thereafter for blood alcohol concentration, from the subject (psychiatric outpatient) in various locations in his community. To avoid possible bias in measurement, the subject was not informed as to specific times that probe measures were to be taken. In fact, these times were randomized in all phases to control for measurement bias.

During baseline (A phase), eight probe measures were obtained. During contingent reinforcement (B), the subject was awarded $3.00 in canteen booklets (redeemable at the hospital commissary for material goods) whenever a negative blood alcohol sample was obtained. In the non-contingent reinforcement phase (C), reinforcement ($3.00 in canteen booklets) was administered regardless of blood alcohol concentration. In the final phase, contingent reinforcement was reinstituted.

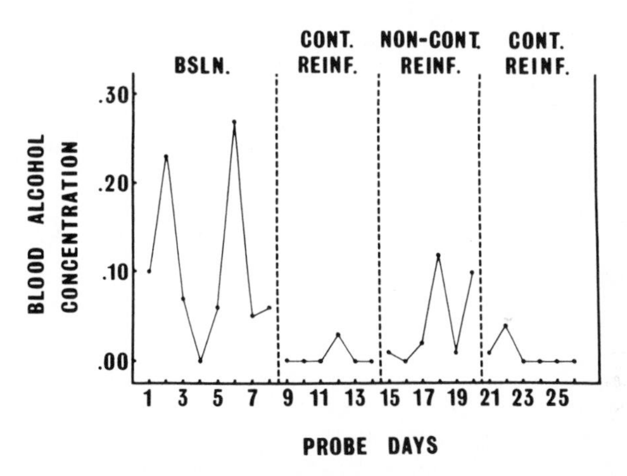

Fig. 5-13. Bi-weekly blood/alcohol concentrations for each phase. (Fig. 1, p. 262, from: Miller, P.M., Hersen, M., Eisler, R.M., and Watts, J.G. Contingent reinforcement of lowered blood/alcohol levels in an outpatient chronic alcoholic. *Behaviour Research and Therapy*, 1974, **12**, 261-263. Reproduced by permission.)

Inspection of Fig. 5-13 reveals a variable baseline pattern ranging from a ·00 to ·27 level of blood alcohol. In contingent reinforcement, five of the six probe measures attained a ·00 level. During non-contingent reinforcement, blood alcohol concentration measures rose, but to lower levels than in baseline. When contingent reinforcement was reinstated, four of the six probe measures yielded ·00 levels of blood alcohol. Therefore, it appears that monetary reinforcement resulted in decreases in drinking in this chronic alcoholic while the contingency was in effect.

References

Agras, W. S., Leitenberg, H., and Barlow, D. H. Social reinforcement in the modification of agoraphobia. *Archives of General Psychiatry*, 1968, **19**, 423-427.

Ashem, R. The treatment of a disaster phobia by systematic desensitization. *Behaviour Research and Therapy*, 1963, **1**, 81-84.

Ayllon, T., and Azrin, N. H. The measurement and reinforcement of behavior of psychotics. *Journal of the Experimental Analysis of Behavior*, 1965, **8**, 357-383.

Ayllon, T., and Azrin, N. H. *The token economy: A motivational system for therapy and rehiabilitation.* New York: Appleton-Century-Crofts, 1968.

Bandura, A. *Principles of behavior modification.* New York: Holt, Rinehart and Winston, 1969.

Barlow, D. H., and Hersen, M. Single-case experimental designs: Uses in applied clinical research. *Archives of General Psychiatry*, 1973, **29**, 319-325.

Barlow, D. H., Leitenberg, H., and Agras, W. S. Experimental control of sexual deviation through manipulation of the noxious scene in covert sensitization. *Journal of Abnormal Psychology*, 1969, 74, 596-601.

Beck, A. T., Ward, C. H., Mendelsohn, M., Mock, J., and Erbaugh, J. An inventory for measuring depression. *Archives of General Psychiatry*, 1961, **4**, 561-571.

Browning, R. M., and Stover, D. O. *Behavior modification in child treatment: An experimental and clinical approach.* Chicago and New York: Aldine-Atherton, 1971.

Campbell, D. T. Reforms as experiments. *American Psychologist*, 1969, **24**, 409-429.

Campbell, D. T., and Stanley, J. C. *Experimental and quasi-experimental designs for research.* Chicago: Rand McNally, 1966.

Eisler, R. M., and Hersen, M. The A-B design: Effects of token economy on behavioral and subjective measures in neurotic depression. Paper read at American Psychological Association, Montreal, August 29, 1973.

Epstein, L. H., and Hersen, M. Behavioral control of hysterical gagging. *Journal of Clinical Psychology*, 1974, **30**, 102-104.

Epstein, L. H., Hersen, M., and Hemphill, D. P. Music feedback as a treatment for tension headache: An experimental case study. *Journal of Behavior Therapy and Experimental Psychiatry*, 1974, **5**, 59-63.

Glass, G. V., Willson, V. L., and Gottman, J. M. Design and analysis of time-series experiments. Boulder: Colorado Associated University Press, 1974.

Gottman, J. M. N-of-one and N-of-two research in psychotherapy. *Psychological Bulletin*, 1973, **80**, 93-105.

Gottman, J. M., McFall, R. M., and Barnett, J. T. Design and analysis of research using time series. *Psychological Bulletin*, 1969, **72**, 299-306.

Hall, R. V., Fox, R., Willard, D., Goldsmith, L., Emerson, M., Owen, M., Davis, F., and Porcia, E. The teacher as observer and experimenter in the modification of disputing and talking-out behaviors. *Journal of Applied Behavior Analysis*, 1971, **4**, 141-149.

Harbert, T. L., Barlow, D. H., Hersen, M., and Austin, J. B. Measurement and modification of incestuous behavior: A case study. *Psychological Reports*, 1974, **34**, 79-86.

Hersen, M. Different design options. Paper read at Association for Advancement of Behavior Therapy, Miami, December 7, 1973.

Hersen, M., Eisler, R. M., Alford, G. S., and Agras, W. S. Effects of token economy on neurotic depression: An experimental analysis. *Behavior Therapy*, 1973, **4**, 392-397.

Hersen, M., Miller, P. M., and Eisler, R. M. Interactions between alcoholics and their wives: A descriptive analysis of verbal and non-verbal behavior. *Quarterly Journal of Studies on Alcohol*, 1973, **34**, 516-520.

Kazdin, A. E. The effect of response cost and aversive stimulation in suppressing punished and non-punished speech dysfluencies. *Behavior Therapy*, 1973a, **4**, 73-82.

Kazdin, A. E. Methodological and assessment considerations in evaluating reinforcement programs in applied settings. *Journal of Applied Behavior Analysis*, 1973b, **6**, 517-531.

Lazarus, A. A. The results of behavior therapy in 126 cases of severe neurosis. *Behaviour Research and Therapy*, 1963, **1**, 69-80.

Lazarus, A. A. Multi-modal behavior therapy: Treating the "BASIC ID." *Journal of Nervous and Mental Disease*, 1973, **156**, 404-411.

Lazarus, A. A., and Davison, G. C. Clinical innovation in research and practice. In A. E. Bergin and S. L. Garfield (Eds.), *Handbook of psychotherapy and behavior change: An empirical analysis.* Pp. 196-213. New York: Wiley, 1971.

Leitenberg, H., Agras, W. S., Thomson, L., and Wright, D. E. Feedback in behavior modification: An experimental analysis in two phobic cases. *Journal of Applied Behavior Analysis*, 1968, **1**, 131-137.

Lovaas, O. I., and Simmons, J. Q. Manipulation of self-destruction in three retarded children. *Journal of Applied Behavior Analysis*, 1969, **2**, 143-158.

Mann, R. A., and Moss, G. R. The therapeutic use of a token economy to manage a young and assaultive inpatient population. *Journal of Nervous and Mental Disease*, 1973, **157**, 1-9.

Miller, P. M. An experimental analysis of retention control training in the treatment of nocturnal enuresis in two institutionalized adolescents. *Behavior Therapy*, 1973, **4**, 288-294.

Miller, P. M., Hersen, M., Eisler, R. M., and Watts, J. G. Contingent reinforcement of lowered blood/alcohol levels in an outpatient chronic alcoholic. *Behaviour Research and Therapy*, 1974, **12**, 261-263.

O'Leary, K. D., and Becker, W. C. Behavior modification of an adjustment class: A token reinforcement program. *Exceptional Children*, 1967, **9**, 637-642.

O'Leary, K. D., Becker, W. C., Evans, M. B., and Saudargas, R. A. A token reinforcement program in a public school: A replication and systematic analysis. *Journal of Applied Behavior Analysis*, 1969, **2**, 3-13.

Rickard, H. C., and Saunders, T. R. Control of "clean-up" behavior in a summer camp. *Behavior Therapy,* 1971, **2**, 340-344.

Risley, T. R. The effects and side-effects of punishing the autistic behavior of a deviant child. *Journal of Applied Behavior Analysis*, 1968, **1**, 21-34.

Risley, T. R., and Wolf, M. M. Strategies for analyzing behavioral change over time. In J. Nesselroade and H. Reese (Eds.), *Life-span developmental psychology: Methodological issues.* Pp. 175-183. New York: Academic Press, 1972.

Sajwaj, T., Twardosz, S., and Burke, M. Side effects of extinction procedures in a remedial preschool. *Journal of Applied Behavior Analysis*, 1972, **5**, 163-175.

Truax, C. B., and Carkhuff, R. R. Experimental manipulation of therapeutic conditions. *Journal of Consulting Psychology*, 1965, **29**, 119-124.

Twardosz, S., and Sajwaj, T. Multiple effects of a procedure to increase sitting in a hyperactive retarded boy. *Journal of Applied Behavior Analysis*, 1972, **5**, 73-78.

Ullmann, L. P., and Krasner, L. (Eds.), *Case studies in behavior modification.* New York: Holt, Rinehart, and Winston, 1965.

Walker, H. M., and Buckley, N. K. The use of positive reinforcement in conditioning attending behavior. *Journal of Applied Behavior Analysis*, 1968, **1**, 245-250.

Williams, J. G., Barlow, D. H., and Agras, W. S. Behavioral measurement of severe depression. *Archives of General Psychiatry*, 1972, **27**, 330-333.

Wolf, M. M., and Risley, T. R. Reinforcement: Applied research. In R. Glaser (Ed.), *The nature of reinforcement*. Pp. 310-325. New York: Academic Press, 1971.

Wolpe, J. *Psychotherapy by reciprocal inhibition*. Stanford: Stanford University Press, 1958.

CHAPTER 6

Extensions of the A-B-A Design, Uses in Drug Evaluation, and Interaction Design Strategies

6.1. EXTENSIONS AND VARIATIONS OF THE A-B-A DESIGN

The applied behavioral literature is replete with examples of extensions and variations of the more basic A-B-A experimental design. These designs can be broadly classified into four major categories. The first category consists of designs in which the A-B pattern is replicated several times. Advantages here are: (1) repeated control of the treatment variable is demonstrated, and (2) extended study can be conducted until full clinical treatment has been achieved. An example of this type of strategy appears in Mann's (1972) work, where he used an A-B-A-B-A-B design to study the effects of contingency contracting on weight loss in overweight subjects.

In the second category separate therapeutic variables are compared with baseline performance during the course of experimentation (e.g., Hall, Axelrod, Tyler, Grief, Jones, and Robertson, 1972; Pendergress, 1972; Wincze, Leitenberg, and Agras, 1972). Subsumed under this category are the A-B-A-C-A designs discussed earlier in Section 3.4 of Chapter 3. There it was pointed out that comparison of differential effectiveness of B and C variables is not possible when both variables appear to effect change over baseline levels. However, in the A-B-A-B-A-C-A design the individual controlling effects of B and C variables can be determined. A careful distinction should be made between these kinds of designs and designs where the interactive effects of variables are investigated (e.g., A-B-A-B-BC-B-BC). In the latter design the effects of C above those of B can be assessed experimentally. Once again, in the A-B-A-C-A design the effects of B and C over A can be evaluated. However, the relative efficacy of B and C cannot be determined in this strategy.

In the third category specific variations of the treatment procedure are examined during the course of experimentation (e.g., Bailey, Wolf, and Phillips, 1970; Coleman, 1970; Hopkins, Schutte, and Garton, 1971; Kaufman and

O'Leary, 1972; McLaughlin and Malaby, 1972; Wheeler and Sulzer, 1970). For example, in some operant paradigms the treatment procedure may be faded out (e.g., Bailey, Wolf, and Phillips, 1970). In other paradigms, differing amounts of reinforcement may be assessed experimentally or in graduated progression (Hopkins, Schutte, and Garton, 1971) following demonstration of the controlling effects of variables in the A-B-A-B portion of the design.

In a fourth category, the interaction or additive effects of two or more variables are examined through variations in the basic A-B-A design (e.g., Agras, Barlow, Chapin, Abel, and Leitenberg, 1974; Hersen, Gullick, Matherne, and Harbert, 1972; Leitenberg, Agras, Thomson, and Wright, 1968; Turner, Hersen, and Alford, 1974). Such analysis is accomplished by examining the effects of both variables alone and in combination to determine the interaction. This extends beyond analysis of the separate effects of two therapeutic variables over baseline as represented by the A-B-A-C-A type design described in the second category. It also extends a step beyond merely adding a variation of a therapeutic variable on the end of an A-B-A-B series (e.g., A-B-A-B-BC) since no experimental analysis of the additive effects of BC is performed. Properly run, interaction designs are complex and usually require more than one subject (see Section 6.5).

In the following subsections we will present examples of extensions and variations, with illustrations selected from each of the four major categories.

6.2. A-B-A-B-A-B DESIGN

Mann (1972) repeatedly introduced and withdrew a treatment variable (contingency contracting) during extended study with overweight subjects who had agreed, prior to experimentation, to achieve a designated weight loss within a specified time period. At the beginning of study, each subject entered into a formal contractual arrangement with the experimenter. In each case the subject agreed to surrender a number of his prize possessions (valuables) to the experimenter. During contingency conditions, the subject was able to regain possession of each valuable (one at a time) by evidencing a 2-pound weight loss over his previous low weight. A further 2-pound weight loss over that resulted in the return of still another valuable, etc. Conversely, a 2-pound weight gain over the previous low weight led to the subject's permanently losing one of the valuables. In addition to these short-term contingency arrangements, 2-week and terminal contingencies (using similar principles) were put into effect during treatment phases. Valuables lost by each subject were subsequently disposed of by the experimenter in equitable fashion (i.e., he did not profit from or retain them). During baseline and "reversal" conditions contractual arrangements were temporarily suspended.

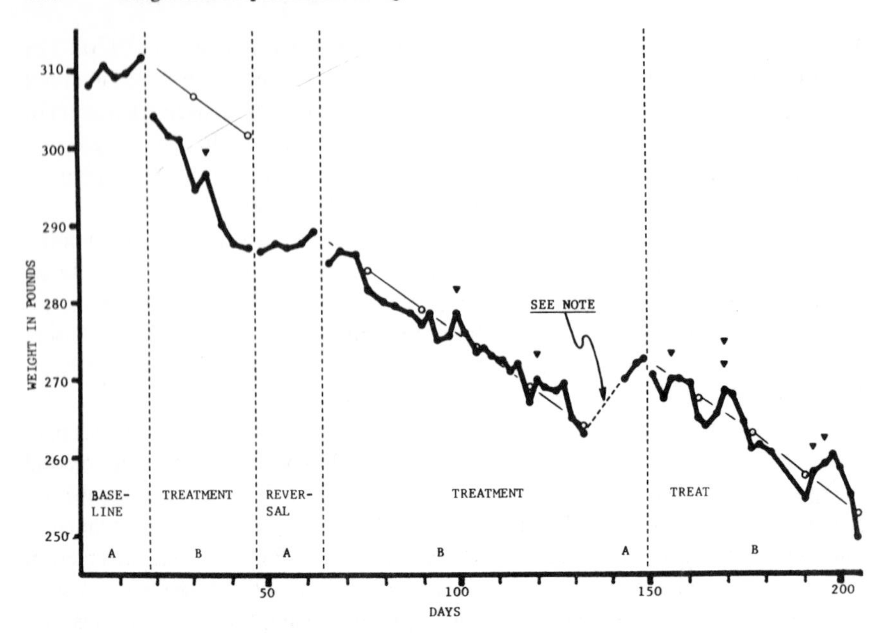

Fig. 6-1. A record of the weight of Subject 1 during all conditions. Each open circle (connected by the thin solid line) represents a 2-week minimum weight loss requirement. Each of the solid dots (connected by the thick solid line) represents the subject's weight on each of the days that he was measured. Each triangle indicates the point at which the subject was penalized by a loss of valuables, either for gaining weight or for not meeting a 2-week minimum weight loss requirement. NOTE: The subject was ordered by his physician to consume at least 2500 calories per day for 10 days, in preparation for medical tests. (Fig. 1a, p. 104, from: Mann, R. A. The behavior-therapeutic use of contingency contracting to control an adult behavior problem: Weight control. *Journal of Applied Behavior Analysis*, 1972, **5**, 99-109. Copyright by Society for the Experimental Analysis of Behavior, Inc. Reproduced by permission.)

The results of this study for a prototypical subject are plotted in Fig. 6-1. Inspection of that figure clearly shows that when contractual arrangements were in force the subject evidenced a steady linear decrease in weight. By contrast, during baseline conditions, weight loss ceased as indicated by a plateau and slightly upward trend in the data. In short, the effects of the treatment variable were repeatedly demonstrated in the alternating increasing and decreasing data trends.

A-B-A-C-A-C′-A design

Wincze, Leitenberg, and Agras (1972) conducted a series of ten experimental single case designs in which the effects of feedback and token reinforcement were examined on the verbal behavior of delusional psychiatric patients. In one of these studies an A-B-A-C-A-C′-A design was used, with B and C representing feedback and token reinforcement phases, respectively. During all phases of study, a delusional patient was questioned daily (15 questions selected randomly from a pool of 105) by his therapist to elicit delusional material. Percentage of responses containing delusional verbalizations was recorded. In addition, percentage of delusional talk on the ward (token economy unit) was monitored by nursing staff on a randomly distributed basis 20 times per day.

During baseline (A), the patient received "free" tokens as no contingencies were placed with respect to delusional verbalizations. During feedback (B), the patient continued to receive tokens non-contingently, but corrective statements in response to delusional verbalizations were offered by the therapist in individual sessions. The third phase (A) consisted of a return to baseline procedures. In phase four (C) a stringent token economy system embracing all aspects of the patient's ward life was instituted. Tokens could be earned by the patient for "talking correctly" (non-delusionally) both in individual sessions and on the ward. Tokens were exchangeable for meals, luxuries, and privileges. Phase five (A) once again involved a return to baseline. In the sixth phase (C′) token bonuses were awarded on a predetermined percentage basis for "talking correctly" (e.g., speaking delusionally less than 10 percent of the time during designated periods). This condition was incorporated to counteract the tendency of the patient to earn tokens merely for increasing frequency of non-delusional talk while still maintaining a high frequency of delusional verbalizations. In the last phase of experimentation (A) baseline conditions were reinstated for the fourth time.

Results for this experimental analysis appear in Fig. 6-2. Percentage of delusional talk in individual sessions and on the ward did not differ substantially during the first three sessions, thus suggesting the ineffectiveness of the feedback variable. Institution of token economy in phase four, however, resulted in a marked decrease of delusional talk in individual sessions. But it failed to effect a change in delusional talk on the ward. Removal of token economy in phase five led to a return to initial levels of delusional talk during individual sessions. Throughout the first five phases, percentage of delusional talk on the ward was consistent, ranging from 0 to 30 percent. Introduction of the token bonus in phase six again resulted in a drop of delusional verbalizations in individual sessions. Additionally, percentage of delusional talk on the ward decreased to zero. In the last phase (baseline) delusional verbalizations rose both on the ward and in individual sessions.

In this case feedback (B) proved to be an ineffective therapeutic agent. However, token economy (C) and token bonuses (C′), respectively, controlled

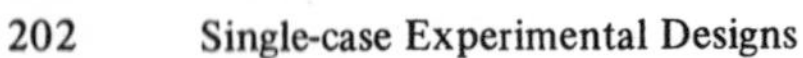

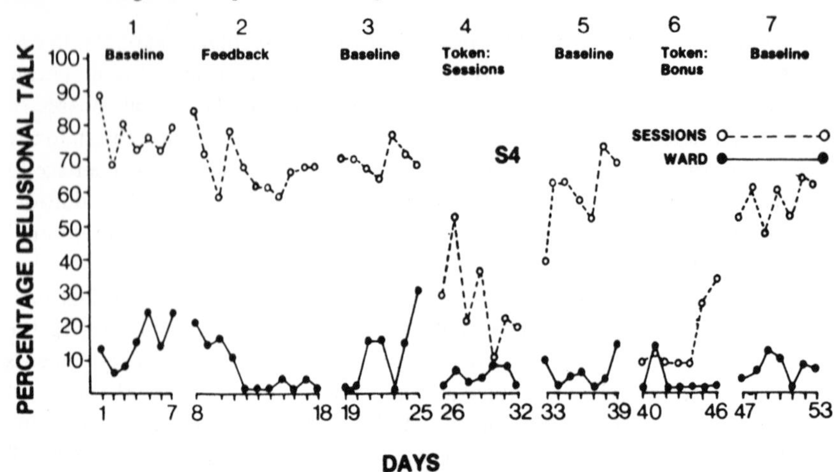

Fig. 6-2. Percentage delusional talk of Subject 4 during therapist sessions and on ward for each experimental day. (Fig. 4, p. 256, from: Wincze, J. P., Leitenberg, H., and Agras, W. S. The effects of token reinforcement and feedback on the delusional verbal behavior of chronic paranoid schizophrenics. *Journal of Applied Behavior Analysis,* 1972, **5**, 247-262. Copyright by Society for the Experimental Analysis of Behavior, Inc. Reproduced by permission.)

percentage of delusional talk in individual sessions and on the ward. Had feedback also effected changes in behavior, the comparative efficacy of feedback and token economy could not be ascertained using this design. Such analysis would require the use of a group comparison design. While this experiment was chosen to illustrate a specific design, it is noteworthy that nine replications ensued. This "direct replication" series will be discussed in greater detail in Chapter 9.

6.3. A-B-A-B-B′-B″-B‴ DESIGN

Our example from the third category of extensions of the A-B-A design is drawn from the child classroom literature. Hopkins, Schutte, and Garton (1971) systematically assessed the effects of access to a playroom on the rate and quality of writing in rural elementary school children. Target measures selected for study were most relevant in that these children came from homes where learning was not a high priority (parents were migrant or seasonal farm workers). Throughout all phases of study, first- and second-grade students were given daily standard written assignments during class periods (class periods were 50 minutes long during the first four phases).

In baseline (A), after each child had completed his assignment, handed it to

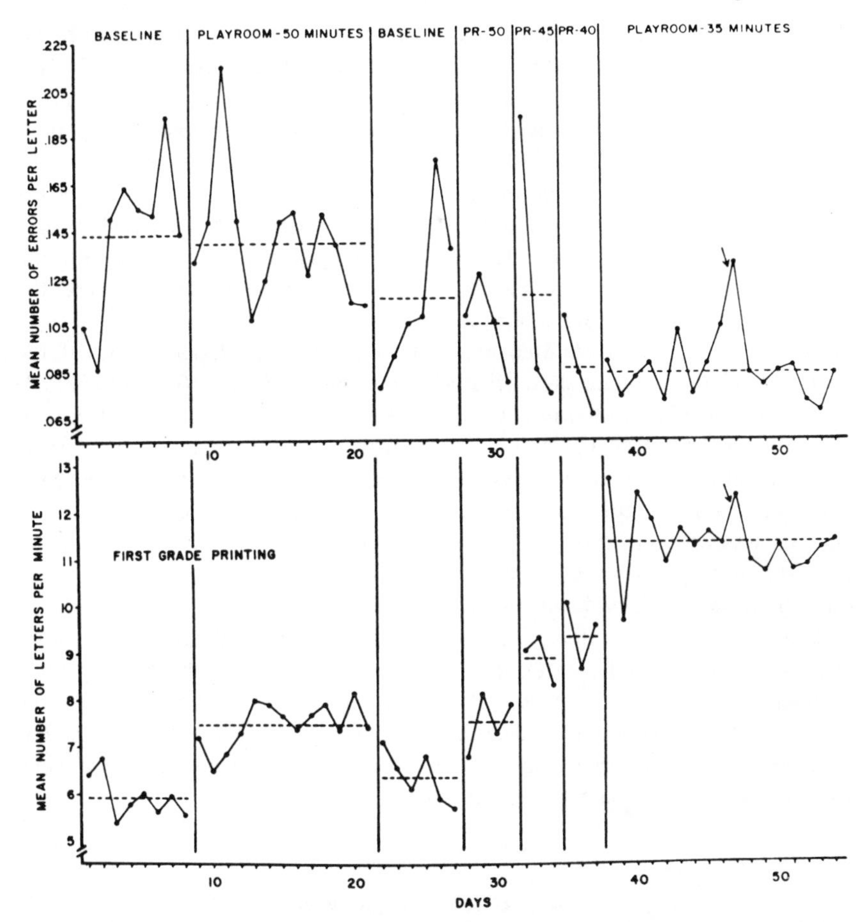

Fig. 6-3. The mean number of letters printed per minute by first-grade children are shown on the lower coordinates and the mean proportion of letters scored as errors are on the upper coordinates. Each data point represents the mean averaged over all children for that day. The horizontal dashed lines are the means of the daily means averaged over all days within the experimental conditions noted by the legends at the top of the figure. (Fig. 1, p. 81, from: Hopkins, B. L., Schutte, R. C., and Garton, K. L. The effects of access to a playroom on the rate and quality of printing and writing of first- and second-grade students. *Journal of Applied Behavior Analysis*, 1971, **4**, 77-87. Copyright by Society for the Experimental Analysis of Behavior, Inc. Reproduced by permission.)

his teacher, and waited for it to be scored, he was expected to return to his seat and remain there quietly until *all* others in class had turned in their papers. In the next phase (B) each child was permitted access to an adjoining playroom, containing attractive toys, after his paper was scored. He was allowed to remain there until the 50-minute period was terminated unless he became too noisy; then he was required to return to his seat. The next two phases (A and B) were identical to the first two. In the last three phases each child was permitted access to the playroom after his paper had been scored, but the length of class periods was gradually decreased (45, 40, 35 minutes). A procedural exception to the aforementioned was made in the last phase on Days 47-54 inasmuch as the teacher noted that a concomitant of increased speed was decreased quality (number of errors) in writing. Therefore, in the last 8 days a quality criterion was imposed before the child gained access to the playroom. In some cases the child was required to recopy a portion of his writing.

Data for first-grade children are plotted in Fig. 6-3. Examination of the bottom half of the figure shows that access to the playroom (50-minute period) increased rate of letter writing over baseline levels. This was confirmed on two occasions in the A-B-A-B portion of study. When total time of classroom periods systematically decreased, a corresponding increase in rate of writing resulted. However, data for the last three phases are correlative as an experimental analysis was not performed. For example, a sequential comparison of 50-, 45-, and 50-minute periods was not made. Therefore, the controlling effects of time differences are not fully documented.

Examination of the top part of the graph shows considerable fluctuation with respect to mean number of errors per letter. However, this did not appear to represent a systematic increase when class periods were shortened. To the contrary, there was a general decrease in error rate from the first to the last phase of study. Nonetheless, the effects of practice cannot be discounted when total length of the investigation is considered.

6.4. DRUG EVALUATIONS

The group comparison approach and the uncontrolled case study predominate in the examination of the effects of drugs on behavior. Scattered throughout the psychological and psychiatric literatures, however, are examples in which the subject has served as his own control in the experimental evaluation of pharmacological agents (e.g., Agras, in press; Bellak and Chassan, 1964; Chassan, 1967; Davis, Sprague, and Werry, 1969; Grinspoon, Ewalt, and Shader, 1967; Liberman, Davis, Moon, and Moore, 1973; Lindsley, 1962; McFarlain and Hersen, 1974; Roxburgh, 1970). Recently, Liberman *et al.* (1973) have encouraged researchers to use the within-subject withdrawal design in assessing drug-environment interactions. In support of their position they contend that "Useful interactions among the drug-patient-environment system can be obtained using this type of methodology. The approach is reliable and rigorous, efficient and inexpensive to mount, and permits sound conclusions and generalizations to other patients with similar behavioral repertoires when systematic replications are performed. . ." (Liberman, Davis, and Moore, 1973, p. 433). There is no doubt that this approach can be of value in the study of both the major forms of psychopathology and those of more exotic origin (Barlow and Hersen, 1973). The single case experimental strategy is especially well suited to the latter as control group analysis in the rarer disorders is obviously not feasible.

Specific issues

It should be pointed out that all procedural issues discussed in Chapter 3 pertain equally to drug evaluation. In addition, there are a number of considerations specific to this area of research. They include the following: (1) nomenclature, (2) carryover effects, and (3) single- and double-blind assessments.

With respect to nomenclature, A is designated as the baseline phase, A_1 as the placebo phase, B as the phase evaluating the first active drug, and C as the phase evaluating the second active drug. The A_1 phase is an intermediary phase between A (baseline) and B (active drug condition) in this schema. This phase controls for the subject's expectancy of improvement associated with mere ingestion of the drug rather than for its contributing pharmacological effects.

Some of the above-mentioned considerations have already been examined in Section 3.4 of Chapter 3 in relation to changing one variable at a time across experimental phases. With regard to this one variable rule, it becomes apparent, then, that A-B, A-B-A, B-A-B, and A-B-A-B designs in drug research involve the manipulation of two variables (expectancy and condition) at one time across phases. However, under certain circumstances where time limitations and clinical considerations prevail, this type of experimental strategy is justified. Of course, when conditions permit, it is preferable to use strategies in which the systematic progression of variables across phases is carefully followed (see Table 6-1, Designs 4, 6, 7, 9-13). For example, this would be the case in the A_1-B-A_1 design strategy, where only one variable at a time is manipulated from phase to phase. Further discussion of these issues will appear in the following section in which the different design options available to drug researchers are to be outlined.

The problem of carryover effects from one phase to the next has already been discussed in Section 3.6 of Chapter 3. There some specific recommendations were made with respect to short-term assessments of drugs and the concurrent monitoring of biochemical changes during different phases of study. In this connection, Barlow and Hersen (1973) have noted that "Since continued measurements are in effect, length of phases can be varied from experiment to experiment to determine precisely the latency of drug effects after beginning the dosage and the residual effects after discontinuing the dosage" (p. 324). This may, at times, necessitate the inequality of phase lengths and suspension of active drug treatment until biochemical measurements (based on blood and urine studies) reach an acceptable level. For example, Roxburgh (1970) examined the effects of a placebo and thiopropazate dihydrochloride on phenothiazine-induced oral dyskinesia in a double-blind crossover in two subjects. In both cases, placebo and active drug treatment were separated by a 1-week interruption during which time no placebo or drug was administered.

A third issue specific to drug evaluation involves the use of single- and double-blind assessments. The double-blind clinical trial is a standard precautionary measure designed to control for possible experimenter bias and patient expectations of improvement under drug conditions when drug and placebo groups are being contrasted. "This is performed by an appropriate method of assigning patients to drugs such that neither the patient nor the investigator observing him knows which medication a patient is receiving at any point along the course of treatment" (Chassan, 1967, pp. 80-81). In these studies, placebos and active drugs are identical in size, shape, markings, and color.

While the double-blind procedure is readily adaptable to group comparison research, it is difficult to engineer for some of the single case strategies and impossible for others. Moreover, in some cases (see Table 6-1, Designs 1, 2, 4, 5, 8) even the single-blind strategy (where only the subject remains unaware of differences in drug and placebo manipulations) is not applicable. In these designs the changes from baseline observation to either placebo or drug conditions obviously cannot be disguised in any manner.

A major difficulty in obtaining a "true" double-blind trial in single case research is related to the experimenter's monitoring of data (i.e., making decisions as to when baseline observation is to be concluded and when various phases are to be introduced and withdrawn) throughout the course of investigation. It is possible to program phase lengths on an *a priori* basis, but then one of the major advantages of the single case strategy (i.e., its flexibility) is lost. However, even though the experimenter is fully aware of treatment changes, the spirit of the double-blind trial can be maintained by keeping the observer (often a research assistant or nursing staff member) unaware of drug and placebo changes (Barlow and Hersen, 1973). We might note here additionally that despite the use of the double-blind procedure, the side effects of drugs in some cases (e.g., Parkinsonism following administration of large doses of phenothiazines) and the marked changes in behavior resulting from removal of active drug therapy in other cases often betray to nursing personnel whether a placebo or drug condition is currently in operation. This problem is equally troublesome for

TABLE 6-1. Single Case Experimental Drug Strategies

No.	Design	Type	Blind Possible
1.	A-A_1	quasi-experimental	none
2.	A-B	quasi-experimental	none
3.	A_1-B	quasi-experimental	single or double
4.	A-A_1-A	experimental	none
5.	A-B-A	experimental	none
6.	A_1-B-A_1	experimental	single or double
7.	A_1-A-A_1	experimental	single or double
8.	B-A-B	experimental	none
9.	B-A_1-B	experimental	single or double
10.	A-A_1-A-A_1	experimental	single or double
11.	A-B-A-B	experimental	none
12.	A_1-B-A_1-B	experimental	single or double
13.	A-A_1-B-A_1-B	experimental	single or double
14.	A-A_1-A-A_1-B-A_1-B	experimental	single or double
15.	A_1-B-A_1-C-A_1-C	experimental	single or double

Note: A = no drug; A_1 = placebo; B = drug 1; C = drug 2.

the researcher concerned with group comparison designs (see Chassan, 1967, Chapter 4).

Different design options

In some of the investigations in which the subject has served as his own control, the standard experimental analysis method of study, where the treatment variable is introduced, withdrawn, and reintroduced following initial measurement, has not been followed rigorously. Thus, the controlling effects of the drug under evaluation have not been fully documented. For example, Davis, Sprague, and Werry (1969) used the following sequence of drug and no-drug conditions in studying rate of stereotypic and nonstereotypic behavior in severe retardates: (1) methylphenidate, (2) thioridazine, (3) placebo, and (4) no drug. Despite the fact that thioridazine significantly (at the statistical level) decreased the rate of stereotypic responses, failure to reintroduce the drug in a final phase weakens the conclusions to some extent from an experimental analysis standpoint.

A careful survey of the experimental analysis of behavior literature reveals relatively little discussion with regard to procedural and design issues in the assessment of drugs. Therefore, in light of the unique problems faced by the drug researcher and in consideration of the relative newness of this area, we will outline the basic quasi-experimental and experimental analysis design strategies for evaluating singular application of drugs. Specific advantages and disadvantages of each design option will be considered. Where possible, we will illustrate with actual examples selected from the research literature. However, to date, most of these strategies have not yet been implemented.

A number of possible single case strategies suitable for drug evaluation are presented in Table 6-1. The first three strategies fall into the A-B category and are really "quasi-experimental" designs in that the controlling effects of the treatment variable (placebo or active drug) cannot be determined. Indeed, it was noted in Section 5.2 of Chapter 5 that changes observed in B might possibly result from the action of a correlated but uncontrolled variable (e.g., time, maturational changes, expectancy of improvement, etc.). These quasi-experimental designs can best be applied in settings (e.g., consulting room practice) where limited time and facilities preclude more formal experimentation. In the first design the effects of placebo over baseline conditions are *suggested*; in the second the effects of active drug over baseline conditions are *suggested*; in the third the effects of an active drug over placebo are *suggested*.

Examination of Strategies 4-6 indicate that they are basically A-B-A designs in which the controlling effects of the treatment variable can be ascertained. In Design No. 4, the controlling effects of a placebo manipulation over no treatment can be assessed experimentally. This design has great potential in the study of disorders such as the conversion reactions and hysterical personalities where attentional factors are presumed to play a major role. Also, the use of this type of design in evaluating the therapeutic contribution of placebos in a variety of psychosomatic disorders could be of considerable importance to clinicians. In Design No. 5, the controlling effects of an active drug are determined over baseline conditions. However, as previously noted, two variables are being manipulated here at one time across phases. Design No. 6 corrects for this deficiency as the active drug condition (B) is preceded and followed by placebo (A_1) conditions. In this design the one variable rule across phases is carefully observed.

An example of an A_1-B-A_1 design appears in a series of single case drug evaluations reported by Liberman, Davis, Moon, and Moore (1973). In one of these studies the effects of fluphenazine on eye contact, verbal self-stimulation (unintelligible or jumbled speech), and motor self-stimulation were examined in a double-blind trial for a 29-year-old regressed schizophrenic who had been continuously hospitalized for 13 years. Double-blind analysis was facilitated by the fact that fluphenazine (10 mg., b.i.d.) or the placebo could be administered twice daily in orange juice without being detected (breaking of the double-blind code) by the patient or the nursing staff as the drug cannot be distinguished either by odor or taste. During all phases of study, 18 randomly distributed 1-minute observations of the patient were obtained daily with respect to incidence of verbal and motor self-stimulation. Evidence of eye contact with the

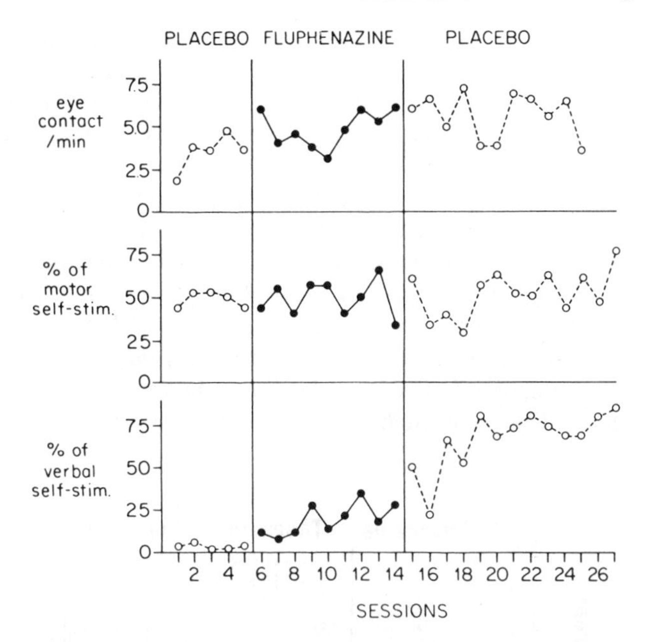

Fig. 6-4. Interpersonal eye contact, motor, and self-stimulation in a schizophrenic young man during placebo and fluphenazine (20 mg. daily) conditions. Each session represents the average of a 2-day block of observations. (Fig. 3, p. 437, from: Liberman, R. P., Davis, J., Moon, W., and Moore, J. Research design for analyzing drug-environment-behavior interactions. *Journal of Nervous and Mental Disease*, 1973, **156,** 432-439. Reproduced by permission.)

patient's therapist was obtained daily in six 10-minute sessions. Each eye contact was reinforced with candy or a puff on a cigarette.

The results of this study are plotted in Fig. 6-4. During the first placebo phase (A_1), stable rates were obtained for each of the target behaviors. Introduction of fluphenazine in the second phase (B) resulted in a very slight increase in eye contact, increased variability in motor self-stimulation, and a linear increase in verbal self-stimulation. Withdrawal of fluphenazine and a return to placebo conditions in the final phase (A_1) failed to yield data trends. On the contrary, eye contact increased slightly while verbal self-stimulation increased dramatically. Motor self-stimulation remained relatively consistent across phases. These data are interpreted by Liberman, Davis, Moon, and Moore (1973) as follows: "The failure to gain a reversal suggests a drug-initiated response facilitation which is seen most clearly in the increase of verbal self-stimulation, and less so in rate of eye contact" (p. 437). It is also suggested that residual phenothiazines during the placebo phase may have contributed to the continued increase in eye contact. However, in the absence of concurrent monitoring of biochemical factors (phenothiazine blood and urine levels) this hypothesis cannot be confirmed. In summary, Liberman, Davis, Moon, and Moore (1973) were not able to confirm the controlling effects of fluphenazine over any of the target behaviors selected for study in this A_1-B-A_1 design.

Let us now continue our examination of drug designs listed in Table 6-1. Strategies 7-9 can be classified as B-A-B designs, and the same advantages and limitations previously outlined in Section 5.5 of Chapter 5 apply here. Strategies 10-12 fall into the general category of A-B-A-B designs and are superior to the A-B-A and B-A-B designs for several reasons: (1) the initial observation period involves baseline or baseline-placebo measurement; (2) there are two occasions in which the controlling effects of the placebo or the treatment variable can be demonstrated; and (3) the concluding phase ends on a treatment variable.

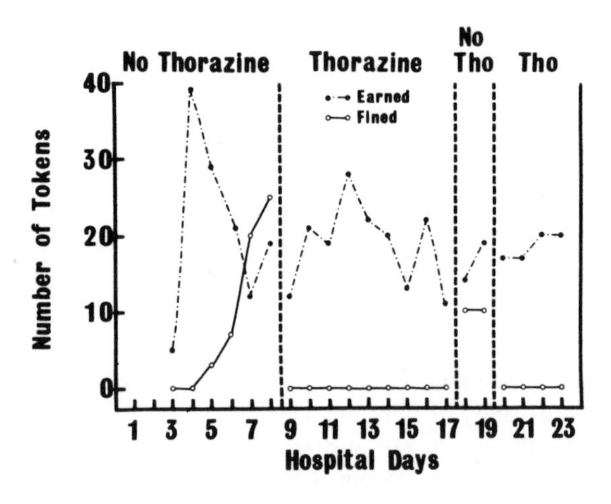

Fig. 6-5. Behavior of an adolescent as indicated by tokens earned or fined in response to chlorpromazine, which was added to token economy. (Fig. 5, from: Agras, W. S. Behavior modification in the general hospital psychiatric unit, appearing in a forthcoming issue of H. Leitenberg (Ed.), *Handbook of behavior modification*. Englewood Cliffs, N.J.: Prentice-Hall. Reproduced by permission.)

Agras (in press) used an A-B-A-B design to assess the effects of chlorpromazine in a 16-year-old, black, brain-damaged, male inpatient who evidenced a wide spectrum of disruptive behaviors on the ward. Included in his repertoire were: temper tantrums, stealing food, eating with his fingers, exposing himself, hallucinations, and begging for money, cigarettes, or food. A specific token economy system was devised for this youth whereby positive behaviors resulted in his earning tokens and inappropriate behaviors resulted in his being penalized with fines. Number of tokens earned and number of tokens fined were the two dependent measures selected for study. The results of this investigation appear in Fig. 6-5. In the first phase (A) no thorazine was administered. Although improvement in appropriate behaviors was noted, the patient's disruptive behaviors continued to increase markedly, resulting in his being fined many times. This occurred in spite of the addition of a timeout contingency. On Hospital Day 9, thorazine (300 mg. per day) was introduced (B phase) in an attempt to control the patient's impulsivity. This dosage was subsequently decreased to 200 mg. per day as he became drowsy. Examination of Fig. 6-5 reveals that fines decreased to a zero level while tokens earned for appropriate behaviors remained at a stable level. In the third phase (A) chlorpromazine was temporarily discontinued, resulting in an increase in fines for disruptive behavior. The no-thorazine condition (A) was only in force for 2 days as the patient's renewal of disruptive activities caused nursing personnel to "demand" reinstatement of his medication. When thorazine was reintroduced in the final phase (B), number of tokens fined once again decreased to a zero level. Thus, the controlling effects of thorazine over disruptive behavior were demonstrated. But Agras (in press) raises the question as to the possible contribution of the token economy program in controlling this patient's behavior. Unfortunately, time considerations did not permit him to systematically tease out the effects of that variable.

We might also note that in the A-B-A-B drug design, where the single- or double-blind trial is not feasible, staff and patient expectations of success during the drug condition are a possible confound with the drug's pharmacological actions. Designs listed in Table 6-1 that show control for these factors are Nos. 12 (A_1-B-A_1-B) and 13 (A-A_1-B-A_1-B). Design No. 13 is particularly useful in this instance. In the event that administration of the placebo fails to lead to behavioral change (A_1 phase of experimentation) over baseline measurement (A), the investigator is in a position to proceed with assessment of the active drug agent in an experimental analysis whereby the drug is twice introduced and once withdrawn (the B-A_1-B portion of study). If, on the other hand, the placebo exerts an effect over behavior, the investigator may wish to show its controlling effects as in Design No. 10 (A-A_1-A-A_1), which then can be followed with a sequential assessment of an active pharmacologic agent (Design No. 14, A-A_1-A-A_1-B-A_1-B). This design, however, does not permit an analysis of the interactive effects of a placebo (A_1) and a drug (B), as this would require the use of an interactive design (see Section 6.5).

An example of the A-A_1-B-A_1-B strategy appears in the series of drug evaluations conducted by Liberman, Davis, Moon, Moore (1973). In their study, the effects of a placebo and trifluperazine (stelazine) were examined on social interaction and content of conversation in a 21-year-old, withdrawn, male in-patient whose behavior had progressively deteriorated over a 3-year period. At the time the experiment was begun, the patient was receiving stelazine, 20 mg. per day. Two dependent measures were selected for study: (1) willingness to engage in 18 daily, randomly time-sampled, one-half minute chats with a member of the nursing staff, and (2) percentage of the "chats" that contained "sick talk." During the first phase of experimentation (A), the patient's medication was discontinued. In the second phase (A_1) a placebo was introduced, followed by application of stelazine, 60 mg. per day, in the next phase (B). Then, the A_1 and B phases were repeated. A double-blind trial was conducted as the patient and nursing staff were not made aware of placebo and drug alternations.

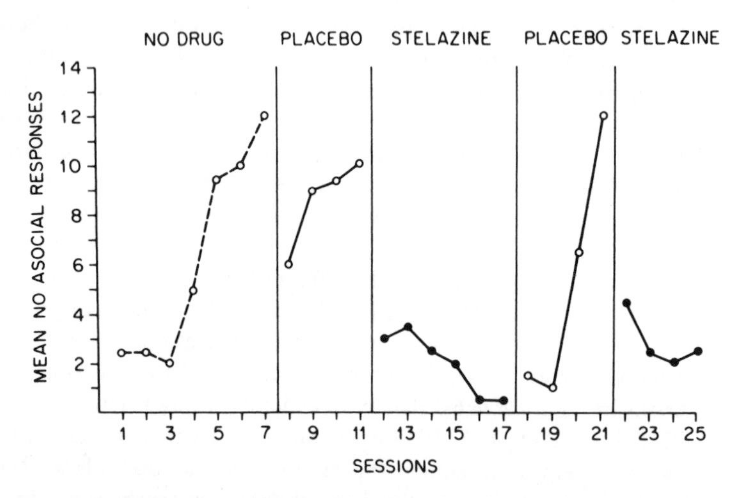

Fig. 6-6. Average number of refusals to engage in a brief conversation. (Fig. 2, p. 435, from: Liberman, R. P., Davis, J., Moon, W., and Moore, J. Research design for analyzing drug-environment-behavior interactions. *Journal of Nervous and Mental Disease*, 1973, **156**, 432-439. Reproduced by permission.)

Results of this study with regard to the patient's willingness to partake in brief conversations appear in Fig. 6-6. In the no-drug condition (A), a marked linear increase in number of asocial responses was observed. Institution of the placebo in phase two (A_1) first led to a decrease followed by a renewed increase in asocial responses, suggesting the overall ineffectiveness of the placebo condition. In phase three (B), administration of stelazine (60 mg. per day) resulted in a substantial decrease in asocial responses. However, a return to placebo conditions (A_1) again led to an increase in refusals to chat. In the final phase (B), reintroduction of stelazine effected a decrease in refusals. To summarize, in this experimental analysis, the effects of an active pharmacological agent were documented twice as indicated by the decreasing data trends in the stelazine phases. Data with respect to content of conversation are not presented graphically, but the authors indicate that under stelazine conditions, rational speech increased. However, administration of stelazine did not appear to modify frequency of delusional and hypochondriacal statements in that they remained at a constant level across all phases of study.

Let us now return to and conclude our examination of drug designs in Table 6-1. In Design 15 (A_1-B-A_1-C-A_1-C) the controlling effects of two drugs (B and C) over placebo conditions (A_1) can be assessed. However, as in the A-B-A-C-A design, cited in Section 6.1 of this chapter, the comparative efficacy of variables B and C are not subject to direct analysis as a group comparison design would be required.

We should point out here that many extensions of these fifteen basic drug designs are possible, including those in which differing levels of the drug are examined. This can be done within the structure of these fifteen designs during active drug treatment or in separate experimental analyses where dosages are systematically varied (e.g., low-high-low-high).

6.5. STRATEGIES FOR STUDYING INTERACTION EFFECTS

Most treatments contain a number of therapeutic components. One task of the clinical researcher is to experimentally analyze these components to determine which are effective and which can be discarded, resulting in a more efficient treatment. Analyzing the separate effects of single therapeutic variables is a necessary way to begin to build therapeutic programs, but it is obvious that these variables may have different effects when interacting with other treatment variables. In advanced stages of the construction of complex treatments it becomes necessary to determine the nature of these interactions. Within the group comparison approach, statistical techniques, such as analysis of variance, are quite valuable in determining the presence of interaction. These techniques are not capable, however, of determining the nature of the interaction or the relative contribution of a given variable to the total effect in a given patient.

To evaluate the interaction of two (or more) variables, one must analyze the effects of both variables separately and in combination in a series of cases. However, one must be careful to adhere to the basic rule of not changing more than one variable at a time (see Chapter 3, Section 3.4).

Before providing examples of strategies for studying interaction, it will be helpful to examine some examples of designs containing two or more variables that are *not* capable of isolating interactive or additive effects. The first example is one where variations of a treatment are added to the end of a successful A-B-A-B (e.g., A-B-A-B^1-B^2-B^3 described above or an A-B-A-B-BC design in which C is a different therapeutic variable). If the BC variable produced an effect over and above the previous B phase, this provides a clue that an interaction may exist but the controlling effects of the BC phase have not been demonstrated. To do this, one would have to return to the B phase and reintroduce the BC phase once again.

A second design, containing two or more variables where analysis of interaction is not possible, occurs if one performs an experimental analysis of one variable against a background of one or more variables already present in the therapeutic situation. For example, O'Leary, Becker, Evans, and Saudargas (1969) measured the disruptive behavior of seven children in a classroom. Three variables (rules, educational structure, and praising appropriate behavior while ignoring disruptive behavior) were introduced sequentially. At this point, we have an A-B-BC-BCD design, where B is rules, C is structure, and D is praise and ignoring. With the exception of one child, these procedures had no effect on disruptive behavior. A fourth treatment–token economy–was then added. In five of six children this was effective, and withdrawal and reinstatement of the token economy confirmed its effectiveness. The last part of the design can be represented as BCD-BCDE-BCD-BCDE, where E is token economy. While this experiment demonstrated that token economy works in this setting, the role of the first three variables is not clear. It is possible that any one of the variables or all three are necessary for the effectiveness of the token program or at least to enhance its effect. On the other hand, the initial three variables may not contribute to the therapeutic effect. Thus, we know that a token program works in this situation against the background of these three variables, but we cannot ascertain the nature of the interaction, if any, since the token program was not analyzed separately.

A third example, where analysis of interaction is not possible, occurs if one is testing the effects of a composite treatment package. Two examples of this strategy were presented in Chapter 3, Section 3.4. In one example (see Fig. 3-13) the effects of covert sensitization on pedophilic interest was examined (Barlow, Leitenberg, and Agras, 1969). Covert sensitization, where a patient is instructed to imagine both unwanted arousing scenes in conjunction with aversive scenes, contains a number of variables such as therapeutic instruction, muscle relaxation, and instructions to imagine each of the two scenes. In this experiment, the whole package was introduced after baseline, followed by withdrawal and reinstatement of one component–the aversive scene. The design can be represented as A-BC-B-BC, where BC is the treatment package and C is the aversive

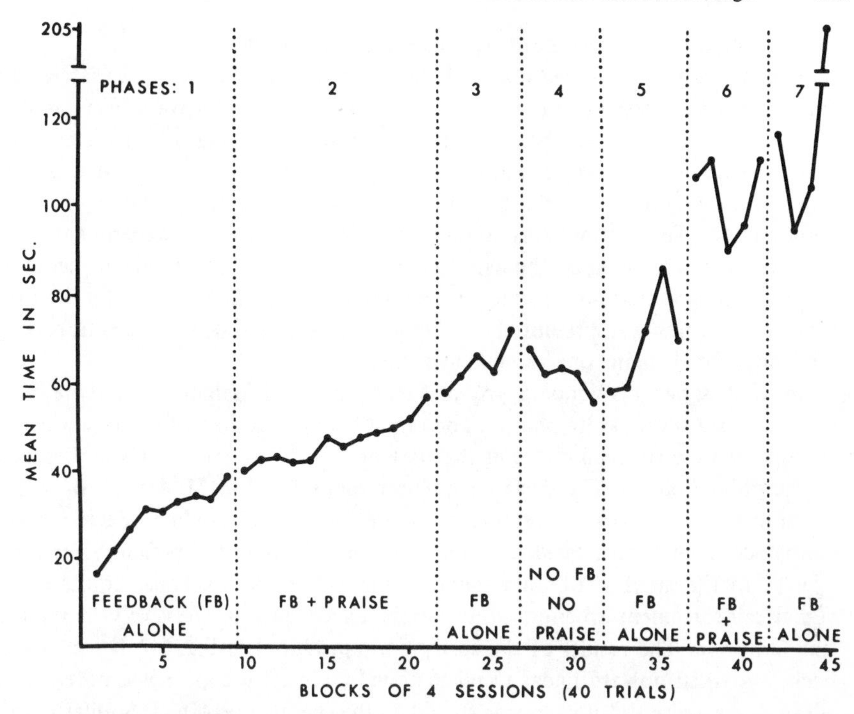

Fig. 6-7. Time in which a knife was kept exposed by a phobic patient as a function of feedback, feedback plus praise, and no feedback or praise conditions. (Fig. 2, p. 136, from: Leitenberg, H., Agras, W. S., Thomson, L. E., and Wright, D. E. Feedback in behavior modification: An experimental analysis in two phobic cases. *Journal of Applied Behavior Analysis*, 1968, **1**, 131-137. Copyright by Society for the Experimental Analysis of Behavior, Inc. Reproduced by permission.)

scene. (Notice that more than one variable was changed during the transition from A-BC. This is in accordance with an exception to the guidelines outlined in Chapter 3, Section 3.4.)

Figure 3-13 demonstrates that pedophilic interest dropped during the treatment package, rose when the aversive scene was removed, and dropped again after reinstatement of the aversive scene. Once again, these data indicate that the noxious scene is important against the background of the other variables present in covert sensitization. The contribution of each of the other variables and the nature of these interactions with the aversive scene, however, has not been demonstrated (nor was this the purpose of the study). In this case, it would seem that an interaction is present since it is hard to conceive of the aversive scene alone producing these decreases in pedophilic interest. The nature of the interaction, however, awaits further experimental inquiry.

The preceding examples outlined designs where two or more variables are simultaneously present but analysis of interactive or additive effects is not possible. While these designs can hint at interaction and set the stage for further experimentation, a thorough analysis of interaction as noted above requires an experimental analysis of two or more variables, separately and in combination. To illustrate this complex process, two series of experiments will be presented that analyze the same variables–feedback and reinforcement–in two separate populations (phobics and anorexics). One experiment from the first series of phobics was presented in Chapter 3, Section 3.4, in connection with guidelines for changing one variable at a time.

In that series (Leitenberg, Agras, Thomson, and Wright, 1968) the first subject was a severe knife phobic. The target behavior selected for study was the amount of time (in seconds) that the patient was able to remain in the presence of the phobic object. The design can be represented as B-BC-B-A-B-BC-B, where B represents feedback, C represents praise, and A is baseline. Each session consisted of ten trials. Feedback consisted of informing the patient after each trial as to the amount of time spent looking at the knife. Praise consisted of verbal reinforcement whenever the patient exceeded a progressively increasing time criterion. The results of the study are reproduced in Fig. 6-7. During feedback, a marked upward linear trend in time looking at the knife was noted. The addition of praise did not appear to add to the therapeutic effect. Similarly, the removal of praise in the next phase did not subtract from the progress. At this point, it appeared that feedback was responsible for the therapeutic gains. Withdrawal and reinstatement of feedback in the next two phases confirmed the controlling effects of feedback. Addition and removal of praise in the remaining two phases replicated the beginning of the experiment in that praise did not demonstrate any additive effect.

This experiment alone does not entirely elucidate the nature of the interaction. At this point, two tentative conclusions are possible. Either praise has no effect on phobic behavior or praise does have an effect, which was masked or overridden by the powerful feedback effect. In other words, this patient may have been progressing at an optimal rate allowing no opportunity for a praise effect to appear. In accordance with the general guidelines of analyzing both variables separately as well as in combination, the next experiment reversed the order of the introduction of variables in a second knife phobic patient (Leitenberg, 1973).

Once again, the target behavior was the amount of time the subject was able to remain in the presence of the knife. The design replicated the first experiment with the exception of the elimination of the last phase. Thus, the design can be

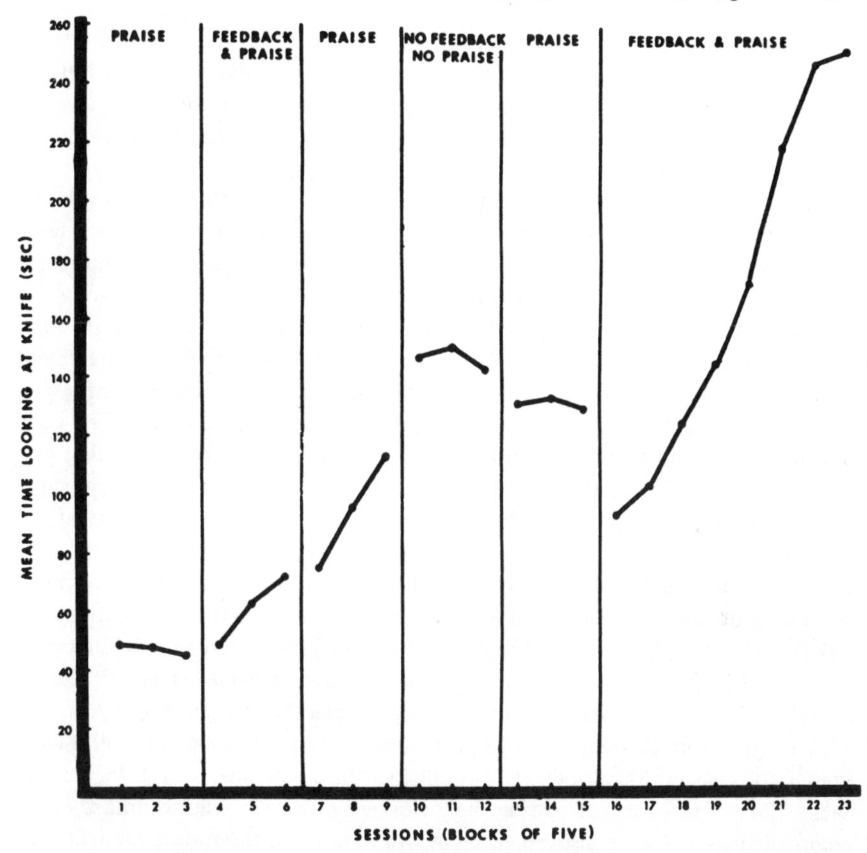

Fig. 6-8. (Fig. 1, from: Leitenberg, H. Interaction designs. Paper read at American Psychological Association, Montreal, August 1973. Reproduced by permission.)

represented as B-BC-B-A-B-BC. In this experiment, however, B refers to praise or verbal reinforcement and C represents feedback of amount of time looking at the knife, which is just the reverse of the last experiment.

In this subject, little progress was observed during the first verbal reinforcement phase (see Fig. 6-8). However, when feedback was added to praise in the second phase, performance increased steadily. Interestingly, this rate of improvement was maintained when feedback was removed. After a sharp gain, performance stabilized when both feedback and praise were removed. Once again, the introduction of praise alone did not produce any further improvement. The addition of feedback to praise for the second time in the experiment resulted in marked improvement in the knife phobic. Direct replication of this experiment in four additional subjects, each with a different phobia, produced similar results. That is, praise did not produce improvement when initially introduced but the addition of feedback resulted in marked improvement. In several cases, however, progress seemed to be maintained by praise after feedback was withdrawn from the package, as in Fig. 6-8.

The overall results of the interaction analysis indicate that feedback is the most active component since marked improvement occurred during both feedback alone and feedback plus praise phases. Praise alone had little or no effect although it was capable of maintaining progress begun in a prior feedback phase in some cases. Similarly, praise did not add to the therapeutic effect when combined with feedback in the first subject. Accordingly, a more efficient treatment package for phobics would emphasize the feedback or knowledge-of-results aspect and de-emphasize or possibly eliminate the social reinforcement component. These results have implications for treatments of phobics by other procedures such as systematic desensitization, where knowledge of results provided by self-observation of progress through a discrete hierarchy of phobic situations is a major component.

The interaction of reinforcement and feedback was also tested in a series of subjects with anorexia nervosa (Agras, Barlow, Chapin, Abel, and Leitenberg, 1974). From the perspective of interaction designs, the experiment is interesting because the contribution of a third therapeutic variable–labeled "size of meals"–was also analyzed. To illustrate the interaction design strategy, several experiments from this series will be presented. All patients were hospitalized and presented with 6000 calories per day, divided into four meals of 1500 calories each. Two measures of eating behavior–weight and caloric intake–were recorded. Patients were also asked to record number of mouthfuls eaten at each meal. Reinforcement consisted of granting privileges based on increases in weight. If weight gain exceeded a certain criterion, the patient could leave her room, watch television, play table games with the nurses, etc. Feedback consisted of providing precise information on weight, caloric intake, and number of mouthfuls eaten. Specifically, the patient plotted on a graph information that was provided by hospital staff.

In one experiment the effect of reinforcement was examined against a background of feedback. The design can be represented as B-BC-BC^1-BC, where B is feedback, C is reinforcement, and C^1 is non-contingent reinforcement. During the first feedback phase (labeled Baseline on the graph), slight gains in caloric intake and weight were noted (see Fig. 6-9). When reinforcement was added to feedback, caloric intake and weight increased sharply. Non-contingent reinforcement produced a drop in caloric intake and a slowing of weight gain, while reintroduction of reinforcement once again produced sharp gains in both measures.

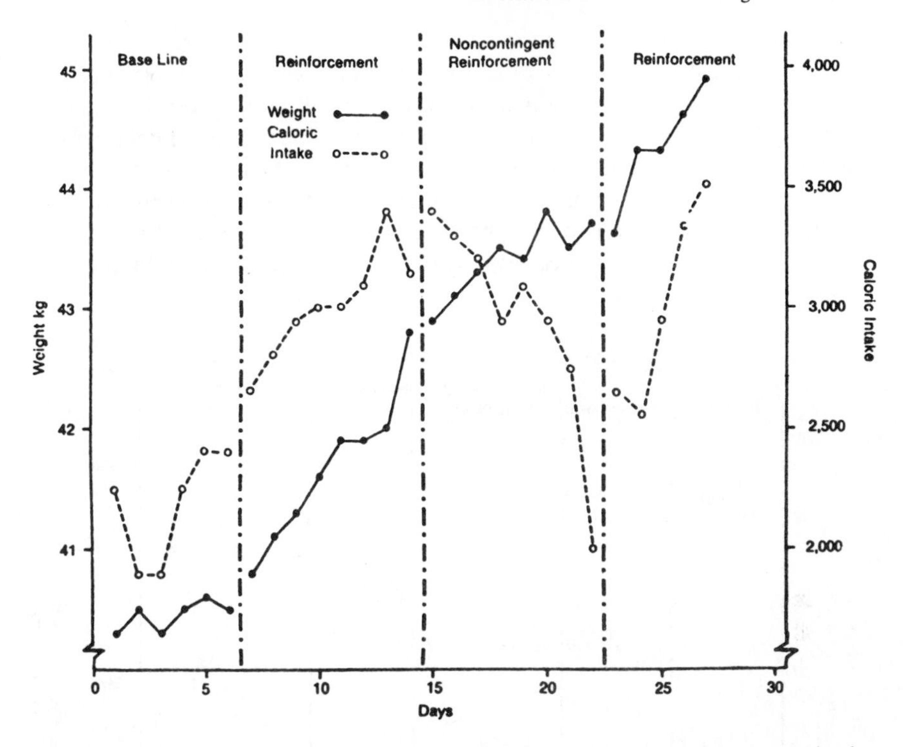

Fig. 6-9. Data from an experiment examining the effect of positive reinforcement in the absence of negative reinforcement (Patient 3). (Fig. 2, p. 281, from: Agras, W. S., Barlow, D. H., Chapin, H. N., Abel, G. G., and Leitenberg, H. Behavior modification of anorexia nervosa. *Archives of General Psychiatry*, 1974, **30,** 279-286. Copyright 1974 American Medical Association. Reproduced by permission.)

These data contain hints of an interaction in that caloric intake and weight rose slightly during the first feedback phase, a finding that replicated two earlier experiments. The addition of reinforcement, however, produced increases over and above those for feedback alone. The drop and subsequent rise of caloric intake and rate of weight gain during the next two phases demonstrated that reinforcement is a controlling variable *when combined with feedback*.

These data only hint at the role of feedback in this study in that some improvement occurred during the initial phase when feedback alone was in effect. Similarly, we cannot know from this experiment the independent effects of reinforcement since this was not analyzed separately. To accomplish this, two experiments were conducted where feedback was introduced against a background of reinforcement. Only one experiment will be presented, although both sets of data are very similar. The design can be represented as A-B-BC-B-BC, where A is baseline, B is reinforcement, and C is feedback (see Fig. 6-10). It should be noted that the patient continued to be presented with 6000 calories throughout the experiment, a point to which we will return later. During baseline, in which no reinforcement or feedback was present, caloric intake actually declined. The introduction of reinforcement did not result in any increases; in fact, a slight decline continued. Adding feedback to reinforcement, however, produced increases in weight and caloric intake. Withdrawal of feedback stopped this increase, which began once again when feedback was reintroduced in the last phase.

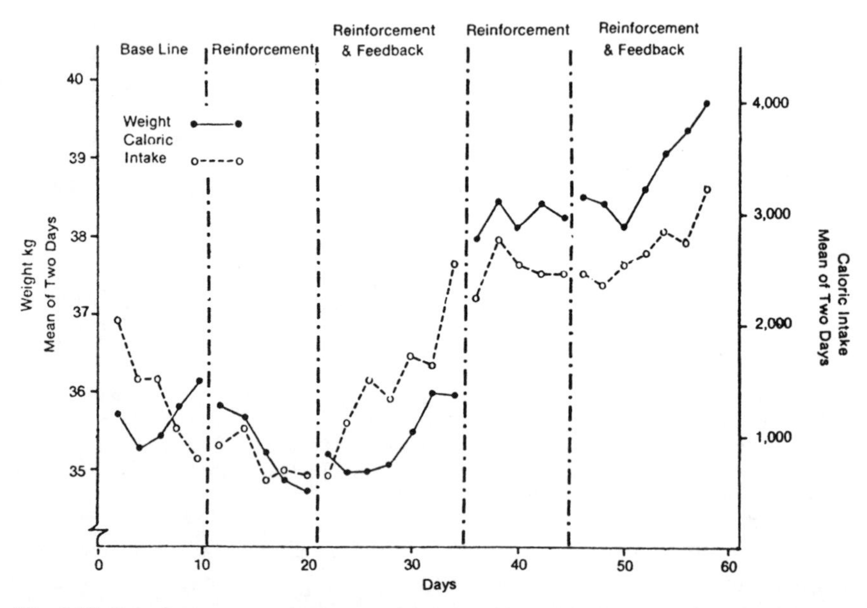

Fig. 6-10. Data from an experiment examining the effect of feedback on the eating behavior of a patient with anorexia nervosa (Patient 5). (Fig. 4, p. 283, from: Agras, W. S., Barlow, D. H., Chapin, H. N., Abel, G. G., and Leitenberg, H. Behavior modification of anorexia nervosa. *Archives of General Psychiatry*, 1974, **30**, 279-286. Copyright 1974 American Medical Association. Reproduced by permission.)

With this experiment (and its replications) it becomes possible to draw conclusions on the nature of what is in this case a complex interaction. When both variables were presented alone, as in the initial phases in the respective experiments, reinforcement produced no increases but feedback produced some increase. When presented in combination, reinforcement added to the feedback effect and, against a background of feedback, became the controlling variable in

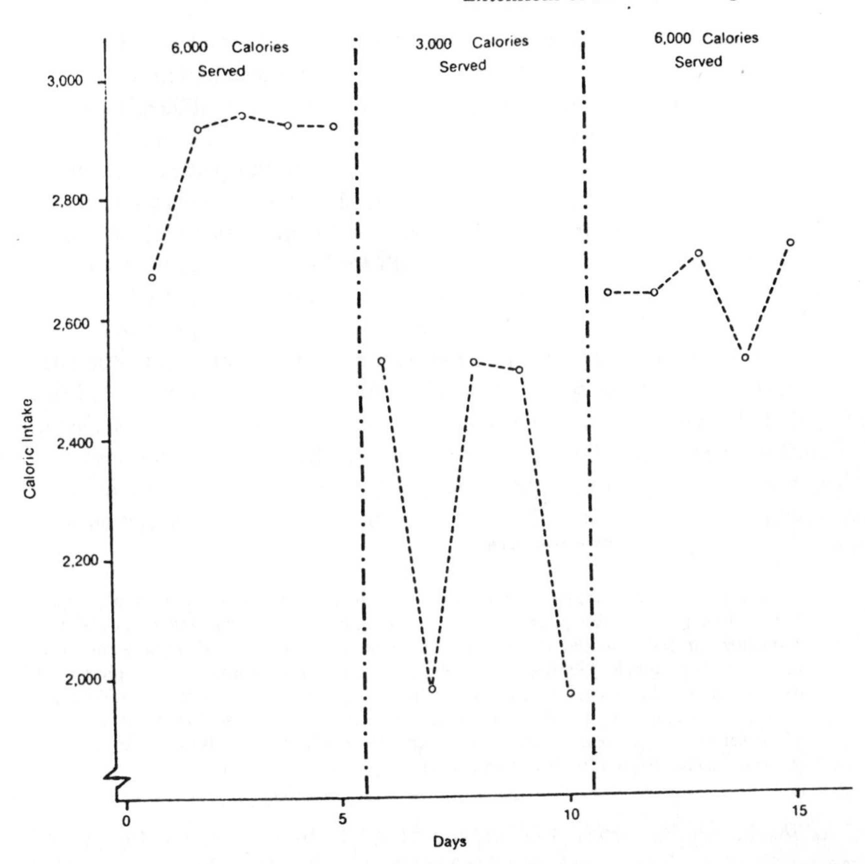

Fig. 6-11. The effect of varying the size of meals upon the caloric intake of a patient with anorexia nervosa (Patient 5). (Fig. 5, p. 285, from: Agras, W. S., Barlow, D. H., Chapin, H. N., Abel, G. G., and Leitenberg, H. Behavior modification of anorexia nervosa. *Archives of General Psychiatry*, 1974, **30,** 279-286. Copyright 1974 American Medical Association. Reproduced by permission.)

that caloric intake decreased when contingent reinforcement was removed. Feedback, however, also exerted a controlling effect when it was removed and reintroduced against a background of reinforcement. Thus, it seems that feedback can maximize the effectiveness of reinforcement to the point where it is a controlling variable. Feedback alone, however, is capable of producing therapeutic results, which is not the case with reinforcement. Feedback, thus, is the more important of the two variables although both contribute to treatment outcome.

It was noted above that the contribution of a third variable–size of meals–was also examined within the context of this interaction. In keeping with the guidelines of analyzing each variable separately and in combination with other variables, phases were examined when the large amount of 6000 calories was presented without the presence of either feedback or reinforcement. The baseline phase of Fig. 6-10 represents one such instance. In this phase caloric intake declined steadily. Examination of other baseline phases in the replications of this experiment revealed similar results. To complete the interaction analysis, size of meal was varied against a background of both feedback and reinforcement. The design can be represented as ABC-ABC^1-ABC, where A is feedback, B is reinforcement, C is 6000 calories per day, and C^1 is 3000 calories per day.

Under this condition, size of meal did have an effect in that more was eaten when 6000 calories were served than when 3000 calories were presented (see Fig. 6-11). In terms of treatment, however, even large meals were incapable of producing weight gain in those phases where it was the only therapeutic variable. Thus, this variable is not as strong as feedback. The authors conclude this series by summarizing the effects of the three variables alone and in combination across five patients:

> Thus large meals and reinforcement were combined in four experimental phases and weight was lost in each phase. On the other hand, large meals and feedback were combined in eight phases and weight was gained in all but one. Finally, all three variables (large meals, feedback, and reinforcement) were combined in 12 phases and weight was gained in each phase. These findings suggest that informational feedback is more important in the treatment of anorexia nervosa than positive reinforcement, while serving large meals is least important. However, the combination of all three variables seems most effective. (Agras, Barlow, Chapin, Abel, and Leitenberg, 1974, p. 285)

As in the phobic series, the juxtaposition of variables within the general framework of analyzing each variable separately and in combination provided information on the interaction of these variables.

References

Agras, W. S. Behavior modification in the general hospital psychiatric unit. In H. Leitenberg (Ed.), *Handbook of behavior modification*. Englewood Cliffs, N.J.: Prentice-Hall, in press.

Agras, W. S., Barlow, D. H., Chapin, H. N., Abel, G. G., and Leitenberg, H. Behavior modification of anorexia nervosa. *Archives of General Psychiatry*, 1974, **30**, 279-286.

Bailey, J. S., Wolf, M. M., and Phillips, E. L. Home-based reinforcement and the modification of pre-delinquents' classroom behavior. *Journal of Applied Behavior Analysis*, 1970, **3**, 223-233.

Barlow, D. H., and Hersen, M. Single-case experimental designs: Uses in applied clinical research. *Archives of General Psychiatry*, 1973, **29**, 319-325.

Barlow, D. H., Leitenberg, H., and Agras, W. S. Experimental control of sexual deviation through manipulation of the noxious scene in covert sensitization. *Journal of Abnormal Psychology*, 1969, **74**, 596-601.

Bellak, L., and Chassan, J. B. An approach to the evaluation of drug effect during psychotherapy: A double-blind study of a single case. *Journal of Nervous and Mental Disease*, 1964, **139**, 20-30.

Chassan, J. B. *Research design in clinical psychology and psychiatry.* New York: Appleton-Century-Crofts, 1967.

Coleman, R. A conditioning technique applicable to elementary school classrooms. *Journal of Applied Behavior Analysis*, 1970, **3**, 293-297.

Davis, K. V., Sprague, R. L., and Werry, J. S. *American Journal of Mental Deficiency*, 1969, **73**, 721-727.

Grinspoon, L., Ewalt, J., and Shader, R. Long term treatment of chronic schizophrenia. *International Journal of Psychiatry*, 1967, **4**, 116-128.

Hall, R. V., Axelrod, S., Tyler, L., Grief, E., Jones, F. C., and Robertson, R. Modification of behavior problems in the home with a parent as observer and experimenter. *Journal of Applied Behavior Analysis*, 1972, **5**, 53-74.

Hersen, M., Gullick, E. L., Matherne, P. M., and Harbert, T. L. Instructions and reinforcement in the modification of a conversion reaction. *Psycholgical Reports*, 1972, **31**, 719-722.

Hopkins, B. L., Schutte, R. C., and Garton, K. L. The effects of access to a playroom on the rate and quality of printing and writing of first and second-grade students. *Journal of Applied Behavior Analysis*, 1971, **4**, 77-87.

Kaufman, K. F., and O'Leary, K. D. Reward cost, and self-evaluation procedures for disrupting adolescents in a psychiatric hospital school. *Journal of Applied Behavior Analysis*, 1972, **5**, 293-309.

Leitenberg, H. Interaction designs. Paper read at American Psychological Association, Montreal, August 1973.

Leitenberg, H., Agras, W. S., Thomson, L., and Wright, D. E. Feedback in behavior modification: An experimental analysis in two phobic cases. *Journal of Applied Behavior Analysis*, 1968, **1**, 131-137.

Liberman, R. P., Davis, J., Moon, W., and Moore, J. Research design for analyzing drug-environment-behavior interactions. *Journal of Nervous and Mental Disease*, 1973, **156**, 432-439.

Lindsley, O. R. Operant conditioning techniques in the measurement of psychopharmacologic response. In J. H. Nodine and J. H. Moyer (Eds.), *Psychosomatic medicine: The first Hahnemann symposium on psychosomatic medicine.* Pp. 373-383. Philadelphia: Lea and Febiger, 1962.

Mann, R. A. The behavior-therapeutic use of contingency contracting to control an adult behavior problem: Weight control. *Journal of Applied Behavior Analysis*, 1972, **5**, 99-109.

McFarlain, R. A., and Hersen, M. Continuous measurement of activity level in psychiatric patients. *Journal of Clinical Psychology*, 1974, **30**, 37-39.

McLaughlin, T. F., and Malaby, J. Intrinsic reinforcers in a classroom token economy. *Journal of Applied Behavior Analysis*, 1972, **5**, 263-270.

O'Leary, K. D., Becker, W. C., Evans, M. B., and Saudargas, R. A. A token reinforcement program in a public school: A replication and systematic analysis. *Journal of Applied Behavior Analysis*, 1969, **2**, 3-13.

Pendergrass, V. E. Timeout from positive reinforcement following persistent, high-rate behavior in retardates. *Journal of Applied Behavior Analysis*, 1972, **5**, 85-91.

Roxburgh, P. A. Treatment of persistent phenothiazine-induced oraldyskinesia. *British Journal of Psychiatry*, 1970, **116**, 277-280.

Turner, S. M., Hersen, M., and Alford, H. Effects of massed practice and meprobamate on spasmodic torticollis: An experimental analysis. *Behaviour Research and Therapy*, 1974, **12**, 259-260.

Wheeler, A. J., and Sulzer, B. Operant training and generalization of a verbal response form in a speech-deficient child. *Journal of Applied Behavior Analysis*, 1970, **3**, 139-147.

Wincze, J. P., Leitenberg, H., and Agras, W. S. The effects of token reinforcement and feedback on the delusional verbal behavior of chronic paranoid schizophrenics. *Journal of Applied Behavior Analysis*, 1972, **5**, 247-262.

CHAPTER 7

Multiple Baseline, Multiple Schedule, and Concurrent Schedule Designs

7.1. INTRODUCTION

The use of sequential withdrawal or reversal designs is inappropriate when treatment variables cannot be withdrawn or reversed due to practical limitations, ethical considerations, or problems in staff cooperation (Baer, Wolf, and Risley, 1968; Barlow and Hersen, 1973; Johnston, 1972; Kazdin, 1973; Kazdin and Bootzin, 1972; Leitenberg, 1973; Risley and Wolf, 1972; Thoresen, 1972; Wolf and Risley, 1971). Practical limitations arise when carryover effects appear across adjacent phases of study, particularly in the case of therapeutic instructions (Barlow and Hersen, 1973). A similar problem may occur when drugs with known long-lasting effects are evaluated in single case withdrawal designs. Despite discontinuation of medication in the withdrawal (placebo) phase, active agents persist physiologically and, with the phenothiazines, traces have been found in body tissues 6 months later (Ban, 1969).

Ethical considerations are of paramount importance when the treatment variable is effective in reducing self- or other-destructive behaviors in subjects. Here the withdrawal of treatment is obviously unwarranted, even for brief periods of time. Related to the problem of undesirable behavior is the matter of environmental cooperation. Even if the behavior in question does not have immediate destructive effects on the environment, if it is considered to be aversive (i.e.; by teachers, parents, or hospital staff) the experimenter will not obtain sufficient cooperation in order to carry out withdrawal or reversal of treatment procedures. Under these circumstances, it is clear that the applied clinical researcher must pursue his study using different experimental strategies. In still other instances withdrawal of treatment, despite absence of harm to the subject or others in his environment, may be undesirable because of the severity of the disorder. Here the importance of preserving therapeutic gains is given priority, especially when a disorder has a lengthy history and previous efforts at remediation have failed.

Three types of experimental strategies have been used by applied clinical researchers when withdrawals and reversals were not feasible. They include the multiple baseline, multiple schedule, and concurrent schedule designs. Although none of these designs has yet been used extensively, the multiple baseline strategy has been applied most frequently. In our review of the applied literature we were able to find three published examples of the multiple schedule design (Agras, Leitenberg, Barlow, and Thomson, 1969; O'Brien, Azrin, and Henson, 1969; Wahler, 1969) and only one of the concurrent schedule design (Browning and Stover, 1971). Of course, in the basic operant research area all of these designs are well known and have been used for years (Catania, 1968; Herrnstein, 1970; Honig, 1966; Reynolds, 1968; Sidman, 1960).

In this chapter we will examine in detail the rationale and procedures for multiple baseline, multiple schedule, and concurrent schedule designs. Examples of each will be presented for illustrative purposes, including the three different varieties of multiple baseline strategies. In addition, the possibility of using multiple baseline designs across subjects in drug evaluations will be considered.

7.2. MULTIPLE BASELINE DESIGNS

The rationale for the multiple baseline design first appeared in the applied behavioral literature in 1968 (Baer, Wolf, and Risley, 1968), although a within-subject multiple baseline strategy previously had been used by Marks and Gelder (1967) in their assessment of electrical aversion therapy for a sexual deviate. Baer, Wolf, and Risely (1968) point out that "In the multiple-baseline technique, a number of responses are identified and measured over time to provide baselines against which changes can be evaluated. With these baselines established, the experimenter then applies an experimental variable to one of the behaviors, produces a change in it, and perhaps notes little or no change in the other baselines" (p. 94). Subsequently, the experimenter applies the same experimental variable to a second behavior and notes rate changes in that behavior. This procedure is continued in sequence until the experimental variable has been applied to all of the target behaviors under study. In each case the treatment variable is usually not applied until baseline stability has been achieved.

Baseline and subsequent treatment interventions for each targeted behavior can be conceptualized as separate A-B designs, with the A phase further extended for each of the succeeding behaviors until the treatment variable is finally applied. The experimenter is assured that his treatment variable is effective when a change in rate appears after its application while the rate of concurrent (untreated) behaviors remains relatively constant. A basic assumption is that the targeted behaviors are independent from one another. If they should happen to covary, then the controlling effects of the treatment variable are subject to question and limitations of the A-B analysis fully apply (see Chapter 5, Section 5.2).

The issue of independence of behaviors within a single subject raises some interesting problems from an experimental standpoint, particularly if the experimenter is involved in a new area of study where no precedents apply. He is then placed in the position where an *a priori* assumption of independence cannot be made, thus leaving an empirical test of the proposition. Leitenberg (1973) argues that "If general effects on multiple behaviors were observed after treatment had been applied to only one, there would be no way to clearly interpret the results. Such results may reflect a specific therapeutic effect and subsequent response generalization, or they may simply reflect non-specific therapeutic effects having little to do with the specific treatment procedure under investigation" (p. 95). In some cases, when independence of behaviors is not found, application of either the multiple schedule or the concurrent schedule design is recommended (see Sections 7.3 and 7.4). In other cases, application of the multiple baseline design across different subjects might yield useful information. Surprisingly, however, in the available published reports the problem of independence has not been insurmountable (Leitenberg, 1973).

The multiple baseline design is considerably weaker than the withdrawal design as the controlling effects of the treatment on each of the target behaviors is not directly demonstrated (e.g., as in the A-B-A design). As noted earlier, the effects of the treatment variable are inferred from the untreated behaviors. This raises an issue, then, as to how many baselines are needed before the experimenter is able to establish confidence in the controlling effects of his treatment. A number of interpretations have appeared in the literature. Baer, Wolf, and Risley (1968) initially considered this issue to be an "audience variable" and were reluctant to specify the minimum number of baselines required. Although theoretically only a minimum of two baselines is needed to derive useful information, Barlow and Hersen (1973) argued that "... the controlling effects of that technique over at least three target behaviors would appear to be a minimum requirement" (p. 323). Similarly, Wolf and Risley (1971) contend that "While a study involving two baselines can be very suggestive, a set of replications across three or four baselines may be almost completely convincing" (p. 316). At this point, we would recommend a minimum of three to four baselines if practical and experimental considerations permit.

Although demonstration of the controlling effects of a treatment variable is obviously weaker in the multiple baseline design, a major advantage of this strategy is that it fosters the simultaneous measurement of several concurrent target behaviors. This is most important for at least two major reasons. First, the monitoring of concurrent behaviors allows for a closer approximation to naturalistic conditions where a variety of responses are occurring at the same time. Second, examination of concurrent behaviors leads to an analysis of covariation among the targeted behaviors. Basic researchers have been concerned with the measurement of concurrent behaviors for some time (Catania, 1968; Herrnstein, 1970; Honig, 1966; Reynold, 1968; Sidman, 1960), but only recently have applied behavioral researchers evidenced a similar interest (Kazdin, 1973; Sajwaj, Twardosz, and Burke, 1972; Twardosz and Sajwaj, 1972). Kazdin (1973) underscores the importance of measuring concurrent (untreated) behaviors when assessing the efficacy of reinforcement paradigms in applied settings. He states that "While changes in target behaviors are the *raison d'être* for undertaking treatment or training programs, concomitant changes may take place as well. If so, they should be assessed. It is one thing to assess and evaluate changes in a target behavior, but quite another to insist on excluding non-target measures. It may be that investigators are short-changing themselves in evaluating the programs" (Kazdin, 1973, p. 527).

As mentioned earlier, there are three basic types of multiple baseline designs. In the first–the multiple baseline design across behaviors–the same treatment variable is applied sequentially to separate (independent) target behaviors in a single subject. A possible variation of this strategy, of course, involves the sequential application of a treatment variable to targeted behaviors for an entire group of subjects. In this connection, Hall, Cristler, Cranston, and Tucker (1970) note that ". . . these multiple baseline designs apply equally well to the behavior of groups if the behavior of the group members is summed or averaged, and the group is treated as a single organism" (p. 253). However, in this case the experimenter would also be expected to present data for individual subjects, demonstrating that sequential treatment applications to independent behaviors affected most subjects in the same direction.

In the second design–the multiple baseline design across subjects–a particular treatment is applied in sequence across *matched* subjects presumably exposed to "identical" environmental conditions. Thus, as the same treatment variable is applied to succeeding subjects, the baseline for each subject increases in length. In contrast to the multiple baseline design across behaviors (the within-subject multiple baseline design), in the multiple baseline design across subjects a single targeted behavior serves as the primary focus of inquiry. However, there is no experimental contraindication to monitoring concurrent (untreated) behaviors as well. Indeed, it is quite likely that the monitoring of concurrent behaviors will lead to additional findings of merit.

As in the multiple baseline design across behaviors, a possible variation of the multiple baseline design across subjects involves the sequential application of the treatment variable across entire groups of subjects. But here, too, it behooves the experimenter to show that a large majority of individual subjects for each group evidenced the same effects of treatment.

We might note that the multiple baseline design across subjects has also been labeled a "time-lagged control" design (Gottman, 1973; Gottman, McFall, and Barnett, 1969). In fact, this strategy was followed by Hilgard (1933) over 40 years ago, in a study in which she examined the effects of early and delayed practice on memory and motoric functions in a set of twins (method of co-twin control).

In the third design–the multiple baseline design across settings–a particular treatment is applied sequentially to a single subject or a group of subjects across independent situations. For example, in a classroom situation, one might apply timeout contingenices for unruly behavior in sequence across different classroom periods. The baseline period for each succeeding classroom period, then, increases in length before application of the treatment. As in the across-subject design, assessment of treatment is usually based on rate changes observed in a selected target behavior. However, once again the monitoring of concurrent behaviors might prove to be of value and should be encouraged where possible.

To recapitulate, in the multiple baseline design across behaviors, a treatment variable is applied sequentially to independent behaviors within the same subject. In the multiple baseline design across subjects, a treatment variable is applied sequentially to the same behavior across different but matched subjects sharing the same environmental conditions. Finally, in the multiple baseline design across settings, a treatment variable is applied sequentially to the same behavior across different and independent settings in the same subject. Published examples of the three basic types of multiple baseline strategies are categorized in Table 7-1 with respect to design type and subject characteristics.

In the following three subsections we will illustrate the use of basic multiple baseline strategies in addition to presenting examples of variations selected from both the clinical and child literatures.

Multiple baseline across behaviors

Hall, Cristler, Cranston, and Tucker (1970) used a multiple baseline strategy (across behaviors) to assess the effects of an "early" bedtime contingency on the amount of time spent in extra-curricular activities for a normal, 10-year-old, fourth-grade girl. The mother of this child, who was enrolled in a course on contingency management in the classroom, served as experimenter and observer, and was interested in systematically increasing the length of time her child spent working in extra-curricular activities at home. Three target behaviors were

TABLE 7-1. Multiple Baseline Designs

Study	Design	Subjects
Allen (1973)	across settings	minimally brain damaged
Bailey *et al.* (1971)	across behaviors	preschool children
Barlow *et al.* (1973)	across behaviors	transsexual
Barrish *et al.* (1969)	across settings	fourth graders
Barton *et al.* (1970)	across behaviors (group)	retardates
Christopherson *et al.* (1972)	across behaviors	three mildly undisciplined siblings
”	across subjects	two normal siblings
Clark *et al.* (1973)	across behaviors	retarded child
Eisler *et al.* (1974)	across behaviors	unassertive inpatients
Epstein and Hersen (1974)	across behaviors	schizophrenic
Hall *et al.* (1970)	across settings	fifth graders
”	across subjects	failing tenth graders
”	across behaviors	normal fourth graders
Hilgard (1933)	across subjects	normal identical twins
Liberman and Smith (1972)	across behaviors	phobic
Liberman *et al.* (1973)	across subjects	schizophrenics
Long and Williams (1973)	across settings (groups)	disruptive seventh graders
Lovaas and Simmons (1969)	across subjects	retardates
	across behaviors	retardates
Marks and Gelder (1967)	across behaviors	sexual deviate
Moore and Bailey (1973)	across behaviors	autistic child
Morganstern (1974)	across behaviors	obese female
Panyon *et al.* (1970)	across subjects (groups)	attendants working with retardates
Patterson and Teigen (1973)	across behaviors	schizophrenic
Rachman (1974)	across behaviors	obsessional
Risley and Hart (1968)	across behaviors	preschool children
Wilson and Hopkins (1973)	across subjects (groups)	seventh and eighth graders

selected for study: time spent daily in clarinet practice, time spent daily working on a Campfire Girl project, and time involved daily reading in preparation for six book reports. Prior to baseline measurement, the girl's mother requested that she spend a minimum of 30 minutes per day in each of these activities. During baseline, time spent in the three activities was monitored by the mother (reliability checks were obtained surreptitiously by an independent observer). At the beginning of the second week of study, a daily bedtime contingency was initiated with respect to time spent in clarinet practice. The child was informed that if she practiced her clarinet less than 30 minutes per day, each minute under 30 would result in her bedtime curfew being decreased by a corresponding 1 minute loss. The same bedtime contingency was instituted sequentially during the third and fourth weeks of the study with regard to amount of time spent on the Campfire Project and time spent reading.

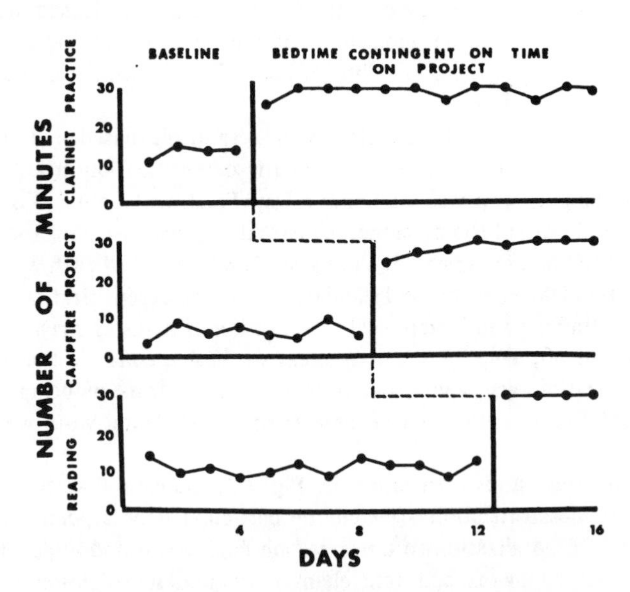

Fig. 7-1. A record of time spent in clarinet practice, camp honors project work, and reading for book reports by a 10-year-old girl. *Baseline*–before experimental procedures; *Early bedtime contingent on less than 30 minutes of behavior*–1 minute earlier bedtime for each minute less than 30 engaged in an activity. (Fig. 3, p. 252, from: Hall, R. V., Cristler, C., Cranston, S. S., and Tucker, B. Teachers and parents as researchers using multiple baseline designs. *Journal of Applied Behavior Analysis*, 1970, **3**, 247-255. Copyright by Society for the Experimental Analysis of Behavior, Inc. Reproduced by permission.)

The results of this investigation are presented in Fig. 7-1. Examination of the graph reveals stable baseline rates for each of the target behaviors. Institution of the bedtime contingency for clarinet practice resulted in a marked increase in time spent practicing (baseline mean = 13·5 minutes per day; contingency mean = 29·0 minutes per day), but rates for the amount of time spent on the Campfire Project and reading remained constant. Institution of the bedtime contingency for the Campfire Project similarly increased the amount of time spent in that activity (baseline mean = 3·5 minutes per day; contingency mean = 29·75 minutes per day), while the rate for reading remained consistent. When the bedtime contingency was applied to reading itself, the rate for that activity increased from an average of 11·0 minutes per day in baseline to 30·0 minutes per day during the contingency condition. Thus, it is clear that the bedtime contingency was effective in increasing the rates of the three target behaviors, but only when the contingency was directly applied to each. Independence of the three behaviors and absence of generalization effects from one behavior to the next facilitate interpretation of these data. On the other hand, had non-treated behaviors covaried following application of the bedtime contingency, unequivocal conclusions as to the controlling effects of the treatment could not have been reached.

Liberman and Smith (1972) also used a multiple baseline design across behaviors in studying the effects of systematic desensitization in a 28-year-old, multiphobic female, who was attending a Day Treatment Center. Four specific phobias were identified (being alone, menstruation, chewing hard foods, dental work), and baseline assessment of the patient's self-report of each was taken for 4 weeks. Subsequently, *in vivo* and standard systematic desensitization (consisting of relaxation training and hierarchical presentation of items in imagination) were administered in sequence to the four areas of phobic concern. Specifically, *in vivo* desensitization was administered in relation to fears of being alone and chewing hard foods, while fears of menstruation and dental work were treated imaginally.

Results of this study, presented in Fig. 7-2, indicate that the sequential application of desensitization affected the particular phobia being treated, but no evidence of generalization to untreated phobias was noted. Independence of the four target behaviors and rate changes when desensitization was finally applied to each support the conclusion that treatment was effective and that it exerted control over the dependent measures (self-reports of degrees of fear). Although the authors argue that a positive set for improvement was maintained throughout all phases of study, the possibility that expectancy of improvement and actual treatment effects were confounded cannot be discounted, especially in light of the primary reliance on self-report data. However, casually conducted behavioral observations corroborate self-report data.

Despite the above-mentioned limitations, Liberman and Smith's (1972) investigation is of interest from a number of standpoints. First, as most multiple baseline studies emanate from the operant framework, this study lends credence to the notion that non-operant procedures (e.g., systematic desensitization) can be assessed in this paradigm. Second, as the particular dependent measure (ratings of subjective fear on the Target Complaint Scale) is based on the patient's self-report, it would appear that this type of single case research might easily be carried out in outpatient facilities and even in consulting room practice (see Chapter 3, Section 3.2). Finally, the treatment was fully implemented by a mental health paraprofessional who had only 1 year's training in psychiatry.

In our next example of a multiple baseline design across behaviors, a

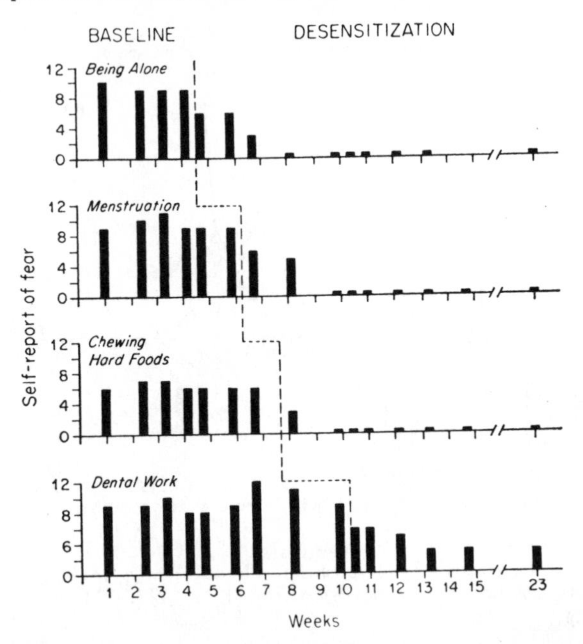

Fig. 7-2. Multiple baseline evaluation of desensitization in a single case with four phobias. (Fig. 1, p. 600, from: Liberman, R. P., and Smith, V. A multiple baseline study of systematic desensitization in a patient with multiple phobias. *Behavior Therapy*, 1972, **3**, 597-603. Reproduced by permission.)

physiological measure (erectile strength as assessed via a penile transducer) was used to determine efficacy of electrical aversion procedures in the treatment of a 21-year-old, male transvestite (Marks and Gelder, 1967). During baseline assessment, several stimuli (panties, slip, skirt, woman's pajamas), previously employed by this patient in cross-dressing episodes, resulted in maximum sexual arousal when he touched or observed them. In addition, maximum erectile strength was obtained in response to a photograph of a nude female. Electrical aversion therapy consisted of the pairing of painful faradic shocks with target stimuli (panties, slip, skirt, woman's pajamas) both *in vivo* and imaginally. Treatment was applied in relation to each of the sexual stimuli in sequence, with about 20 pairings per stimulus. However, erectile responses to the photograph of the nude remained untreated as this behavior served as a control for the four treated behaviors.

The results of this study are plotted in Fig. 7-3. Examination of the top portion of the figure shows that erectile latency to deviant stimuli increased as treatment progressed for each. Erectile strength following exposure to the five

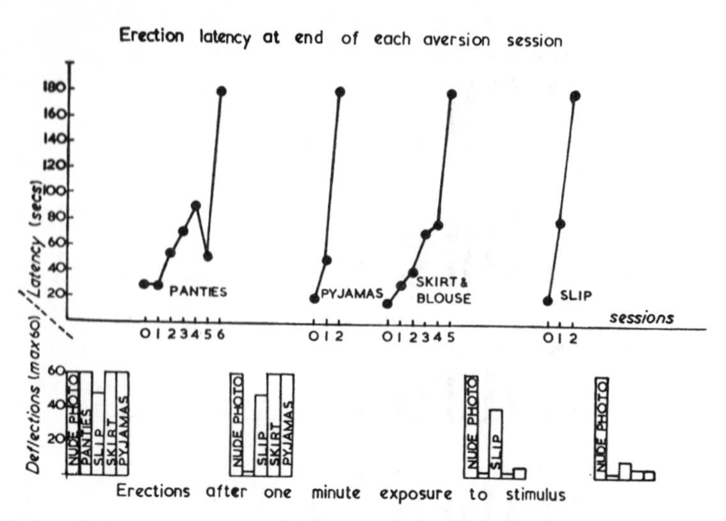

Fig. 7-3. Specificity of autonomic changes (Patient B). (Fig. 1, p. 714, from: Marks, I. M., and Gelder, M. G. Transvestism and fetishism: Clinical and psychological changes during faradic aversion. *British Journal of Psychiatry*, 1967, **113**, 711-729. Reproduced by permission.)

stimuli appears in the bottom part of the graph. Baseline responding is presented in the first block of five stimuli. Maximum responding to each was obtained with the exception of the "slip" stimulus. Erectile responses to stimuli are then presented in the second block following pairing of shock with "panties." Results indicate that minimal responding to "panties" was obtained whereas untreated responses to stimuli maintained their peviously high levels. In the third block erectile strength to sexual stimuli are presented following electrical aversion directed toward responses to "pajamas" and "skirt and blouse." Here, too, the effects of treatment are specific, but little change is noted in the untreated behaviors. In the fourth block treatment specificity is further demonstrated as evidenced by a significant decrease in responding to the "slip" stimulus. However, no changes in erectile strength toward the photographed nude appeared throughout the study. Thus, specificity of electrical aversion therapy was demonstrated in regard to the particular stimulus paired with faradic shock. Barlow and Hersen (1973) point out that "In this study erectile responses to the nude photo may be conceptualized as a control for the other four stimuli. If erections to untreated articles of clothing and the nude photo had decreased, then one would either conclude that electrical aversion had some general effect on sexual arousal or that a correlated therapeutic variable (e.g., expectancy) was the crucial therapeutic agent" (p. 324). Once again, independence of target behaviors in a multiple baseline strategy was evidenced.

The study by Barton, Guess, Garcia, and Baer (1970) illustrates the use of a multiple baseline design in which treatment was applied sequentially to separate targeted behaviors for an entire group of subjects. Sixteen severely and profoundly retarded males served as subjects in an experiment designed to improve their mealtime behaviors through the use of timeout procedures. Several undesirable mealtime behaviors were selected as targets for study during preliminary observations. They included *stealing* (taking food from another resident's tray), *fingers* (eating food with fingers that required the use of utensils), *messy utensils* (e.g., using utensil to push food off the dish, spilling food, etc.), and *pigging* (eating spilled food from floor, tray, etc., placing mouth directly over food without the use of a utensil). Observations of these behaviors were made 5 days per week during the noon and evening meals by using a time-sampling procedure. Independent observations were also obtained as reliability checks. The treatment–timeout–involved removing the subject (cottage resident) from the dining area for the remainder of a meal or for a designated time period contingent upon his evidencing an undesirable mealtime behavior.

The full timeout contingency (removal from the dining area for the entire meal) was initially applied to *stealing* following 6 days of baseline recording. Timeout contingencies for *fingers, messy utensils,* and *pigging* were then applied in sequence, each time maintaining the contingency in force for the previously treated behavior. During the application of timeout for *fingers*, the contingency involved timeout from the entire meal for 11 subjects, but only for 15 seconds for five of the subjects. This differentiation was made in response to nursing staff's concerns that a complete timeout contingency for the five subjects might jeopardize their health. Timeout procedures for *messy utensils* and *pigging* were limited to 15 seconds per infraction for all 16 subjects.

The results of this study are presented in Fig. 7-4. Examination of the graph indicates that when timeout was applied to *stealing* and *fingers*, rates for these behaviors decreased. However, application of timeout to *fingers* also resulted in a concurrent increase in the rate for *messy utensils*. But subsequent application of timeout for *messy utensils* effected a decrease in rate for that behavior. Finally, application of timeout for *pigging* proved successful in reducing its rate.

Independence of the target behaviors was observed with the exception of *messy utensils,* which increased in rate when the timeout contingency was applied to *fingers.* Although group data for the 16 subjects are presented, it would have been desirable if the authors had presented data for individual subjects. Unfortunately, the time-sampling procedure used by Barton, Guess, Garcia, and Baer (1970) precluded obtaining such information. However, this factor should not overshadow the clinical and social significance of this study in that: (1) mealtime behaviors improved significantly; (2) a result of improved

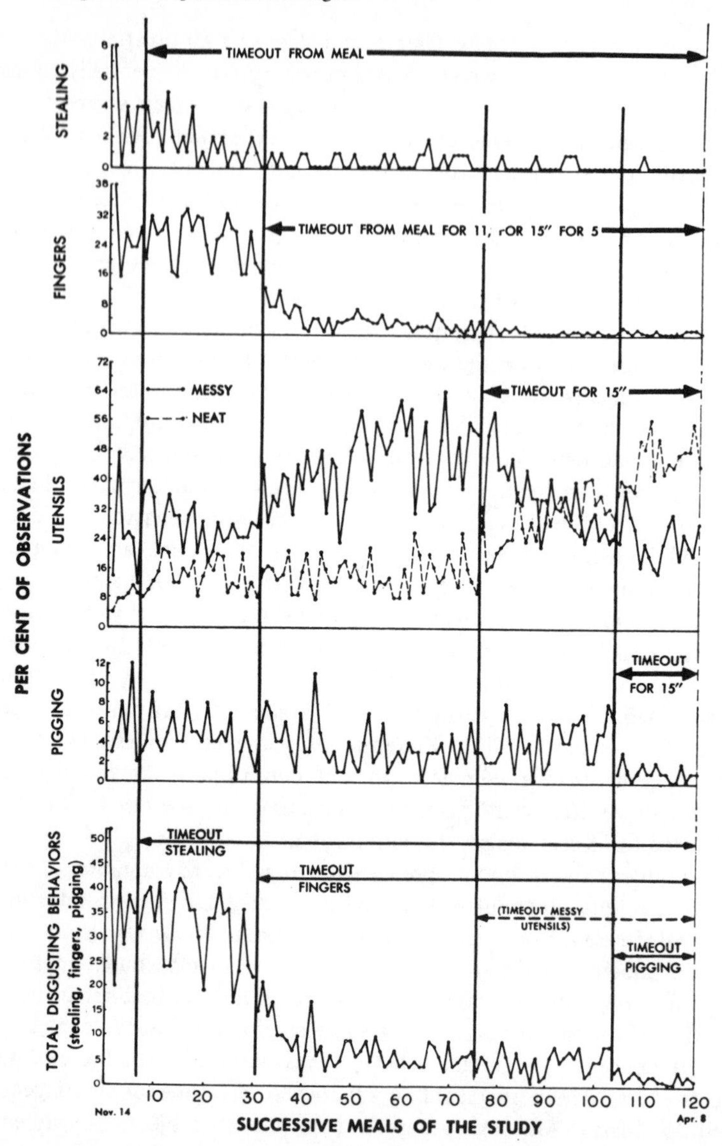

Fig. 7-4. Concurrent group rates of Stealing, Fingers, Utensils, and Pigging behaviors, and the sum of Stealing, Fingers, and Pigging (Total Disgusting Behaviors), through the baseline and experimental phases of the study. (Fig. 1, p. 80, from: Barton, E. S., Guess, D., Garcia, E., and Baer, D. M. Improvement of retardates' mealtime behaviors by timeout procedures using multiple baseline techniques. *Journal of Applied Behavior Analysis*, 1970, **3**, 77-84. Copyright by Society for the Experimental Analysis of Behavior, Inc. Reproduced by permission.)

mealtime behaviors was a concomitant improvement in staff morale, thereby facilitating more favorable interactions with the subjects; and (3) staff in other cottages were sufficiently impressed with the results of this study so that they began to implement similar mealtime programs for their own resident retardates.

Although the multiple baseline design is frequently used in clinical research when withdrawal of treatment is considered to be detrimental to the patient, on occasion, withdrawal procedures have been instituted following the sequential administration of treatment to target behaviors, particularly when reinforcement techniques are being evaluated (e.g., Patterson and Teigen, 1973). If treatment is reintroduced after withdrawal, a powerful demonstration of its controlling effects can be documented: first, in sequence across behaviors, and second, in terms of rate changes for all behaviors in accordance with withdrawal and reinstatement procedures. This type of multiple baseline strategy was used by Patterson and Teigen (1973) in their assessment of prompting and reinforcement on delusional verbalizations in a 60-year-old, female, psychotic patient, residing on a token economy unit. This patient invariably gave delusional responses when questioned specifically by staff as to her name, age, place of birth, record of hospitalizations, and names of family members.

During baseline, this patient was relieved from partaking in work assignments on the token economy unit and was given a "free" daily supply of tokens based on an average of previous earnings. Prior to each of three daily meals, the patient was asked to respond to the five questions concerning her identity, and these were scored on a dichotomous correct or incorrect basis. In the treatment phase a prompting and reinforcement procedure was applied sequentially to the patient's responses. During this contingency phase, the patient was instructed that she could earn tokens necessary for meals and privileges by responding correctly to questions posed prior to each meal. One-third the amount of tokens needed for daily expenditures could be earned in *each session* if the five questions were correctly answered. If the patient responded incorrectly, the staff member prompted her; if the same questions were subsequently answered correctly, she was given credit and awarded tokens. During Trials 45-59, a partial schedule of reinforcement was introduced, followed by 24 sessions in which tokens were awarded non-contingently prior to questioning. Beginning in Trial 85, tokens were given to the patient following her responses regardless of their accuracy (she was given instructions to that effect). Finally, the full reinforcement contingency was reinstated until the patient was discharged from inpatient status.

The results of this investigation are plotted in Fig. 7-5, and show that during baseline none of the questions was answered correctly. When prompting and reinforcement procedures were applied sequentially, correct responses to each

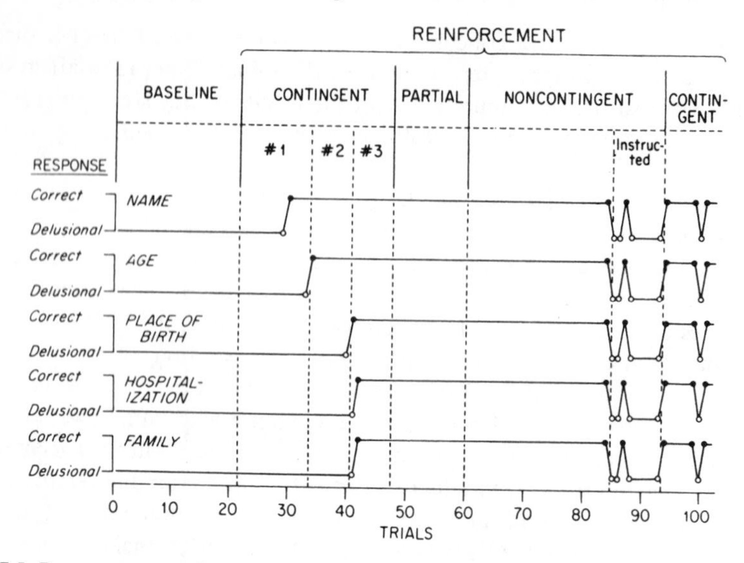

Fig. 7-5. Responses to each of the standard interview questions as a function of trials and experimental conditions. (Fig. 1, p. 67, from: Patterson, R. L., and Teigen, J. R. Conditioning and post-hospital generalization of nondelusional responses in a chronic psychotic patient. *Journal of Applied Behavior Analysis,* 1973, **6,** 65-70. Copyright by Society for the Experimental Analysis of Behavior, Inc. Reproduced by permission.)

question were obtained. Correct responses were maintained throughout the partial reinforcement condition. Surprisingly, the non-contingent allotment of tokens prior to question sessions did not result in a return to delusional responding. However, when the patient was given tokens after each response, regardless of its accuracy, and also informed that the "correctness" contingency was lifted, she then responded delusionally. Reintroduction of the contingency in the final phase of study led to correct responding, with the exception of one trial in which all questions were answered delusionally.

In summary, this study illustrates the use of the multiple baseline design across behaviors in a single subject, demonstrating independence of target behaviors. Sequential application of a reinforcement contingency to individual responses showed the controlling effects of the contingency. Finally, additional experimental manipulations (withdrawal and reintroduction of the contingency), introduced after completion of the multiple baseline portion of study, further confirm the controlling effects of the treatment variable.

Multiple baseline across subjects

Our first example of multiple baseline strategy across subjects is taken from the classroom literature. Hall, Cristler, Cranston, and Tucker (1970) examined the sequential administration of an after-school tutoring contingency on three tenth-grade students who were obtaining grades of D and F on quizzes given three to four times a week in their French class. Prior to this study, the French teacher, who was enrolled in a course in classroom management, attempted to use verbal praise contingently on successful classroom performance evidenced by these students. However, this manipulation failed to effect an increase in their quiz grades.

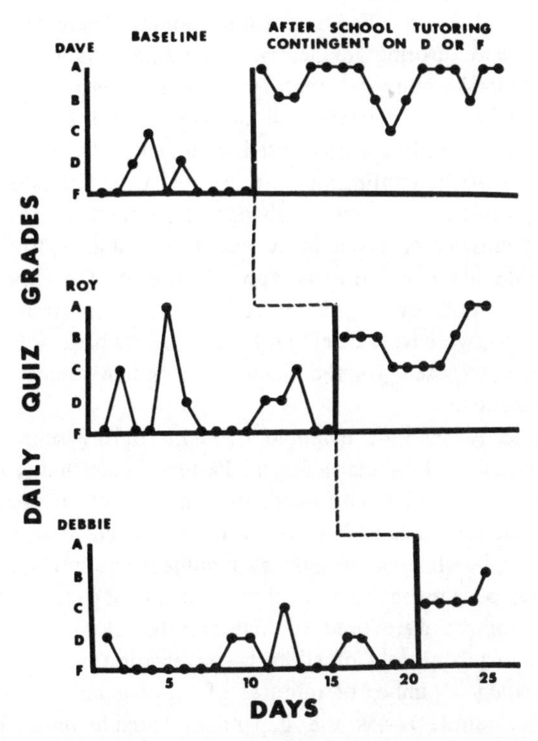

Fig. 7-6. A record of quiz score grades for three high school French-class students. *Baseline*–before experimental procedures. *After school tutoring contingent on D and F grades*–pupils required to stay after school for tutoring if they score D and F on daily quizzes. (Fig. 2, p. 251, from: Hall, R. V., Cristler, C., Cranston, S. S., and Tucker, B. Teachers and parents as researchers using multiple baseline designs. *Journal of Applied Behavior Analysis*, 1970, **3**, 247-255. Copyright by Society for the Experimental Analysis of Behavior, Inc. Reproduced by permission.)

During baseline, quizzes given throughout the week were scored independently by the teacher and an outstanding student (100 percent agreement was achieved). Starting on Day 10, the first student (Dave) began receiving after-school tutoring whenever he obtained a D or F grade on a quiz. Baseline conditions were maintained for the other two students (Roy and Debbie). On Day 15, Roy began his after-school tutoring contingency, whereas Debbie's tutoring contingency began on Day 20.

The results of this study appear in Fig. 7-6, and show that grades improved only after each student was exposed to the after-school tutoring contingency. For example, the median grade (when scored on a 0-4 point scale) obtained in baseline by Dave was 0·4; the median grade obtained during the tutoring phase was 3·6. Similarly, median grades for Roy and Debbie rose from 0·7 and 0·4 in baseline to 2·8 and 2·2 during the tutoring phases. Therefore, the beneficial effects of contingent tutoring are clearly documented. As in the multiple baseline across behaviors, baseline and treatment phases for each subject in this study can be conceptualized as separate A-B designs, with the length of baselines increased for each succeeding subject used in the multiple baseline analysis. The controlling effects of the contingency are inferred from the rate changes in the treated subject, while rates remain unchanged in untreated subjects. When rate changes are sequentially observed in at least three subjects, but only after the treatment variable has been directly applied to each, the experimenter gains confidence in the efficacy of his procedure (i.e., the after-school tutoring contingency). Thus, we have a direct replication of the basic A-B design in three matched subjects exposed to the same environment under "time-lagged" contingency conditions.

A more clinically relevant example of a multiple baseline design across subjects is presented by Liberman, Teigen, Patterson, and Baker (1973). In this study the effects of a social reinforcement contingency on rational speech in four chronic paranoid schizophrenics were examined under "time-lagged" conditions. During baseline assessment, each subject was interviewed four times daily, 10 minutes per session, by a member of the nursing staff. When necessary, questions and prompts were used to maintain the flow of conversation. In addition, nurses surreptitiously timed each conversation until the patient began speaking delusionally. Number of minutes of rational talk per day (maximum possible was 40 minutes) was the dependent variable under evaluation. In baseline each patient also participated in a 30-minute nightly chat with his therapist, during which time snacks were made available. During these chats, delusional statements made by the patient were acknowledged by the therapist but no comment was made as to specific content. When the contingency condition was enforced, each of the four daily 10 minute interviews was terminated by the therapist as soon as the patient spoke delusionally. Also, time allotted for the evening session was now made proportional (one-to-one ratio) to the amount of rational talk accumulated in daily interviews. Subjects were specifically apprised of this contingency.

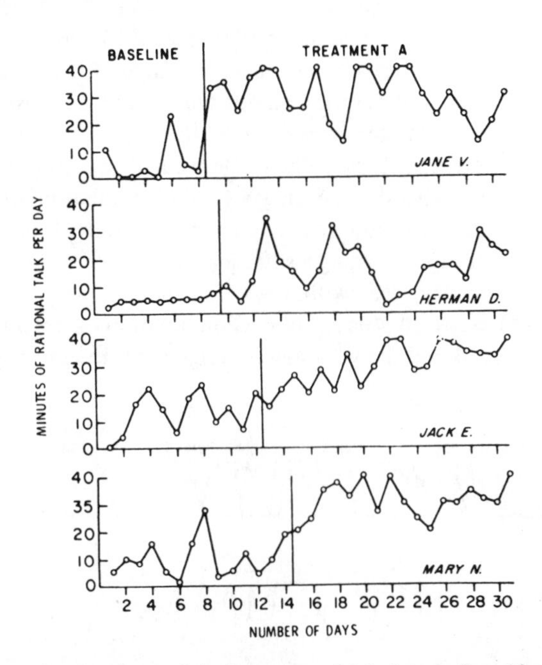

Fig. 7-7. Duration of rational speech before onset of delusions in four 10-minute interviews under baseline and contingent social reinforcement (Treatment A) conditions. (Fig. 1, p. 61, from: Liberman, R. P., Teigen, J., Patterson, R., and Baker, V. Reducing delusional speech in chronic paranoid schizophrenics. *Journal of Applied Behavior Analysis*, 1973, **6**, 57-64. Copyright by Society for the Experimental Analysis of Behavior, Inc. Reproduced by permission.)

Data for the four subjects are plotted in Fig. 7-7. Examination of this figure shows that when the social contingency was put into operation for each subject the amount of rational talk subsequently increased. Specifically, a 200 percent increase for Jack, a 300 percent increase for Mary and Harry, and a 600 percent increase for Jane were noted. "These differences between the Baseline and Phase A are statistically significant at beyond the 0·05 level (Rn Test, Revusky, 1967)" (Liberman, Teigen, Patterson, and Baker, 1973, p. 61).

Christopherson, Arnold, Hill, and Quilitch (1972) used an interesting variation of the multiple baseline strategy across subjects in their assessment of feedback and reinforcement techniques in two siblings (7-year-old boy and 10-year-old girl) whose parents were unable to maintain discipline. The specific problem involved the children's unwillingness to carry out simple household chores. During baseline, a list of three specific chores for each sibling was established, with the expectation that all chores would be completed by 9:00 p.m. each day. Baseline was maintained for 9 and 13 days, respectively, for Robin and Teresa. The next phase, feedback, consisted of the mother's report to each sibling of chores satisfactorily completed as per her inspection. In the following condition a gold star was glued to the child's report for each chore satisfactorily completed. This was then followed by a bonus phase in which additional money was awarded contingently on completion of chores for 7 consecutive days. A point economy was then instituted, with privileges for each child made contingent upon earning a sufficient number of points by completing chores. This contingency was withdrawn (no points) and finally reinstated (points) in the last phase of study. Thus, a combination of the multiple baseline design across behaviors and the withdrawal design (last three phases) is provided.

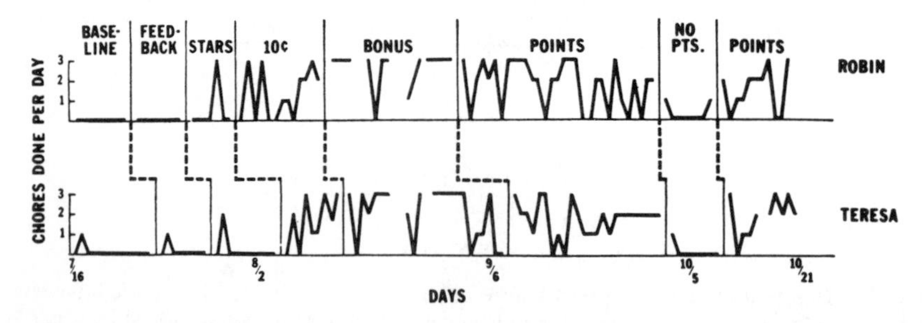

Fig. 7-8. Multiple baseline analysis across two siblings (7-year-old boy and 10-year-old girl) of the effects of various conditions on baseline maintenance behaviors (second family). (Fig. 5, p. 496, from: Christopherson, E. R., Arnold, C. M., Hill, D. W., and Quilitch, H. R. The home point system: Token reinforcement procedures for application by parents of children with behavior problems. *Journal of Applied Behavior Analysis*, 1972, **5**, 485-497. Copyright by Society for the Experimental Analysis of Behavior, Inc. Reproduced by permission.)

The results of this study are plotted in Fig. 7-8, and clearly show the effects of "time-lagged" contingencies. That is, changes in rate for each child only took place as the feedback or contingency condition in force was applied. Mean percentages of chores successfully completed for Robin and Teresa in the different phases are presented in Table 7-2. Examination of this table indicates that feedback and stars effected little change for both children. However, institution of the ten cents contingency resulted in approximately one-half of

TABLE 7-2. Mean Percentage of Chores Completed by Robin and Teresa

Condition	Robin	Teresa
	%	%
baseline	0	2
feedback	0	4
starts	12	6
ten cents	47	54
bonus	87	68
points	59	45
no points	8	4
points	47	60

Note: Based on data taken from Christopherson, Arnold, and Quilitch (1970).

the chores completed for each child. Addition of a bonus further increased the number of chores completed by each child, and appeared to be the most effective condition. However, this condition was not withdrawn and reinstated as was the point contingency in the last three phases. Not only are the "time-lagged" effects of points and no points evidenced across subjects in the last three phases, but the controlling effects over percentage of chores completed are available for each child in separate B-A-B analyses (points, no points, points).

A three-group application of the multiple baseline strategy across subjects (groups) is provided by Wilson and Hopkins (1973) in addition to their presenting data in a withdrawal design (A-A-B-C-B) for a fourth group of subjects. The effects of contingent music (radio-operated by voice-activated-relay) on acceptable levels of classroom noise were examined in four seventh- and eighth-grade home economics classes. Whenever classroom noise exceeded a predetermined decibel level, the radio was automatically turned off for a minimum of 20 seconds. Conversely, whenever noise was below an acceptable decibel level (70 for Groups B, C, and D; 76 for Group A), music continued to be played. This condition was labeled "Radio-Quiet."

Baseline assessment (No Radio) for Groups B, C, and D consisted of measuring the percentage of time that the noise level rose above threshold. Such measurement was facilitated by coupling the voice-operated-relay with timing equipment. In the next phase (Radio-Quiet), the music contingency was applied in sequence to Groups D, C, and B under "time-lagged" conditions.

The results of this portion of the study are presented for Groups B, C, and D (unfortunately in reversed order) in Fig. 7-9. Examination of the three bottom parts of the graph, from bottom to top, indicate that as soon as the Radio-Quiet contingency was applied to each group, a significant decrease in noise level was

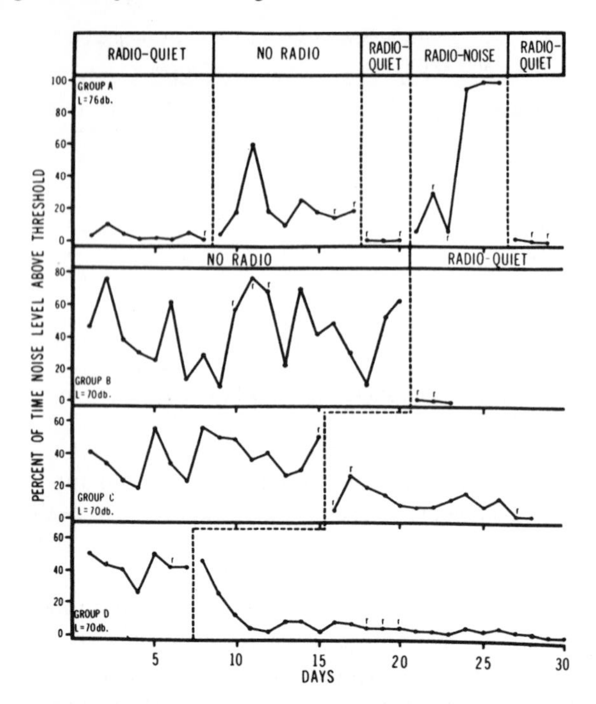

Fig. 7-9. The percent of time that the intensity of noise exceeded the teacher-determined threshold for all four classes. (Fig. 1, p. 273, from: Wilson, C. W., and Hopkins, B. L. The effects of contingent music on the intensity of noise in junior high home economics classes. *Journal of Applied Behavior Analysis*, 1973, **6**, 269-275. Copyright by Society for the Experimental Analysis of Behavior, Inc. Reproduced by permission.)

obtained, thus showing the controlling effects of the contigency. (We might note that these results could have been presented more clearly by plotting data with respect to the actual sequence of application of the Radio-Quiet contingency—i.e., D first, C second, and B third.)

Date for Group A clearly show that when the Radio-Quiet contingency was removed in the second phase of study an increase in noise level resulted. However, when the Radio-Quiet contingency was reinstated in Phase 3, noise level above threshold markedly decreased to a near zero percent. To further demonstrate the effectiveness of the music contingency in controlling the behavior of Group A, a Radio-Noise contingency was put into effect in Phase 4 whereby the radio was turned on automatically only when the noise level *exceeded* the threshold continuously for at least 20 seconds. In the last phase the Radio-Quiet contingency was once again reinstated. Results for these two final phases show that the Radio-Noise contingency markedly increased noise level above threshold whereas the Radio-Quiet contingency effected a significant decrease in noise level.

In summary, Wilson and Hopkins (1973) present a powerful demonstration of the Radio-Quiet contingency's controlling effects over decibel levels of classroom noise, both across groups D, C, and B and within Group A in a withdrawal strategy.

Panyan, Boozer, and Morris (1970) also used a multiple baseline strategy across subjects (groups) in their assessment of feedback procedures in reinforcing attendants to apply operant techniques with severely and profoundly retarded residents. Prior to this investigation, attendants from 11 living units (halls) in a state institution for retarded children were given a 4-week course in operant techniques, with emphasis directed toward using such techniques in shaping self-help skills (e.g., self-feeding, hand-washing, dressing, bathing, and toileting) in residents. The investigation involved staff and residents from four halls: Hall E housed 32 residents with a staff of seven attendants; Hall O housed 26 residents with a staff of nine attendants; Hall C housed 26 residents with a staff of nine attendants; and Hall R housed 28 residents with a staff of nine attendants.

During baseline and feedback conditions, staff members from each of the four halls were instructed to conduct prescribed numbers of sessions to develop self-help skills in residents. Also, in accordance with their prior training, they were asked to maintain daily performance records for those residents involved in training. In baseline no contingencies were attached to percentage of sessions conducted by hall staff. However, during feedback, percentage of sessions conducted, including names of staff members and residents involved, were posted weekly in the halls. Halls were ranked with respect to percentage of sessions conducted, and these data were also posted.

Feedback was provided to Hall E after a 4-week baseline, Hall O after an 8-week baseline, and Hall C after a 38-week baseline period. Hall R was given feedback throughout the entire study.

The results of this study are plotted in Fig. 7-10. Examination of these data indicates that for Halls E, O, and C, performance deteriorated during baseline, but subsequent institution of feedback under "time-lagged" conditions resulted in marked improvements in percentage of sessions conducted in each hall. Data also show that the hall (Hall C) with the longest baseline period required the most time for the feedback contingency to effect a change in rate. By contrast, "The data from Hall R suggest that initiating the feedback procedure as soon as possible after formal training is an effective means of maintaining high performance levels" (Panyan, Boozer, and Morris, 1970, p. 3). The percentage of sessions completed by Hall R personnel ranged from 69 to 100.

Although Panyan, Boozer, and Morris (1970) demonstrated the controlling

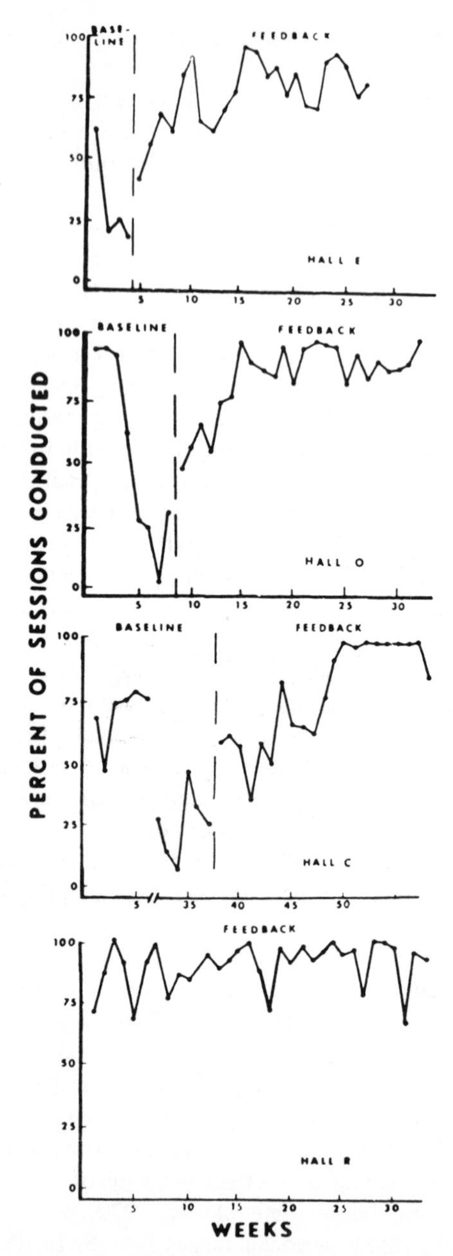

Fig. 7-10. Percent of requested training sessions conducted by the staff on Halls E, O, C, and R. (Fig. 1, p. 3, from: Panyan, M., Boozer, H., and Morris, N. Feedback to attendants as a reinforcer for applying operant techniques. *Journal of Applied Behavior Analysis*, 1970, **3**, 1-4. Copyright by Society for the Experimental Analysis of Behavior, Inc. Reproduced by permission.)

effects of their feedback contingency, they failed to present data showing its effectiveness for individual attendants. Such data are important as it is conceivable (as frequently occurs when a group comparison design is used) that some subjects (attendants) were unaffected by the contingency in force. Therefore, once again we recommend that investigators employing group variations of multiple baseline strategies provide data showing efficacy of their procedures in a majority of individual subjects in each respective group.

Multiple baseline across settings

Our first example of a multiple baseline strategy across settings was conducted in a summer camp and illustrates well how both treatment and research considerations can be combined under naturalistic conditions (Allen, 1973). The subject was an 8-year-old boy who had been given a diagnosis of minimal brain damage. This child evidenced a high frequency of bizarre verbalizations, primarily concerned with "penguins" (his parents noted that he spent a considerable amount of time at home fantasizing about penguins). During camping activities, the child's high rate of bizarre verbalizations interfered with his developing good interpersonal relations both with peers and adults. However, it was observed that camp counselors frequently reinforced these verbalizations by their paying attention to them.

Allen's (1973) study was directed to systematically reduce the child's high frequency of bizarre verbalizations in four separate camp activities (walking on trail evening activity, dining hall, cabin, education). During the first 6 days of baseline assessment, counselors were asked to record the child's rate of bizarre verbalizations in the four designated settings, but they were not given specific instructions as to how they were to respond. Beginning on Day 7, a treatment procedure (Ignore), consisting of inattention to bizarre responses and attention to the child's positive initiatives, was implemented during the evening activity. The highest rate of bizarre verbalizations for this setting had been recorded during the baseline period. Treatment (Ignore) was then applied in sequence under "time-lagged" conditions to the remaining three settings until it was simultaneously in force for all four. Inter-observer reliability for the 27 days of observation ranged from 81 percent to 100 percent (mean = 94 percent).

Fig. 7-11 shows that the sequential application of treatment procedures resulted in a near zero rate of bizarre responses in all four settings. A careful examination of the graph reveals that the baselines for the first two settings (evening activity, dining hall) appear to be independent, but that a decreasing rate for the cabin setting coincided with application of treatment in the dining hall and that a slight decrease in rate for the education setting coincided with

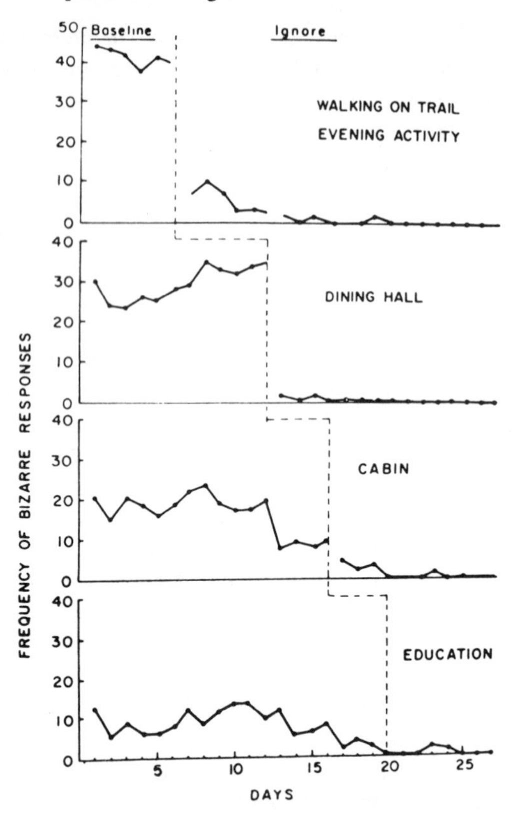

Fig. 7-11. Daily number of bizarre verbalizations in specific camp settings. (Fig. 1, p. 573, from: Allen, G. J. Case study: Implementation of behavior modification techniques in summer camp settings. *Behavior Therapy*, 1973, **4,** 570-575. Reproduced by permission.)

application of treatment in the cabin setting. However, despite covariation among some of the target measures (settings), data generally support the contention that the application of treatment (Ignore) in each setting was the primary variable responsible for change in frequency of the child's emission of bizarre verbalizations.

Barrish, Saunders, and Wolf (1969) combined elements of the multiple baseline strategy across settings and the withdrawal design in a group classroom application. The subjects were 24 fourth-grade students, seven of whom had been referred to the principal on several occasions as behavior problems. Two target behaviors (talking-out, out-of-seat) were selected for study during separate math and reading periods. Presence of target behaviors per consecutive 1-minute

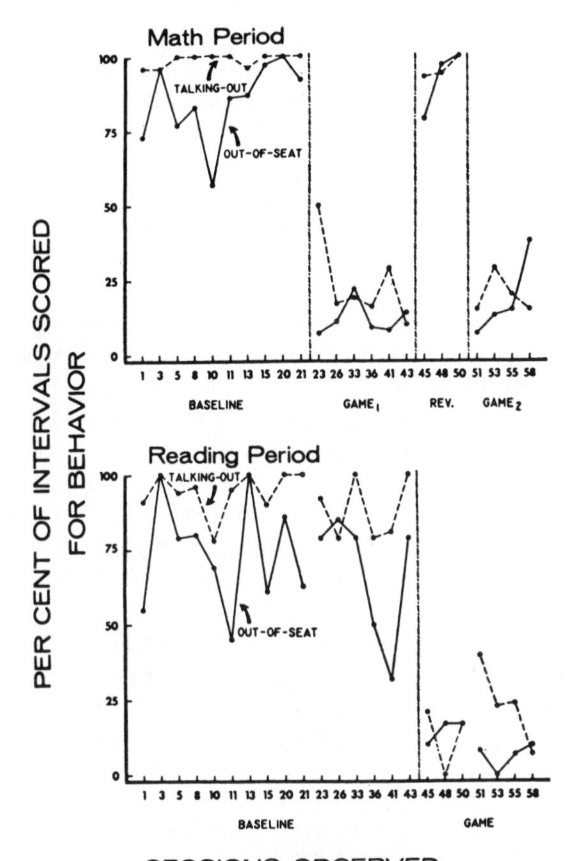

Fig. 7-12. Percent of 1-minute intervals scored by an observer as containing talking-out and out-of-seat behaviors occurring in a classroom of 24 fourth-grade school children during math and reading periods. In the baseline conditions the teacher attempted to manage the disruptive classroom behavior in her usual manner. During the game conditions, out-of-seat and talking-out responses by a student resulted in a possible loss of privileges for the student and his team. (Fig. 1, p. 122, from: Barrish, H. H., Saunders, M., and Wolf, M. M. Good behavior game: Effects of individual contingencies for group consequences on disruptive behavior in a classroom. *Journal of Applied Behavior Analysis*, 1969, **2**, 119-124. Copyright by Society for the Experimental Analysis of Behavior, Inc. Reproduced by permission.)

time periods was scored by an observer throughout all phases of the investigation.

Following a 21-session baseline assessment, a game was introduced during the math period. This game involved dividing the class into two competing teams. During this competition, talking-out and out-of-seat behaviors for an individual member of the team could result in possible loss of privileges (marks) for the team as a whole. A variety of special privileges were made available (e.g., wearing victory tags, lining up first for lunch, 30 minutes of free time at the end of the school day) to the winning team or to both teams if they obtained fewer than five marks in one week. After 44 sessions, the game was discontinued during the math period but applied during the reading period. Beginning in Session 51, the game was reintroduced during the math period and maintained during the reading period.

The results of this study are plotted in Fig. 7-12. Examination of the figure indicates that during the baseline math period the median number of intervals in which talking-out and out-of-seat behavior occurred was 96 percent and 82 percent, respectively. Application of the game during the math period resulted in a decrease to 19 percent talking-out and 9 percent out-of-seat behaviors. During this phase, talking-out and out-of-seat behaviors remained at baseline levels for the reading period. When the game was withdrawn from the math period, target behaviors returned to baseline levels. By contrast, when the game was applied during the reading period, a dramatic decrease in disruptive behaviors was noted. Finally, when the game was reinstated in the math period, disruptive behaviors were once again substantially reduced.

The investigation clearly shows that target behaviors decreased in frequency *only* when the game was introduced in the particular period under study, thus demonstrating the sequential effectiveness of the contingency across settings (math and reading periods). Further confirmation of the game's controlling effects over target behaviors was obtained in the withdrawal strategy (A-B-A-B design) applied during the math period portion of the study.

Hall, Cristler, Cranston, and Tucker (1970) also combined aspects of the multiple baseline strategy across settings and the withdrawal strategy in a group classroom application. The subjects were 25 fifth-grade students, a number of whom were returning late to class following their noon, morning, and afternoon recesses. The number of pupils returning late was monitored during baseline for each of the three recess periods. After 13 days of baseline recording for the noon period, the teacher, who was enrolled in a course on classroom contingency management, introduced a "game-like" procedure in which students returning from recess on time would have their names placed on a "Patriots' Chart." This was considered to be a likely reinforcing event as the class had previously responded with great enthusiasm to reading books relating to patriotism in American history. "One suggestion was that if they obeyed class rules they were good citizens and, therefore, patriots" (Hall, Cristler, Cranston, and Tucker, 1970, p. 248).

The Patriots' Chart contingency was then instituted in Session 22 for the

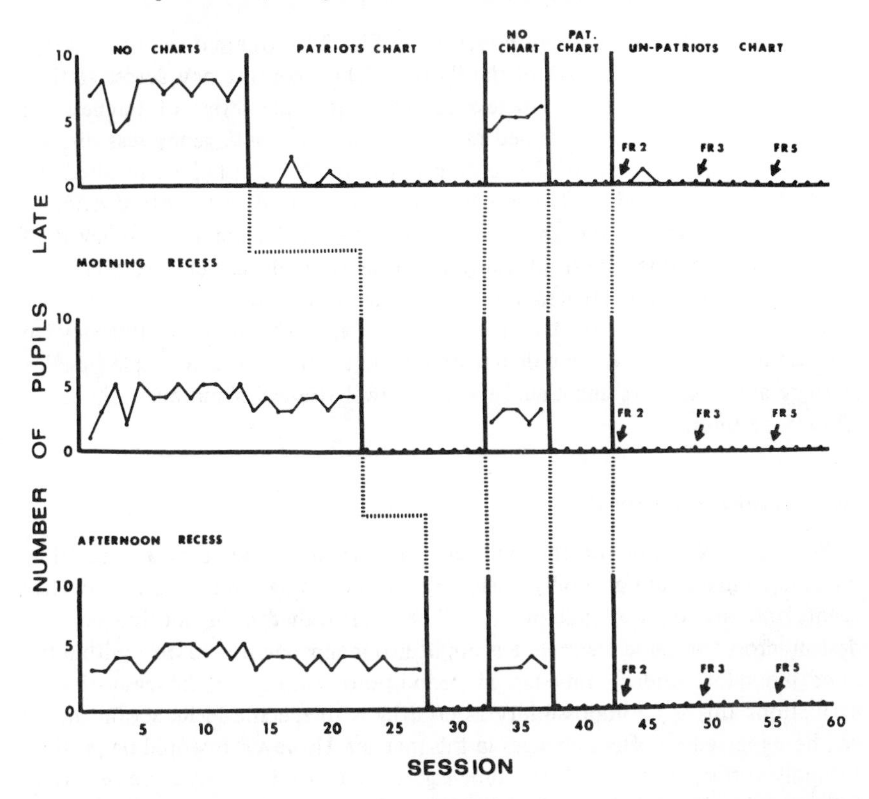

Fig. 7-13. A record of the number of pupils late in returning to their fifth-grade classroom after noon, morning, and afernoon recess. *No Charts*—baseline, before experimental procedures. *Patriots' Chart*—posting of pupil names on "Today's Patriots" chart contingent on entering class on time after recess. *No Chart*—posting of names discontinued. *Patriots' Chart*—return to Patrots' Chart conditions. *Un-Patrots' Chart*—posting of names on "Un-patriots'" chart contingent on being late after recess (FR2) every 2 days, (FR3) every 3 days, and (FR5) every 5 days. (Fig. 1, p. 249, from: Hall, R. V., Cristler, C., Cranston, S. S., and Tucker, B. Teachers and parents as researchers using multiple baseline designs. *Journal of Applied Behavior Analysis*, 1970, **3**, 247-255. Copyright by Society for the Experimental Analysis of Behavior, Inc. Reproduced by permission.)

morning recess while it was maintained in force for the noon recess. Subsequently, the Patriots' Chart contingency was applied to the afternoon recess period in Session 28 while keeping the contingency in force for the noon and morning recesses. This concluded the multiple baseline portion of the study.

The withdrawal strategy was introduced in Session 33 when the Patriots' Chart contingency was removed. It was then reinstated in Session 38. Beginning in Session 43, FR2, FR3, and FR5 schedules were applied, during which time latenesses were posted (Un-patriots' Chart).

The results of this study are presented in Fig. 7-13, where it can be seen that the "time-lagged" application of the Patriots' Chart contingency across settings resulted in a zero level of lateness for each. Baseline levels of lateness for untreated recess periods remained unaltered until the contingency was directly applied to each. Removal of the contingency in the No Chart phase resulted in a return to baseline levels of lateness for each recess period, but reintroduction of the Patriots' Chart in the fourth phase effected a decline to zero levels of lateness. In the final phase of study a partial schedule of reinforcement was effective in controlling latenesses for all three recess periods.

In summary, the effectiveness of the Patriots' Chart contingency in eliminating lateness was first demonstrated sequentially in a multiple baseline strategy across settings and then in a withdrawal strategy simultaneously across all three settings.

Issues in drug evaluations

With the exception of the multiple baseline across subjects, the multiple baseline strategies are generally unsuitable for the evaluation of pharmacological agents on behavior. For example, it will be recalled that in the multiple baseline design across the same treatment is applied to independent behaviors within the same individual under "time-lagged" conditions. Clearly in the case of drug evaluations this is an impossibility as no drug is so specific in its action that it can be expected to effect changes in this manner. However, it would be possible to apply *different drugs* under "time-lagged" conditions to *separate behaviors* following baseline placebo administrations for each. But this kind of design would involve a radical departure from the basic assumptions underlying the multiple baseline strategy across behaviors and would only permit very tentative conclusions based on separate A_1-B designs for each targeted behavior. In addition, the possible interactive effects of drugs might obfuscate specific results. Indeed, the interaction design (see Chapter 6) is better suited for evaluation of combined effects of therapeutic strategies.

Similarly, the use of the multiple baseline across different settings in drug evaluations would prove difficult unless the particular drug applied worked immediately, had extremely short-term effects, and could be rapidly eliminated from body tissues. However, as most drugs used in controlling behavior disorders do not meet these three requirements, this kind of design strategy is not useful in drug research.

Of the three types of multiple baseline strategies currently in use, the multiple baseline across subjects is most readily adaptable to drug evaluations. Using this type of strategy across matched subjects, baseline administration of a placebo (A_1) could be followed by the sequential administration (under "time-lagged" conditions) of an active drug (B). Thus, a series of A_1-B (quasi-experimental) designs would result, with inferences made in accordance with changes observed when the B (drug) condition is applied. Although an approximation of a double-blind procedure is feasible (observer and patient blind to conditions in force) it is more likely that single-blind (patient only) conditions will prevail.

The application of the multiple baseline design across subjects in drug evaluations could be most useful when withdrawal procedures (return to A_1-baseline placebo) are unwarranted either for ethical or clinical considerations. However, to date, the evaluation of drug effects in this design has not been reported in the archival literature.

7.3. MULTIPLE SCHEDULE DESIGNS

Whereas in the multiple baseline design across behaviors independent behaviors are treated individually in sequence, in the multiple schedule design the same behavior is treated differentially under varying stimulus conditions. In short, response discrimination is involved in the multiple baseline strategy across behaviors while stimulus discrimination is involved in the multiple schedule strategy. In an analogous situation from the operant animal laboratory, consider a stimulus discrimination condition whereby a rat is provided reinforcement when bar-pressing in the presence of a buzzer and not reinforced for bar-pressing in its absence. Leitenberg (1973) points out that the multiple schedule design ". . . is based on discrimination learning principles; that is, if the same behavior is treated differently in the presence of different physical or social stimuli, it will exhibit different characteristics in the presence of these stimuli" (p. 93). He further indicates that in the therapeutic context different situations or different discriminative stimuli might involve time differences, separate physical locations, different therapists, or different family members.

As in the case of the reversal design (illustrated in Section 3.5 of Chapter 3), the multiple schedule design is somewhat cumbersome in that: (1) considerable control over staff in settings is required; (2) the ability to obtain discrimination between settings or stimuli is difficult; and (3) it consists of a rather limited (artificially structured) approximation of naturalistic conditions. However, when this strategy can be implemented, it obviates the necessity for withdrawing active treatment procedures. Moreover, it is particularly useful if independence of behaviors cannot be achieved under multiple baseline conditions, especially if the investigator is still intent on demonstrating the efficacy of his procedures in an experimental analysis paradigm.

Examples of multiple schedule designs

An excellent example of a multiple schedule design appears in a study by O'Brien, Azrin, and Henson (1969) in which the communications of chronic mental patients were examined. Thirteen patients (all evidencing the passivity associated with chronic institutionalization) housed on a token economy unit served as subjects. During all phases of study, these patients met daily with one of two group leaders (A or B). In each meeting the group leader directed three

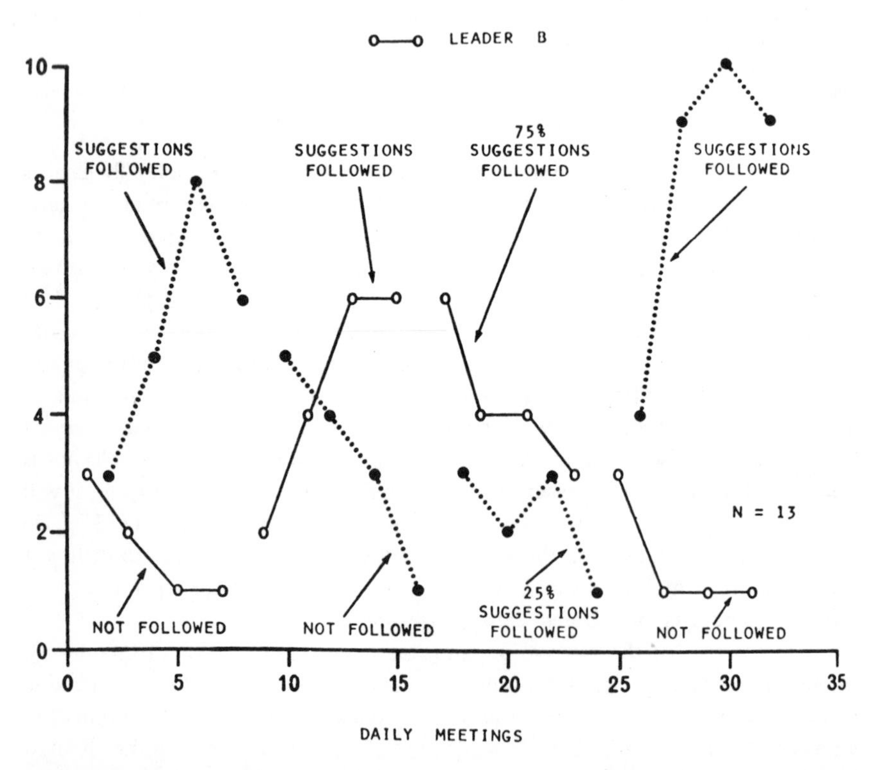

Fig. 7-14. The number of suggestions made by 13 mental patients daily, during the multiple schedule of Exp. III. The schedule stimuli were the two meeting leaders, A and B, who conducted the meetings and reacted to the suggestions. On Days 25 to 33, a third person conducted the meetings, during which group leaders A and B reacted to the suggestions as noted. (Fig. 2, p. 26, from: O'Brien, F., Azrin, N. H., and Henson, K. Increased communications of chronic mental patients by reinforcement and by response priming. *Journal of Applied Behavior Analysis*, 1969, **2**, 23-29. Copyright by Society for the Experimental Analysis of Behavior, Inc. Reproduced by permission.)

questions separately to each patient in attendance in order to prompt comments and suggestions. During the different phases of study, the two group leaders alternated in either following or not following suggestions made by the patients. The dependent measure was the number of suggestions made by patients at daily meetings as a function of the group leader's responsiveness.

Data presented in Fig. 7-14 indicate that initially an equal number of suggestions were made in the presence of the two leaders. However, the number of suggestions subsequently made in the presence of Leader A (followed 100 percent of the time) increased markedly, whereas the number of suggestions offered in the presence of Leader B (followed 0 percent of the time) decreased dramatically. In the second phase, where the role of the two leaders was reversed, the number of suggestions now made in the presence of Leader B (followed 100 percent of the time) increased, but those made in the presence of Leader A (followed 0 percent) decreased. In the third phase the schedule was modified slightly, with Leader B following 75 percent of the suggestions and Leader A following 25 percent of the suggestions. Data for these eight meetings reflect the change in each leader's responsiveness, as evidenced by a decrease in the number of suggestions made in the presence of Leader A throughout the phase. By contrast, the number of suggestions made in the presence of Leader B generally increased when compared with the last data point in the previous phase. In the final phase a naive leader conducted the meetings, but "Suggestions gradually increased in the presence of leader A who granted 100 percent instead of 25 percent, and gradually decreased in the presence of leader B who now granted 0 percent instead of 75 percent of the suggestions" (O'Brien, Azrin, and Henson, 1969, p. 27).

To summarize, this study demonstrates that when a group leader follows suggestions made by patients at a unit meeting, his responsiveness to suggestions may serve to increase (reinforce) the number of suggestions presented by these same patients at subsequent meetings. In addition, it shows that chronic mental patients not only have the capacity to make discriminations between socially reinforcing and non-reinforcing group leaders, but that they are able to shift their response patterns when such leaders reverse their roles.

Agras, Leitenberg, Barlow, and Thomson (1969) also used a multiple schedule design in their examination of the effects of social reinforcement in a severely claustrophobic patient. Their subject was a 50-year-old female, whose symptoms intensified following the death of her husband some 7 years before her being admitted as a research patient on a psychiatric unit. When admitted, the patient was unable to remain in a closed room for longer than 1 minute without experiencing considerable anxiety. As a consequence of this disorder, the patient's activities were seriously restricted.

During the study, a small windowless room with a chair facing the door was used to measure the extent of her phobic reaction. During baseline and all three experimental phases of study, the patient was assessed four times daily and asked to remain inside this room until she began feeling anxious. Thus, elapsed time between the closing and reopening of the door by the patient served as the dependent measure.

Two therapists were assigned to work with the patient and alternated sessions with one another. Following baseline assessment, one therapist (reinforcing therapist) administered social reinforcement (consisting of praise) to the patient contingent on her remaining in the room for an increasing time criterion. The second therapist (non-reinforcing therapist) maintained a pleasant relationship with the patient but did not praise her in any way with respect to time spent in the phobic situation. In the next experimental phase the therapists' roles were reversed (reinforcing therapist now became non-reinforcing and *vice versa*), while in the last phase their roles were once again reversed.

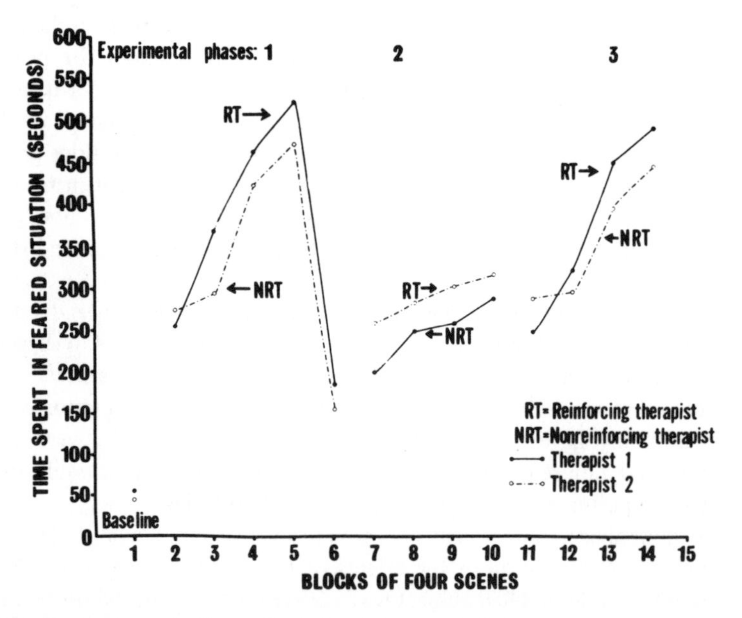

Fig. 7-15. Comparison of effects of reinforcing and non-reinforcing therapists in the modification of claustrophobic behavior. (Fig. 3, p. 1438, redrawn from: Agras, W. S., Leitenberg, H., Barlow, D. H., and Thomson, L. E. Instructions and reinforcement in the modification of neurotic behavior. *American Journal of Psychiatry*, 1969, **125**, 1435-1439. Copyright 1969 the American Psychiatric Association. Reproduced by permission.)

The results for this study are plotted in Fig. 7-15. Examination of these data indicate that during the first experimental phase improvement appeared to be greater in the presence of the reinforcing therapist. In phase two, with the roles of the two therapists now reversed, the new reinforcing therapist seemed to be more effective in eliciting behavioral change. Finally, in the third phase, with therapists' roles reversed a second time, the reinforcing therapist seemed to be more successful than his non-reinforcing counterpart. Confirmation of the superiority of the "reinforcing therapist" was obtained statistically in each phase by comparing data with correlated t-test analyses. However, differences between the two therapists in baseline were not significant. It would appear that the use of social reinforcement was the critical variable accounting for differential efficacy of therapists in each of the three experimental phases.

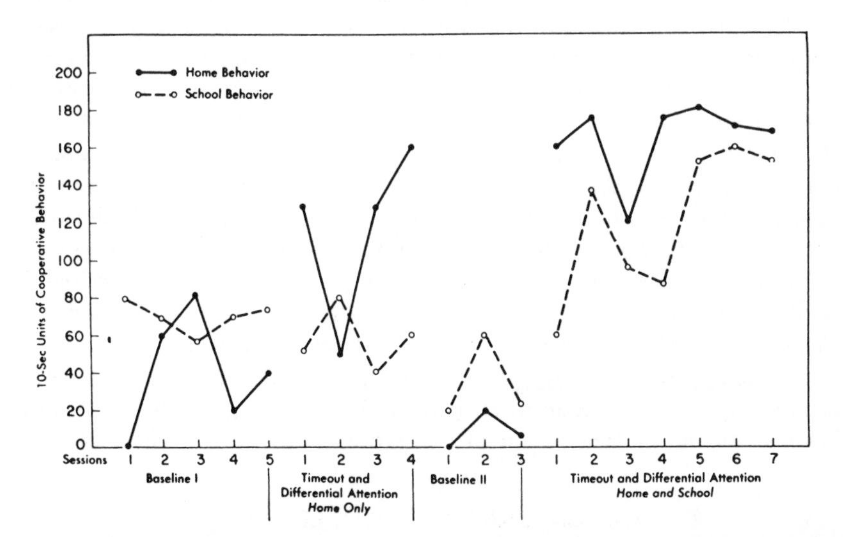

Fig. 7-16. (Fig. 1, p. 243, from: Wahler, R. G. Setting generality: Some specific and general effects of child behavior therapy. *Journal of Applied Behavior Analysis*, 1969, **2**, 239-246. Copyright by Society for the Experimental Analysis of Behavior, Inc. Reproduced by permission.)

In our next example, aspects of the multiple schedule, multiple baseline across settings, and withdrawal designs were incorporated in one study (Wahler, 1969). In this study the effects of timeout and differential attention were examined at home and in school for a 5-year-old boy described as "oppositional." The specific aim of the study was to determine whether changes brought about in one setting might generalize to a second. The target behavior selected for study was cooperative behavior evidenced by the boy in response to directives issued by his parents and the teacher (scored in terms of presence or absence in a 10-second period). Treatment involved positive attention (verbal and physical) administered contingently when cooperative behavior was evidenced. Disruptive behaviors were ignored unless they persisted, at which point timeout procedures were initiated. Prior to the actual study, both the parents and teacher underwent training in applying these specific behavioral techniques.

As can be seen in Fig. 7-16, baseline procedures at home and at school were maintained until a relatively stable pattern emerged. In the second phase, treatment was applied at home and resulted in a significant improvement in cooperative behavior. This portion of the study represents the multiple schedule aspect (discrimination in settings), but could also be conceptualized within the framework of the multiple baseline strategy across settings (two independent baselines with delayed application of treatment in one while maintaining the second untreated as a control). However, in the third phase, treatment was discontinued (withdrawal aspect of the design), resulting in a marked decrease in cooperation at home (independent examination of all home data reveals an A-B-A-B strategy). If, in the third phase, treatment had been applied in school and discontinued at home, then a more complete version of the multiple schedule strategy would have been apparent. In the final phase, treatment was applied simultaneously in both settings, resulting in rate changes for both. With respect to the specific aim of the study (the effects of setting generality), data permit conclusions only as to the effects of change in school as a function of treatment at home (no change). The effects of treatment in school on the home setting cannot be determined as this condition would have required treatment at school while maintaining baseline assessment at home. Wahler (1969) indicates that the boy's parents resisted implementation of this procedure in the last phase of study.

7.4. CONCURRENT SCHEDULE DESIGN

Of the complex experimental analysis designs currently used by applied clinical researchers, the concurrent schedule strategy best approximates the conditions of the natural environment. Whereas in the multiple schedule strategy the same behavior is treated differently under different stimulus conditions (stimuli presented individually in alternation), in the concurrent schedule strategy the subject is *simultaneously* exposed to different stimulus conditions. An advantage of this latter strategy is that proportional efficacy of varying schedules (treatments) can be ascertained. In addition, this kind of design strategy requires considerably less time to conduct than the withdrawal, reversal, multiple baseline, or multiple schedule designs. "For this reason, designs which utilize simultaneous comparison of several hypothesized treatment techniques are clearly preferred to designs using successive treatment conditions. The simultaneous design clearly requires greater skill on the part of the investigator and considerably more planning and organization for the . . . staff" (Browning and Stover, 1971, p. 110).

Although we were able to find only one published example of a concurrent schedule design in the applied literature (Browning and Stover, 1971, pp.105-110), the use of concurrent scheduling in basic laboratory research has gained in popularity (e.g., Catania, 1968; Herrnstein, 1970; Honig, 1966; Reynolds, 1968). This fact suggests that more cross-fertilization of ideas between basic researchers and applied researchers is needed, particularly if the trend toward increased experimental sophistication is to continue in the applied areas.

Browning and Stover (1971) present an elegant example of the concurrent schedule design in their comparative assessment of three techniques (praise, verbal admonishment, and purposely ignoring) used in reducing "grandiose bragging" in a 10-year-old boy, who was admitted for residential treatment with a diagnosis of Anxiety Reaction. Bragging consisted of the boy's concocting untrue tales about himself and presenting them to staff members in order to gain their attention. A casual examination of staff members' reactions to bragging suggested that they primarily fell into two broad categories: (1) praise and attention, and (2) verbal admonishments. Staff members rarely ignored (extinguished) the boy's responses.

During the first 4 weeks of the study, members were asked to record the frequency of bragging responses. In the fifth week baseline assessment procedures were continued, but now staff members were requested to ignore the boy's bragging. During Weeks 6-8, staff members were divided into three teams with two members in each. This constituted the beginning of the concurrent schedule design, during which time the three reinforcement conditions (praise, admonish, and ignore) were administered simultaneously and successively for 3 weeks by the three teams. In accordance with counterbalancing principles, each of the three teams administered all treatments but in different orders for each of the 3 weeks (i.e., *Week One*: Team 1–praise, Team 2–admonish, Team 3–ignore; *Week Two*: Team 1–admonish, Team 2–ignore, Team 3–praise; *Week Three*: Team 1–ignore, Team 2–praise, Team 3–admonish). We might underscore that all conditions (schedules) were actually in force at the same time as the child was exposed daily to the six members comprising the three teams. During Weeks 6-8, bragging responses were timed (length in seconds) in addition to their being monitored as to frequency.

The results of this study appear in Fig. 7-17, and show that during the

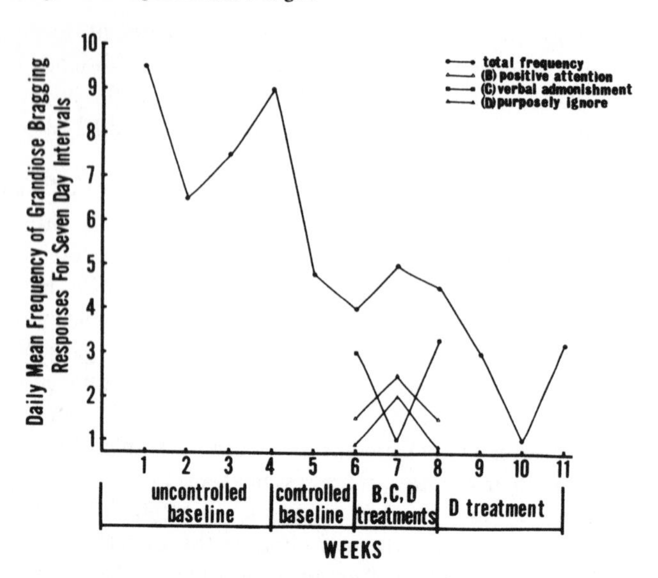

Fig. 7-17. Total mean frequency of grandiose bragging responses throughout study and for each reinforcement contingency during experimental period. (Fig. 4-14, p. 107, redrawn from: Browning, R. M., and Stover, D. O. *Behavior modification in child treatment: An experimental and clinical approach*. Chicago and New York: Aldine-Atherton, 1971. Copyright 1971 by Robert M. Browning and Donald O. Stover. Reproduced by permission.)

uncontrolled baseline bragging ranged from six to ten responses. During the controlled baseline (ignore), bragging dropped to less than five responses. In Weeks 6-8, total number of bragging responses ranged from 4·5 to 5·5. Separate data are also plotted for responses to the three schedules. These data were examined in a Latin square design with the subject as his own control. Data analyzed with respect to frequency of bragging under the three conditions failed to yield significant differences. However, a significant effect was found when data were examined in terms of length of the response (second for each bragging response). Analysis of mean differences for the three schedules indicates that ignoring the child's bragging was more successful than either praising or admonishing him ($p < \cdot 01$, $p < \cdot 01$). Interestingly, significantly shorter responses were obtained under the praise than the admonishment condition ($p < \cdot 05$).

As the ignore condition appeared to be the most effective in reducing length of bragging responses, this schedule was applied by all three teams during Weeks 9-11. Examination of the graph shows that bragging responses ranged from one to three during this last phase of treatment. Browning and Stover (1971) note that "Although it was not an objective of this study, the design and statistical analysis would enable an investigator to identify which staff would serve as the most effective behavior therapists to administer the selected treatment technique with the patient" (p. 108).

References

Agras, W. S., Leitenberg, H., Barlow, D. H., and Thomson, L. E. Instruction and reinforcement in the modification of neurotic behavior. *American Journal of Psychiatry*, 1969, **125,** 1435-1439.

Allen, G. J. Case study: Implementation of behavior modification techniques in summer camp settings. *Behavior Therapy*, 1973, **4,** 570-575.

Baer, D. M., Wolf, M. M., and Risley, T. R. Some current dimensions of applied behavior analysis. *Journal of Applied Behavior Analysis*, 1968, **1,** 91-97.

Bailey, J. S., Timbers, G. D., Phillips, E. L., and Wolf, M. M. Modification of articulation errors of pre-delinquents by their peers. *Journal of Applied Behavior Analysis*, 1971, **4,** 265-281.

Ban T. *Psychopharmacology.* Baltimore: Williams and Wilkins, 1969.

Barlow, D. H., and Hersen, M. Single-case experimental designs: Uses in applied clinical research. *Archives of General Psychiatry*, 1973, **29,** 319-325.

Barlow, D. H., Reynolds, J., and Agras, W. S. Gender identity change in a transsexual. *Archives of General Psychiatry*, 1973, **28,** 569-576.

Barrish, H. H., Saunders, M., and Wolf, M. M. Good behavior game: Effects of individual contigencies for group consequences on disruptive behavior in a classroom. *Journal of Applied Behavior Analysis*, 1969, **2,** 119-124.

Barton, E. S., Guess, D., Garcia, E., and Baer, D. M. Improvement of retardates' mealtime behaviors by timeout procedures using multiple baseline techniques. *Journal of Applied Behavior Analysis*, 1970, **3,** 77-84.

Browning, R. M., and Stover, D. O. *Behavior modification in child treatment: An experimental and clinical approach.* Chicago and New York: Aldine-Atherton, 1971.

Catania, A. C. (Ed.) *Contemporary research in operant behavior*. Glenview, Ill. Scott, Foresman, 1968.

Christopherson, E. R., Arnold, C. M., Hill, D. W., and Quilitch, H. R. The home point system: Token reinforcement procedures for application by parents of children with behavior problems. *Journal of Applied Behavior Analysis*, 1972, **5,** 485-497.

Clark, H. B., Rowbury, T., Baer, A. M., and Baer, D. M. Timeout as a punishing stimulus in continuous and intermittent schedules. *Journal of Applied Behavior Analysis*, 1973, **6,** 443-455.

Eisler, R. M., Hersen, M., and Miller, P. M. Shaping components of assertive behavior with instructions and feedback. *American Journal of Psychiatry*, 1974, **131,** 1344-1347.

Epstein, L. H., and Hersen, M. A multiple baseline analysis of coverant control. *Journal of Behavior Therapy and Experimental Psychiatry*, 1974, **5,** 7-12.

Gottman, J. M. N-of-one and N-of-two research in psychotherapy. *Psychological Bulletin*, 1973, **80,** 93-105.

Gottman, J. M., McFall, R. M., and Barnett, J. T. Design and analysis of research using time series. *Psychological Bulletin*, 1969, **72,** 299-306.

Hall, R. V., Cristler, C., Cranston, S. S., and Tucker, B. Teachers and parents as researchers using multiple baseline designs. *Journal of Applied Behavior Analysis*, 1970, **3,** 247-255.

Hart, B. M., and Risley, T. R. Establishing use of descriptive adjectives in the spontaneous speech of disadvantaged preschool children. *Journal of Applied Behavior Analysis*, 1968, **1**, 109-120.

Herrnstein, R. J. On the law of effect. *Journal of the Experimental Analysis of Behavior*, 1970, **13**, 243-266.

Hilgard, J. R. The effect of early and delayed practice on memory and motor performances studied by the method of co-twin control. *Genetic Psychology Minographs*, 1933, **14**, 493-567.

Honig, W. K. (Ed.) *Operant behavior: Areas of research and application.* New York: Appleton-Century-Crofts, 1966.

Johnston, J. M. Punishment of human behavior. *American Psychologist*, 1972, **27**, 1033-1054.

Kazdin, A. E. Methodological and assessment considerations in evaluating reinforcement programs in applied settings. *Journal of Applied Behavior Analysis*, 1973, **6**, 517-531.

Kazdin, A. E., and Bootzin, R. R. The token economy: An evaluative review. *Journal of Applied Behavior Analysis*, 1972, **5**, 343-372.

Leitenberg, H. The use of single-case methodology in psychotherapy research. *Journal of Abnormal Psychology*, 1973, **82**, 87-101.

Liberman, R. P., and Smith, V. A multiple baseline study of systematic desensitization in a patient with multiple phobias. *Behavior Therapy*, 1972, **3**, 597-603.

Liberman, R. P., Teigen, J., Patterson, R., and Baker, V. Reducing delusional speech in chronic paranoid schizophrenics. *Journal of Applied Behavior Analysis*, 1973, **6**, 57-64.

Long, J. D., and Williams, R. L. The comparative effectiveness of group and individually contingent free time with inner-city junior high school students. *Journal of Applied Behavior Analysis*, 1973, **6**, 465-474.

Lovaas, O. L., and Simmons, J. Q. Manipulation of self-destruction in three retarded children. *Journal of Applied Behavior Analysis*, 1969, **2**, 143-158.

Marks, I. M., and Gelder, M. G. Transvestism and fetishism: Clinical and psychological changes during faradic aversion. *British Journal of Psychiatry*, 1967, **113**, 711-729.

Moore, B. L., and Bailey, J. S. Social punishment in the modification of a pre-school child's "autistic-like" behavior with a mother as therapist. *Journal of Applied Behavior Analysis*, 1973, **6**, 497-507.

Morganstern, K. P. Cigarette smoke as a noxious stimulus in self-managed aversion therapy for compulsive eating: Techniques and case illustration. *Behavior Therapy*, 1974, **5**, 255-260.

O'Brien, F., Azrin, N. H., and Henson, K. Increased communications of chronic mental patients by reinforcement and by response priming. *Journal of Applied Behavior Analysis*, 1969, **2**, 23-29.

Panyan, M., Boozer, H., and Morris, N. Feedback to attendants as a reinforcer for applying operant techniques. *Journal of Applied Behavior Analysis*, 1970, **3**, 1-4.

Patterson, R. L., and Teigen, J. R. Conditioning and post-hospital generalization of non-delusional responses in a chronic psychotic patient. *Journal of Applied Behavior Analysis*, 1973, **6**, 65-70.

Rachman, S. Primary obsessional slowness. *Behaviour Research and Therapy*, 1974, **12**, 9-18.

Reynolds, G. S. *A primer of operant conditioning.* Glenview, Ill. Scott, Foresman, 1968.

Risley, T. R., and Hart, B. M. Developing correspondence between the nonverbal and verbal behavior of preschool children. *Journal of Applied Behavior Analysis*, 1968, **1**, 267-281.

Risley, T. R., and Wolf, M. M. Strategies for analyzing behavioral change over time. In J. Nesselroade and H. Reese (Eds.), *Life-span developmental psychology: Methodological issues.* Pp. 175-183. New York: Academic Press, 1972.

Sajwaj, T., Twardosz, S., and Burke, M. Side effects of extinction procedures in a remedial preschool. *Journal of Applied Behavior Analysis*, 1972, **5**, 163-175.

Sidman, M. *Tactics of scientific research: Evaluating experimental data in psychology.* New York: Basic Books, 1960.

Thoresen, C. E. The intensive design: An intimate approach to counseling research. Paper read at American Educational Research Association, Chicago, April 1972.

Twardosz, S., and Sajwaj, T. Multiple effects of a procedure to increase sitting in a hyperactive retarded boy. *Journal of Applied Behavior Analysis*, 1972, **5**, 73-78.

Wahler, R. G. Setting generality: Some specific and general effects of child behavior therapy. *Journal of Applied Behavior Analysis*, 1969, **2**, 239-246.

Wilson, C. W., and Hopkins, B. L. The effects of contingent music on the intensity of noise in junior high home economics classes. *Journal of Applied Behavior Analysis*, 1973, **6**, 269-275.

Wolf, M. M., and Risley, T. R. Reinforcement: Applied research. In R. Glaser (Ed.), *The nature of reinforcement.* Pp. 310-325. New York: Academic Press, 1971.

CHAPTER 8

Statistical Analyses for Single-case Experimental Designs[1]

ALAN E. KAZDIN

The Pennsylvania State University

8.1. INTRODUCTION

Statistical analyses have been reported relatively infrequently in investigations employing intra-subject replication designs. Recently, increased attention has been devoted to the problems attendant upon applying conventional analyses to N=1 research and to alternative analyses aimed at resolving these problems (e.g., Edgar and Billingsley, 1974; Gentile, Roden, and Klein, 1972; Hartmann, 1974; Kelly, McNeil, and Newman, 1973; Keselman and Leventhal, 1974; Kratochwill, Alden, Demuth, Dawson, Panicucci, Arntson, McMurray, Hempstead, and Levin, 1974; Namboodiri, 1972; Shine and Bower, 1971; Thoresen and Elashoff, 1974). The increased attention is noteworthy because the development of statistical analyses for the single case has lagged behind analyses in the tradition of R. A. Fisher. However, the limited use of statistics for N=1 research has not been simply a matter of developing powerful statistical techniques. Whether statistical analyses *should* be used to evaluate change in intra-subject replication designs has been a major point of contention (cf. Michael, 1974).

A salient issue in the controversy over the use of statistics pertains to the criteria for evaluating change, including the notions of clinical and statistical significance, discussed earlier. In the present chapter, the main points in this controversy will be outlined. However, the primary purpose of the chapter is to provide a nontechnical description and evaluation of different statistical tests that can be applied to intra-subject designs for the single case. Because most of the tests have not been widely employed, examples and computational procedures will be detailed. In addition, special considerations and limitations that may

[1] Preparation of this chapter was facilitated by a grant from the National Institute of Mental Health (MH 23399). The author wishes to gratefully acknowledge Eugene Edgington, Donald P. Hartmann, Richard R. Jones, and Thomas R. Kratochwill who provided comments on an earlier draft of this chapter.

restrict the use of specific tests will be discussed. In general, this chapter provides a nontechnical introduction to diverse statistics. Although the focus of the chapter is on N=1 research, the statistical tests are not necessarily limited to the single case. Indeed, many N=1 statistical tests, as well as more commonly used statistics (e.g., repeated-measures analysis of variance) can be readily employed in both between-group and single-case experimental designs.

Prior to considering specific statistical tests, the issue of whether statistics should be employed at all will be addressed. No attempt will be made to resolve the controversy surrounding the use of statistics. Of course, application of the tests detailed in this chapter presupposes that statistics can be of some use in applied clinical work. The case for and against the use of statistics will be presented. The case against will be detailed first because it represents the dominant position in laboratory and applied intra-subject replication research.

8.2. THE CASE AGAINST THE USE OF STATISTICS

Arguments against the use of statistics for intra-subject replication research have centered around the criteria for evaluating research in general. In applied research, the criticisms against statistical evaluation become particularly salient, as will be detailed below. Other sources of criticism of statistical analyses pertain to the notion of statistical evaluation in the context of *group* research and the obfuscation of performance of the individual subject. Many of the points pertaining to objections against traditional group research will be mentioned briefly because they have been detailed elsewhere (Kazdin, 1973; Sidman, 1960).

Criteria for evaluation

Clinical and experimental criteria have been posed for evaluating applied behavioral research (Risley, 1970). The clinical criterion refers to the importance of the change achieved, whereas the experimental criterion refers to the reliability of that change.

The clinical criterion for "significance" refers to a comparison between behavior change that has been accomplished and the level of change required for the client's adequate functioning in society (Risley, 1970). This notion of significance departs drastically from statistical significance. Small changes in behavior of one individual or many individuals usually are not considered to be clinically significant. Even if behavior change is reliable and clearly related to the experimental intervention, the change cannot necessarily be regarded as clinically significant. For example, reducing delusional statements of a psychotic patient from 80 percent to 40 percent of the periods observed probably would not be regarded as an important change. Relative to a standard of "normal social interaction," the effect of the intervention leaves a great deal to be desired. The patient could not function in a community without censure from peers for bizarre statements. To make the change of clinical significance, a much greater reduction in bizarre statements would be required.

The precise criterion that makes a given change of clinical significance is difficult to specify because individuals in everyday life (parents, teachers, peers, colleagues, friends) determine which behaviors are deviant, obnoxious, intolerable, or acceptable. When a behavior is altered, as evidenced by objective data, *and* when individuals in contact with the client indicate that the original behavioral goal has been achieved, the program has attained a change of clinical significance.

As an ideal, applied interventions strive for changes that ordinarily surpass statistical significance. Yet, for clinical evaluation, statistical significance is an ancillary and, to many individuals, irrelevant criterion. The concern with clinical rather than statistical significance has led investigators to focus upon independent variables that frequently produce dramatic changes. Variables that have subtle influences on behavior, at least at present, are given lower priority. The reason for this is that in achieving therapeutic results, gross rather than subtle changes in behavior usually are required. While experimental manipulations with weak effects might be interesting in their own right and eventually prove to have practical value, they would not solve the immediate problem of making a clinically significant change for a given client.

Aside from the clinical or therapeutic criterion for evaluating an intervention, there is an experimental criterion. The experimental criterion refers to a comparison between behavior during the intervention with what it would be if the intervention had not been implemented (Risley, 1970). Traditionally, the experimental criterion has been satisfied by comparing performance among groups and evaluating the comparison statistically. Strictly, the use of statistics is not essential to inferring a reliable effect. The intra-subject approach detailed in previous chapters describes the procedures for satisfying the experimental criterion without statistical evaluation.

Baseline data are collected, and serve as a basis for determining the present level of behavior and predicting what behavior would be without the intervention. Implementing a program can reveal a change from this projected performance. The reliability of a finding in the experimental sense is achieved by *replicating* the baseline level of performance and a different level of performance during intervention, as shown in the commonly employed A-B-A-B design. Other intra-subject designs (e.g., multiple baseline designs) make different comparisons to project what performance would be if treatment were not implemented (cf. Kazdin, 1976).

In practice, there are a few ways in which results clearly meet the experimental criterion. First, performance during the treatment intervention when plotted may not overlap with performance during baseline. The data points of baseline may not extend to the levels achieved by the data points during the intervention. If this is replicated either over time with a given subject (A-B-A-B design) or across behaviors (multiple baseline design), there can be little question that the results are reliable. Second, a more typical criterion for experimental evaluation is related to divergent slopes in baseline and treatment phases but is less stringent than completely non-overlapping distributions. This criterion emphasizes the trends or slopes in each phase. Usually, the baseline phase is not terminated if there is a trend (e.g., improvement) in the behavior that is to be changed. Because experimental evaluation depends upon extrapolating how performance would be if no intervention were made, it is important to have a stable rate of behavior during baseline. If there is a trend, it should be opposite from the direction that is to be achieved with the intervention. In any case, baseline usually shows a relatively stable performance rate with no particular trend. When treatment is implemented, usually a definite trend is evident indicating that behavior is changing from baseline. The intervention is continued until the trend clearly departs from the predicted performance of baseline or is at a stable level different from baseline. If baseline conditions are reinstated, the trend is likely to be in the opposite direction of the intervention. By alternating baseline and experimental phases, systematic changes in trend strongly argue for the experimental reliability of the effect.

The experimental criterion for evaluating applied interventions with intra-subject designs is retained by the experimental design (intra-subject replication) rather than by the design and statistical comparisons characteristic of traditional between-group research. The rejection of statistical criteria should be viewed in light of the goals of the evaluating and implementing potent interventions. Seeking unambiguous effects for which statistical analyses are unnecessary is an established research goal rather than a position that derives from research in a specific content area. Campbell (1963) has expressed this point in noting:

> If the more advanced sciences use tests of significance less than do psychology and sociology for example, it is undoubtedly because the magnitude and the clarity of effects with which they deal are such as to render tests of significance unnecessary; were our conventional test of significance to be applied, high degrees of significance would be found. It seems typical of the ecology of the social sciences, however, that they must work the low-grade ore in which tests of significance are necessary. (p. 224)

Of course, whether all areas in social science qualify as "low-grade ore" in the sense referred to by Campbell can only be determined empirically. In any case, many researchers view the search for marked unambiguous effects as a logical priority in psychological research. Thus, statistical evaluation has been rejected as a criterion for establishing the effect of an intervention. In clinical research, one is concerned only with marked changes to effect change in the patient.

Group research and statistical analyses

There have been other criticisms associated with the use of statistics in group research, some of which were noted in previous chapters. Initially, in many statistical tests, inter-subject variability serves as a basis of evaluating the effect of an intervention. Yet, within-group variability is not part of the behavioral processes of the individual subject and, perhaps, should not be included in the evaluation of the processes (Sidman, 1960). A related objection is that statistical tests in group research often hide the performance of the individual subject. Means and variances alone are not useful in deciding the degree to which group performance represents a particular individual who received the intervention. Thus, the effect of an intervention may be obfuscated with statistical analyses (Skinner, 1956).

Some of the criticisms against the use of statistics have been based upon the logic of many statistical tests. For example, Sidman (1960) criticized the circularity of statistics, which makes assumptions (e.g., form of the distribution) about the parent population from which a sample of observations was drawn and subsequently tests the data on the basis of those assumed properties. Other authors (e.g., Edgington, 1969) also have questioned the legitimacy of making assumptions about a population from which samples putatively have been derived.

Another objection against the use of statistics pertains to the notion of replication. Conclusions based on statistical criteria will lead to significant group differences given the null hypothesis on the basis of chance alone on a certain number of occasions. Similarly, if an experimental intervention does have a veridical effect, replication will prove nonsignificant in some instances. A single experiment is subject to both of these problems (cf. Tversky and Kahneman, 1971).

Criticisms of statistical tests because of erroneous conclusions or a failure to meet assumptions are somewhat less fundamental to the overall objection to statistical evaluation in applied research. The thrust of the argument pertains to the criteria that should be employed to evaluate the effect of an intervention.

8.3. THE CASE FOR THE USE OF STATISTICS

Despite the above considerations, there are several situations in intra-subject replication research that many investigators believe are suitable for statistics. The situations include those instances in which stable baseline rates of behavior are not readily established, when "new" variables with unestablished effects are being explored, and where intra-subject variability is relatively large, perhaps due to the lack of experimental control over the situation.

Failure to establish a stable baseline

Statistical evaluation may be helpful when desirable conditions for evaluation cannot be met. For example, it is desirable in an A-B-A-B design to begin the intervention (B) phase only when a stable baseline rate of behavior has been achieved. If there is a trend in baseline, it should not be in the direction of the expected change, partially because this makes evaluation of subsequent phases difficult (and may indicate that the intervention is not needed). Yet it is not always feasible to achieve a stable baseline, as some investigators have lamented (Browning, 1967; Browning and Stover, 1971; White, 1972). Further, sometimes a trend in baseline *is* in the direction of therapeutic change. Yet it may be desirable to accelerate the change. For example, an autistic child may be beating himself severely. During baseline, a slight reduction in behavior may be evident. Although the reduction may be systematic, its therapeutic value may be limited and the change should be accelerated. Although baseline is not stable, treatment can be introduced. Some statistical analyses, to be discussed later, allow for evaluation of an intervention taking into account the trend in baseline. Whereas visual inspection of the data often entails noting distinct changes in trends across phases, statistical analysis can scrutinize continuous shifts across phases where there is no change in trend. Thus, statistical analyses in some cases permit evaluation of an intervention that could not be easily assessed by visual inspection.

Investigations in "new" areas of research

Statistical analyses may be useful at the beginning of a research program when the investigator is not familiar with the relative impact of different interventions. The data may appear to reveal reliable effects, although visual inspection is equivocal. Moreover, the change during the intervention may not be rapid, making evaluation especially difficult. In cases where the results are ambiguous, statistical evaluation can assess whether the effects are reliable. While the intervention may not achieve clinical significance, statistical criteria may direct the investigator to focus on those interventions that have reliable (although not obvious) effects. The investigator can then devote subsequent attention to varying parameters of the interventions to determine whether dramatic changes can be achieved. In any case, at the beginning of research, statistics may provide a preliminary screening tool to aid the investigator in deciding which variables produce reliable changes.

A major point of contention, as indicated in the previous section, is the use of statistics to "tease" out subtle effects in the data. Those individuals favoring statistics note that often reliable but nondramatic effects do occur in the data. Applied clinical research currently focuses on variables that effect large changes. Yet as the technology increases, attention may be directed toward those variables with relatively subtle effects. Visual inspection of the data may not prove adequate in these instances.

It should be noted that an intervention which in itself does not effect dramatic change is not by that fact alone unimportant for applied purposes. Variables that exert subtle control may be added to interventions that are known to alter behavior and may enhance their effects. Also, variables that by themselves exert little control may be shown in subsequent work to interact with other variables to produce dramatic effects. Thus, the search for effective variables that might be evaluated with statistical, rather than visual, criteria should not be thwarted.

Increased intra-subject variability

Often variables in applied research are investigated in the context of fairly well-controlled settings (e.g., classrooms, day-care centers, closed wards). When the setting is well-controlled, the likelihood of producing dramatic treatment effects is enhanced. Extraneous contingencies that otherwise might attenuate the effect of the intervention can be reduced or eliminated. However, increasingly, applied research is extended to open-field settings in society where control over the environment is limited (see Kazdin, 1975a, 1975b). Where there is more uncontrolled variation, the effects of the interventions may be less obvious. Variability within phases and systematic variability due to extraneous factors across phases may make previously dramatic interventions appear less potent. In such cases, statistical evaluation will increase the precision in determining a significant (reliable) effect. Under these circumstances, the value of statistical evaluation may increase.

SUMMARY: CASE FOR AND AGAINST

Individuals who recommend the use of statistics do not advocate subjecting the effect of marked behavior change to statistical analyses (e.g., Jones, Vaught, and Weinrott, 1976). Rather, they pose that a variety of situations may make visual inspection difficult to invoke as the criterion for experimental reliability.

Moreover, they question the value of disregarding reliable but nondramatic effects that might under various conditions prove to be of theoretical or clinical importance.

Individuals who advocate non-statistical criteria for evaluation caution against "teasing out" subtle effects because these effects are least likely to be replicable. Moreover, invoking statistical significance as the only criterion for evaluation does not encourage the investigator to obtain clear unequivocal experimental control over behavior. Finally, many investigators believe that in clinical work, statistical evaluation is simply not relevant for assessing therapeutic change.

The previous discussion does not attempt to resolve the controversy over statistical tests for intra-subject replication research. Independently of the final resolution, specific statistical tests that are available for the single case can be discussed. However, prior to considering specific tests, it is important to describe a salient characteristic of data based upon repeated measurement–i.e., time series data. One characteristic, referred to as serial dependency, has limited the use of conventional statistical analyses for $N=1$ research.

8.4. SERIAL DEPENDENCY AND CONVENTIONAL STATISTICAL ANALYSES

As is well known, there are several assumptions that are made in the use of analysis of variance (ANOVA) models. Simulation studies generally have shown the "robust" nature of ANOVA in handling the violation of most of the assumptions in ordinary circumstances even when different assumptions are simultaneously violated (e.g., Atiqullah, 1967; Boneau, 1960; Cochran, 1947; Glass, Peckham and Sanders, 1972; Scheffé, 1959).

There is one assumption which, if violated, seriously affects ANOVA and makes the t or F test inappropriate. The assumption is the independence of error components. The assumption refers to the correlation between the error (e) components of pairs of observations (within and across conditions) for i and j subjects. The expected value of the correlation for pairs of observations is assumed to be zero (i.e., $r_{e_i e_j} = 0$). Typically, in between-group designs independence of error components is assured by randomly assigning subjects to conditions.

In the case of continuous or repeated measures over time, the assumption of independence of observations usually is not met. Successive observations in a time series tend to be correlated. Thus, knowing the level of performance of a subject at a given time allows one to make predictions about subsequent points in the series. The extent to which there is dependency among successive observations can be assessed by examining autocorrelation (or serial correlation) in the data.[2] Autocorrelation refers to a correlation (r) between data points separated by different time intervals (lags) in the series. An autocorrelation of lag 1 (or r_1) is computed by pairing the initial observation with the second observation, the second with the third, the third with the fourth, and so on throughout the time series. Autocorrelation of lag 1 yields the correlation coefficient that reflects serial dependency. If the correlation is significantly different from zero, this indicates a performance at a given point in time can be predicted from performance on the previous occasion (the direction of the prediction determined by the sign of the autocorrelation).

Generally, autocorrelation of lag 1 (i.e., correlating adjacent data points) is sufficient to reveal serial dependency in the data. However, a finer analysis of dependency may be obtained by computing several autocorrelations with different time lags (e.g., autocorrelations of lags of 2, 3, 4, and so on). For the general case, an autocorrelation of the lag t is computed by pairing observations t data points apart. For example, autocorrelation of lag 2 is computed by pairing the initial observation in the series with the third, the second with the fourth, the third with the fifth, and so on.

Serial dependency throughout the time series is clarified by computing and plotting correlations of different lags.[3] The plot of the autocorrelations is referred to as a *correlogram.* Figure 8-1 provides correlograms (i.e., autocorrelations plotted as a function of different lags) for two hypothetical sets of data. In each correlogram, the point that is plotted reflects the correlation coefficient for observations of a given lag. As can be seen for the data in the upper portion of the figure, the correlations with short lags are positive and relatively high. As the lag (i.e., the distance between the data points) increases, the autocorrelation approaches zero and eventually becomes negative. The hypothetical data in the upper portion of Fig. 8-1 reflect serial dependency because the autocorrelation of lag 1 is likely to be significantly different from zero.[4] Moreover, the correlogram reveals that the dependency continues beyond lag 1 until the autocorrelation approaches 0. In contrast, the lower portion of Fig. 8-1 reveals a

[2]Serial correlation and autocorrelation are used synonymously by many authors. A distinction is sometimes made by designating autocorrelation as the correlation of data points in a single time series and serial correlation as the correlation between values of one time series and lagged values of another series (i.e., two variables) (Ezekiel and Fox, 1959). For a technical discussion of autocorrelation and dependency, the reader is referred to Anderson (1971, Chapter 6).

[3]As the lag increases, the correlation becomes somewhat less stable, in part, because of the decrease in the number of pairs of observations upon which the coefficient can be based (Holtzman, 1963).

[4]While the statistical significance of autocorrelations can be approximated by testing them as correlations in the usual manner, Anderson (1942) has provided tables for the exact test. (See also Anderson, 1971, and Ezekiel and Fox, 1959.)

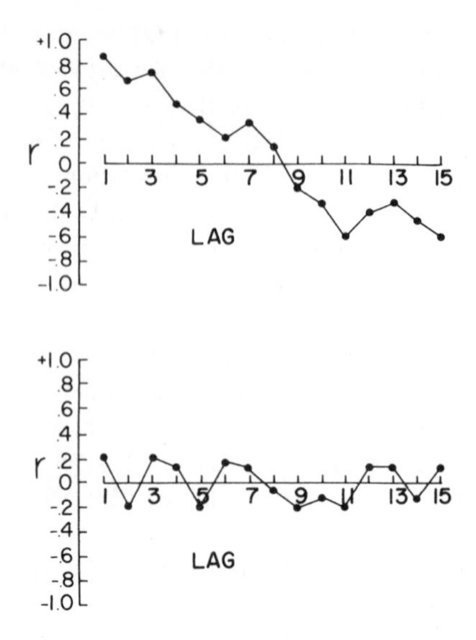

Fig. 8-1. Correlograms for data with (upper portion) and without serial dependency (lower portion).

hypothetical correlogram where the observations in the time series are not dependent. The autocorrelations do not significantly deviate from 0. The lack of dependence signifies that the errors of successive observations are "random"– i.e., a data point below the "average" value is just as likely to be followed by a high value as by another low value. Time series data that reveal this latter pattern can be treated as independent observations and can be subjected to conventional statisitical analyses.

A significant autocorrelation of lag 1 indicates that the assumption of independence of error components is not achieved and that conventional analyses are not appropriate. There are serious problems that occur if conventional analyses are used (cf. Scheffé, 1959). Initially, serial dependency reduces the number of independent sources of information in the data. The degrees of freedom based upon the actual number of observations is inappropriate because it assumes that the observations are independent. Any *F* test is likely to overestimate the true *F* value because of an inappropriate estimate of the degrees of freedom.[5] For the appropriate application of *t* and *F* tests, the degrees of freedom must be independent (uncorrelated) sources of information.

[5] An approximate correction formula has been designed to adjust degrees of freedom for autocorrelation (cf. Ezekiel and Fox, 1959, Chapter 20).

A second and related problem associated with dependency is that the autocorrelation spuriously reduces the variability of the time series data. Thus, error terms derived from the data underestimate the variability that would result from independent observations. The smaller error term inflates or positively biases F. In recommending the use of conventional analyses, various authors (e.g., Chassan, 1967; Kelly, McNeil, and Norman, 1973) have underestimated the impact of serial dependency. The straightforward application of conventional statistical analyses is inappropriate. Certainly, the inapplicability of conventional analyses has thwarted the use of statistics for time series data. Various statistical tests are presented below that address or circumvent the issue of dependency in time series data.

8.5. VARIATIONS OF *t* TEST AND ANOVA

Variations of the t test and ANOVA have been posed to evaluate $N=1$ studies of two or more phases. Performance of a single subject across baseline (A) and intervention (B) phases is compared. A t test assesses whether baseline and intervention means are different. Of course, as noted in the discussion of serial dependency, the usual applications of t and ANOVA are inappropriate.

Variations of t and ANOVA have been suggested for analyzing single case data across different phases (i.e., A-B-A-B design). Essentially, the techniques differ in the way of handling violation of the assumption of independent error components. Consider the case where one subject is exposed to different phases in an A-B-A-B design. Gentile, Roden, and Klein (1972) posed that the different phases can be viewed as the independent variable or treatment effect. The effect of two "treatments" (A vs. B) can be compared. In an A-B-C-A-B-C design, three treatments would be evaluated. The treatment factor is analogous to the between-group factor in the usual (nonrepeated measures) ANOVA. The number of observations of the subject within a phase constitutes "within-group" observations.

The rationale for using within-phase observations in the construction of an error term is as follows. The responses generated by a single individual are assumed to be statistically independent and normally distributed about a central "true" value (Shine and Bower, 1971). The subject's responses are considered to vary about some "true" mean point; the mean of the obtained responses is an unbiased estimate of this point. On an *a priori* basis, Gentile, Roden, and Klein assume that performance of a given response for a given individual is independent of every other response. This assumption is crucial for justifying use of the test.

Computation of the analysis of variance using the method proposed by

Gentile, Roden, and Klein (1972) is straightforward. If there are only two treatments (A and B) in four phases, a simple *t* test across of the combined A and the combined B treatments is proposed. Analysis of variance is only required for three or more treatments.[6,7]

Gentile, Roden, and Klein were aware of the non-independence of observations. They suggested that the high autocorrelation of adjacent observations could be ameliorated by combining non-adjacent phases that employed the same treatment. In an A-B-A-B design (N=1), they recommended combining both A phases ($A_1 + A_2$) and both B phases ($B_1 + B_2$) and making a simple comparison (*t* test). They suggested that the autocorrelation would not be problematic by combining phases. The rationale for combining phases is based on the fact that autocorrelations tend to decrease as the lag between observations increases. Assuming serial dependency in the data, observation 1 in phase A_1 would be more highly correlated with observation 1 in phase B_1 (i.e., the immediately adjacent phase) than with observation 1 in phase A_2 (i.e., a non-adjacent phase). Since the error components of all observations in A_1 are more like the components for the observations in B_1 than in A_2, it is assumed that combining treatments separated in time will reduce the dependency. Combining phases that are not adjacent should make A and B treatments more *dissimilar* due to dependency in the data. The resulting *t* (or *F*) should be reduced because the dependency of adjacent observations will *minimize* treatment differences. Again, the rationale is that data from adjacent phase (A_1 B_1 and A_2 B_2) will be more similar due to serial dependency than data from non-adjacent phases. The combining of non-adjacent phases to make the comparison statistically should utilize dependency to deflate the *t* value.

The model proposed by Gentile, Roden, and Klein (1972) can be extended to a two-way ANOVA with the addition of another subject to the above design

[6]It is possible to evaluate the A-B-A-B design with $N = 1$ as a two-factor design comparing the effect of Treatment (A vs. B) and Time (First phase vs. Second phase) and their interaction.

[7]To determine the number of observations, Gentile, Roden, and Klein (1972) employed the number of recorded intervals of behavior, thus treating separate intervals (for a given day of observation) as separate observations. Thus, if an individual was observed for ten intervals over 10 days in a given phase, the number of observations for that phase would be 100. A more acceptable procedure might be to count each day (mean of each day) as the unit of observation rather than employing separate intervals. There are at least two reasons for this recommendation. First, by counting each interval as an observation, it is possible to arbitrarily employ an exhaustively large number of intervals with inordinately small durations merely to inflate the degrees of freedom. Second, the observations within a given session (day) certainly are not independent since they include systematic error components associated with unique features of that day of observation. The problem of serial dependency across days is compounded by dependency within a session. If the number of observations within a session differs across days or phases, new sources of bias can enter in the calculation of error terms.

(i.e., N=2). Phases remain one factor in the A-B-A-B design and subjects become another factor. The analysis yields an effect for Subjects, Treatments (phases), and the interaction of these terms. The interaction term, of course, can reveal whether subjects were differentially affected by treatment.

Shine and Bower (1971) proposed an analysis somewhat different from the model outlined above. These investigators proposed adding a Trials factor to the one-way ANOVA of treatment. The Trials factor in the single case design replaces the Subject factor that would be present in the ordinary repeated measures design employing several subjects. In the single case design, the error term would become the mean square due to the interaction of Treatment × Trials (instead of Treatment × Subjects in the repeated measures analysis). Employing Trials as a factor means that there is only one observation for each treatment-trial cell, thus eliminating the usual within group (error) term. Shine and Bower derived error terms to test both Treatment and Trials effects. The authors noted that it is not feasible to count Trials as a random factor (as Subjects would be counted in repeated measures design) because the trials represent performance proceeding in a sequential fashion (i.e., are correlated). Thus, special error terms are required to handle Trials as a fixed factor. (See Shine and Bower, 1971, for the formula for computing the error terms.)

Problems with variations of *t* and ANOVA

There are two basic problems with ANOVA procedures that are not well handled by the above variations. The problems are serial dependency and the focus on changes in means across phases.

Serial dependency. The solution for handling dependency in the Gentile, Roden, and Klein (1972) procedure is equivocal (Hartmann, 1974; Kratochwill *et al*., 1974; Thoresen and Elashoff, 1974). Combining phases does not necessarily remove or attenuate serial dependency for at least two reasons. First, combining phases does not at all affect the problem of non-independent data points and the decreased variability among observations *within* phases, two factors that can positively bias F tests. Second, the ability of combining phases to minimize the bias due to serial dependency depends upon the specific structure of the time series (e.g., pattern of autocorrelations with different lags throughout the series) and the length of each phase in relation to this structure. At the present time, it is unclear how different variations of ANOVA could handle diverse patterns of serial dependency in the data. It is generally agreed that ANOVA is not at all "robust" with respect to violation of the independence assumption generally characteristic of time series data. Unless dependency is not a problem in the data, the inadequacy of variations of ANOVA remains largely unaffected (Hartmann, 1974).

Evaluation of means. The ANOVA models evaluate phases or "treatments" by comparing means. A problem in evaluating means of time series data is that trends in the data are ignored. Ignoring trends makes conclusions about mean changes ambiguous, and in some cases, unimportant. For example, a linear slope beginning in baseline and continuing throughout the period of intervention would be statistically significant if the means are examined, even if treatment had *no* effect. Conversely, if there were an increasing slope in baseline and a decreasing slope during the intervention (but no mean difference across phases), an analysis of means would not represent the effect of treatment. In short, a problem with the usual ANOVA models is that they represent only mean changes and ignore trends. The discussion that follows elaborates the variety of situations in which this might be problematic.

8.6. TIME SERIES ANALYSIS

As alluded to earlier, evaluation of mean changes in time series data omits information about trends within and across phases. Omission of trends in the analysis may obfuscate the effect of the intervention. Time series analysis simultaneously examines trends in the data and changes in level of the data across phases. Prior to presenting an overview of the statistical analysis, it is important to amplify the characteristics of time series data that make the analysis especially useful.

Patterns of change in time series data

To describe the characteristics of changes in time, consider as an example the first two phases of an A-B-A-B design. It is important to note three characteristics of changes across phases (Jones, Vaught, and Weinrott, 1976): (1) change in level, (2) change in slope, and (3) presence or absence of drift or slope. The change of level from phase A to B is similar to but not identical with a change in mean values across phases. A change in level usually refers to a change at the point at which the intervention is made (i.e., inception of the B phase). A change in slope refers to a change in trend between (or among) phases. Drift or slope refers to whether or not there is a linear trend in the data in any phase. Graphically, no drift would be represented as a horizontal line or simply as no trend in the data.[8]

[8]Ordinarily, "drift" and "slope" are used interchangeably. These synonyms are employed in separate uses here for the purposes of describing the three properties of changes in time series data. However, the latter property might be referred to accurately as the presence or absence of slope.

Prior to considering different patterns of time series data that are likely to occur in operant investigations or are likely to obfuscate evaluation, a further clarification is required regarding the notion of level. Investigators employing intra-subject designs frequently refer to changes in means across phases. However, sources on time series analysis refer to changes in level (Box and Jenkins, 1970; Glass, Willson, and Gottman, 1974; Jones, Vaught, and Weinrott, 1976). While the notions overlap in some cases, they can be clearly distinguished. A change in *level* across phases usually refers to a discontinuity in the data at the point of intervention. A change in *mean* refers to a change in the "average" values across phases. A change in level across phases does not necessarily entail a change in level. However, a change in one *does* entail a change in the other when there is no drift or slope in the data in either phase.

One reason that the notion of level is not distinguished from the mean in applied research is that an attempt is made during baseline to obtain data with no drift (i.e., a "steady state"). However, baseline rates of behavior do not always conform to this ideal. A second reason that the mean is focused upon in applied research is that it relates directly to the goal of the intervention—namely, changing the overall rate of performance. Whether or not there is a change at the precise point of intervention (beginning of the B phase) is not necessarily crucial. In any case, the distinction between changes in mean and level is important in evaluating the results from time series analysis.

In time series analysis, a "significant" change in the mean across phases will be reflected either as a change in level or change in slope or both. For example, a very gradual change in behavior after the intervention might be detected as a change in slope but no change in level. What is commonly referred to as the mean is detected in a more analytic fashion in time series analysis by distinguishing level from slope. This distinction will become clearer by providing examples of data patterns using intra-subject replication designs.

To understand precisely what is accomplished by time series analysis, it is useful to examine different patterns of time series data over the initial A and B phases of an A-B-A-B design. (Obviously, the remarks made apply to changes across all phases.) Four different patterns are illustrated in Fig. 8-2. Figure 8-2a represents no change in level or slope and the absence of drift. Of course, this pattern of data does not represent a problem of evaluation because there is no hint of an effective intervention during the B phase. Figure 8-2b shows a change in level with no change in slope. There is no drift during baseline so the example is easy to evaluate visually. The pattern of data in Fig. 8-2c shows no change in level although there is a change in slope across phases. Although there is no change at the point of intervention (i.e., level), there is a mean change in performance across phases. The absence of drift during baseline makes evaluation of

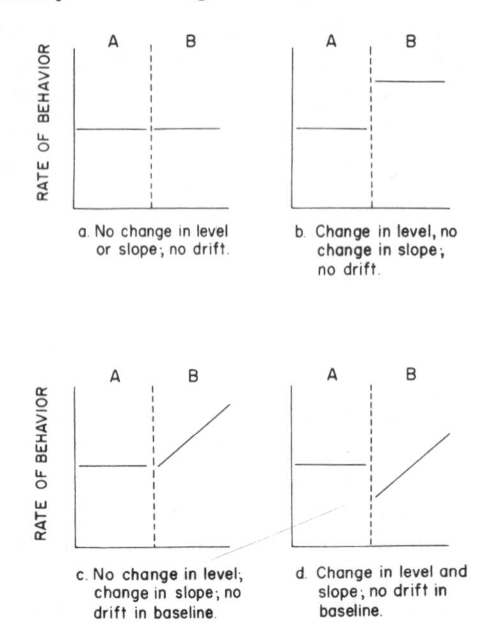

Fig. 8-2. Patterns of data for Baseline (A) and Intervention (B) phases in the intial portions of an A-B-A-B design.

this pattern of data relatively easy through visual inspection. Figure 8-2d indicates an interesting pattern of data. There is no drift in baseline, a change in level across phases (i.e., at the point of intervention), and a change in slope. The pattern is interesting because these data would reveal no "mean" change across phases with a conventional analysis. Yet, there is a change in level accompanied by a return of behavior to the previous level in baseline, a data pattern easily revealed by time series analysis.

Additional patterns of data are depicted in Fig. 8-3. These patterns are somewhat more difficult to evaluate than the previous examples, in part because of the presence of drift during baseline. Figure 8-3a reveals a change in level but no change in slope. Although there is drift, or an overall trend continued throughout the time series, the change in level is independent of this overall trend. Essentially, the intervention appears to have accelerated a behavior change pattern apparent during baseline. Figure 8-3b indicates no change in level or slope. The presence of drift reflects a continuous trend over time. Of course, the means of the two phases differ as distinct from a change in level. The mean differences across phases can be attributed solely to a continuous trend seen in baseline. (As noted earlier, this pattern of data would present a problem in interpreting the

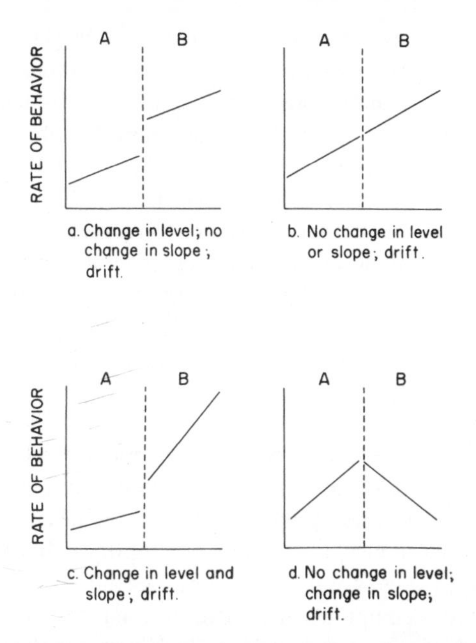

Fig. 8-3. Patterns of data for Baseline (A) and Intervention (B) phases in the initial portions of an A-B-A-B design.

results if a *t* test or ANOVA were used.) Time series analysis protects against interpreting the change in means as a significant change in patterns of data similar to those in Fig. 8-3b. Figure 8-3c shows both changes in level and slope. The presence of drift in baseline may make visual inspection of these data somewhat difficult. During baseline, behavior is changing in a "therapeutic" direction, a pattern that is accelerated during the intervention phase. Of course, in actual practice, the patterns portrayed in Fig. 8-3a, b, and c are unlikely to occur because operant methodology, outlined in earlier chapters, entails withholding an intervention when baseline data show a clear trend in the direction of therapeutic change.

Figure 8-3d shows no change in level (at the point of the intervention) and an obvious change in slope. Although there is drift, there is a change in the slope across phases. This pattern of data is worth noting because the level and mean do not change across phases. However, the change in slope in its own right might be of applied importance and worth evaluating. An example of such a data pattern might be the effects of some intervention to alter crime rates or mortality rates for a given disease. An accelerating trend appears to have been reversed by the intervention, although this is not yet reflected in changes of means.

The last four data patterns, and others that could be generated with the presence of drift and changes in slope, represent those situations in which visual inspection is likely to be ambiguous. These potentially problematic data patterns emphasize what time series analysis can accomplish. Time series analysis can determine changes in level and slope which, as obvious from the above examples, do not necessarily go together. Conventional tests might obfuscate that pattern of change that can be represented by separate analyses of level and slope. Similarly, visual inspection alone might be ambiguous not because of the reliability of the effect but because of under- or overestimating the effect of drift in the data.

Statistical analysis

The statistical analysis using time series techniques will be outlined briefly. The present section alone is not sufficient to employ the analysis. An overview is provided for two reasons. First, there are different models of time series analysis, each of which makes somewhat different assumptions about the nature of the data and provides different formal equations. Second, as explained below, time series analysis requires computational operations that are handled by computers. Several computer programs are available, which obviates the need for discussing computational details.[9] Thus, the present discussion will describe some general features of time series analysis.

The statistical analysis of time series evaluates change in light of the characteristics of the obtained data. Of course, a major characteristic of time series data often is serial dependency, discussed earlier. Time series analysis evaluates changes in the series (level and slope), taking into account the dependency in the data. Indeed, the evaluation of change in the series depends upon determining the precise nature of the dependency. Autocorrelations are employed as part of the analysis to determine the dependency—i.e., how much individual data points are influenced by (or "remember") past observations.

The purpose of time series analysis, of course, is to separate chance fluctuations from intervention effects. The analysis is accomplished by applying a

[9] For a detailed discussion of different time series models, conditions dictating their use, and examples of their application, the reader is referred to Box and Jenkins (1970), Glass (1972), and Glass, Willson, and Gottman (1974). The reader may also wish to consult Anderson (1971) for a detailed development of mathematical properties of time series. Computer programs have been made available by several sources including Glass, Willson, and Gottman (1974), Gottman (1973), Maguire and Glass (1967), Sween and Campbell (1965), and the University of Wisconsin (Univariate Time Series Analysis Using Methods of Box and Jenkins), and Synergy Incorporated in Washington, D. C. (Econometric Software Package).

model to the data to describe the structure of the specific time series. Different models are available that can be applied to the data (Box and Jenkins, 1970; Glass, Willson, and Gottman, 1974). Importantly, the adequacy of the model that is applied can be examined directly based upon how well it fits the data.

The *integrated (or summed) moving average* (IMA) model has served as a general form to represent time series data (Box and Jenkins, 1970; Box and Tiao, 1965). The model refers to the representation of each observation in a time series as a weighted sum of an infinite number of previous observations. There are different IMA models which, depending upon the structure of the obtained data, can represent a given time series. For present purposes, a simple model will be used as an example of time series analysis. The model is referred to as an *exponentially weighted moving averages* (EWMA) model and has been discussed elsewhere (Gottman, 1973; Gottman, McFall, and Barnett, 1969).[10]

The time series models transform the data to remove the serial dependency (e.g., so that the error terms are uncorrelated). As ordinarily is the case with data transformations (e.g., square root, logarithmic), the scores are altered so that they meet the properties required for statistical analysis (Jones, Vaught, and Weinrott, 1976). Similarly, a transformation can adjust for serial dependency so that statistical tests can be appropriately performed. Indeed, once the data are transformed, t tests for changes in level and slope are performed in the time series analysis.

The transformation of time series data is controlled by gamma (γ), a parameter important in the EWMA model.[11] The magnitude of gamma reflects the degree of autocorrelation in the obtained data. If there is no autocorrelation in the data, the observations in the time series are independent, and gamma equals zero. Usually there is serial dependency in the data, so a transformation based upon gamma is required. Values of gamma (which can range from zero to two) for a given time series are generated with the use of computer programs (Glass and Maguire, 1968). By iterating through the entire range of gammas (in increments of ·01), a maximum likelihood estimate is determined (i.e., that value of

[10] The EWMA is equivalent to a first-order IMA (1, 1) model, which indicates that a given observation (x_t) in the time series is influenced most by the previous observation (x_{t-1}); observations further away are less influential.

[11] The basic EWMA model transforms the data as follows:

$$\hat{x}_{t+1} = \gamma x_t + e_t,$$

where $\hat{x}_{t+1}$ = the predicted value of a given observation; γ = gamma, a constant which reflects the degree of dependence in the data; e_t = error of the prediction at time t. (See Gottmann, McFall, and Barnett, 1969.)

gamma which minimizes the residual sum of squares).[12] For each value of gamma that is iterated, t tests for changes in level and slope are provided. The investigator simply selects the value of gamma that minimizes the sum of the squared residuals, sometimes referred to as the maximum likelihood gamma (Gottman and Leiblum, 1974), which provides the separate t values. The t values test the null hypothesis that the series following or during the intervention (i.e., the B phase) can be represented by the baseline series with no statistically significant change in level or slope.

Example. Time series analyses are rarely reported in the behavior modification literature. When they are reported, often they do not provide the reader with great insight about the analysis because once the investigator adopts a given model, a computer handles the model evaluation, parameter estimation, and statistical tests (see Schnelle and Frank, 1974). Notwithstanding, various sources have provided analyses of applied behavior analysis investigations (Glass, Willson, and Gottman, 1974; Jones, Vaught, and Weinrott, 1976).

Jones, Vaught, and Reid (1975) and Jones, Vaught, and Weinrott (1976) reanalyzed several applied behavior analysis investigations using time series analysis. In one experiment (Hall, Fox, Willard, Goldsmith, Emerson, Owen, Davis, and Porcia, 1971, Exp. 6), inappropriate talking of children in a classroom setting was altered by reinforcing appropriate classroom behavior. The daily number of "talk-outs" for the 27 children was recorded throughout an A-B-A-B design. The results are plotted in Fig. 8-4. Considering the first two phases, the portion analyzed by Jones, Vaught, and Reid, it appears that "talk-outs" decreased substantially after baseline. The data appear to show not only a change in level but a decrease (downward slope) over the treatment phase.

These data were analyzed using the time series procedure, mentioned earlier. Initially, to point out the serial dependency in the data, the authors computed autocorrelations with different lags (Jones, Vaught, and Reid, 1975). Autocorrelations for lags 1, 2, 3, and 4 were ·96, ·94, ·92, and ·89, respectively. These large correlations reveal substantial dependency in the data, obviously making conventional analyses inappropriate. Using a computer program, the authors obtained the maximum likelihood gamma for the data, which provided separate t's for level and slope changes. The analyses revealed a significant change in level ($t = 3{\cdot}90$, $df = 39$, $p < {\cdot}01$) but not in slope across the first two phases. Although a change in trend during the B phase may appear upon visual inspection of Fig. 4, there was no statistically significant change. The apparent trend during the treatment phase can be parsimoniously interpreted as a "carry-

[12] The sum of the square of the deviations of the predicted value of the observations from the obtained value $\Sigma(\hat{x}_i - x_i)^2$ is computed for separate values of gamma. The sum of squares can be plotted as a function of gamma. The value of gamma that minimizes the sum of squares is selected because it best fits the data (given the maximum likelihood criterion).

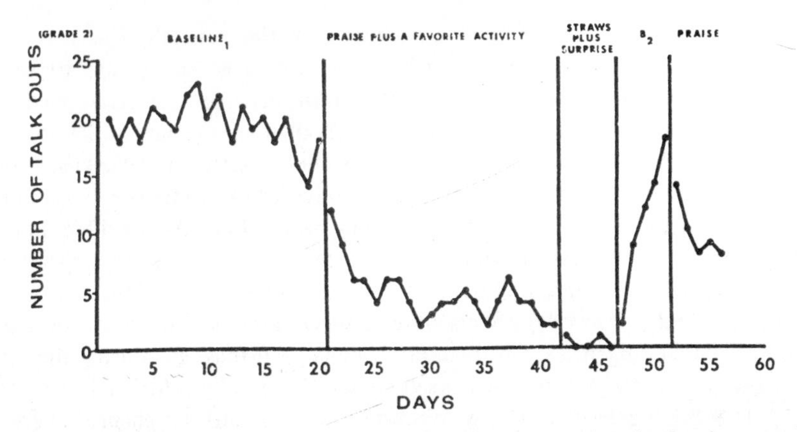

Fig. 8-4. A record of the daily number of "talk outs" in a second-grade class in a poverty area school. Baseline$_1$–before experimental conditions. Praise Plus a Favorite Activity–systematic praise and permission to engage in a favorite classroom activity contingent on not talking out. Straws Plus Surprise–systematic praise plus a token system (straws) backed by the promise of a surprise at the end of the week. B_2–withdrawal of reinforcement and reinstatement of attention to talking out. Praise–systematic teacher attention and praise for handraising and ignoring talking out. (Fig. 6, p. 147, from: Hall, R. V., Fox, R., Willard, D., Goldsmith, L., Emerson, M., Owen, M., Davis, F., and Porcia, E. The teacher as observer and experimenter in the modification of disputing and talking-out behaviors. *Journal of Applied Behavior Analysis*, 1971, **4**, 141-149. Copyright by Society for the Experimental Analysis of Behavior, Inc., reproduced by permission.)

over" effect of the trend during baseline. In any case, an important point is that differences might be obtained in visual and statistical criteria for evaluating the slope during treatment. Of course, the goal of the original study was to effect a change in level (and in means), which was achieved according to both statistical and visual criteria.

Jones, Vaught, and Reid (1975) point out the difference in analyzing the change in level in the above experiment with a "time-series" *t* and with a conventional *t* that does not account for serial dependency in the data. The latter *t* for the change in level of the first two phases of the data plotted in Fig. 8-4 would be 20·64, a value substantially larger than the *t* that accounted for dependency.

An important point from the analysis of the Hall, Fox, Willard, Goldsmith, Emerson, Owen, Davis, and Porcia (1971) experiment is that different conclusions might be reached in relying on visual and statistical criteria for evaluating change in trend. Visually, a change in slope may have been apparent although this was not statistically significant. This suggests that in some situations statistical evaluation might be more reliable than visual inspection as a criterion of change.

While statistical evaluation may be a more reliable criterion with a given pattern of data, there might be disagreement in stating generally that it is more reliable. In some cases, data that have been interpreted as "significant" by visual inspection have not achieved statistical significance in time series analysis. Indeed, in select cases visual inspection appears obvious, yet time series analysis does not reveal statistical significance. For example, Gottman (1973) used time series analysis to analyze data from a program designed to increase social interaction of a withdrawn boy (Harris, Wolf, and Baer, 1964). As is evident in Fig. 8-5 the program, evaluated in an A-B-A-B design, yielded results that appear dramatic by visual inspection. Indeed, across the first three phases of the design, the data points did not overlap. Social interaction clearly was controlled by alteration of the contingencies. Gottman (1973) analyzed these data using the integrated moving average model, discussed earlier. The results, which are presented in Table 8-1, indicate that there were no statistically significant changes in either level or slope across any adjacent phases. The *t* values did not even approach

TABLE 8-1. Time Series Analysis of Harris, Wolf, and Baer: Data presented by Bandura (1969) without statistical analysis showing lack of significance of shifts

Time periods	Student's *t* for shift in level	Student's *t* for shift in slope	*df*
Base line to interaction reinforced	·10	−·09	5
Interaction reinforced to solitary play reinforced	−·01	·00	4
Solitary play reinforced to interaction reinforced	·05	−·05	13

(Table 4, p. 96, from: Gottman, J. M. N-of-one and N-of-two research in psychotherapy. *Psychological Bulletin*, 1973, **80**, 93-105. Reproduced by permission.)

significance. Thus, what appears visually to be dramatic behavior change was not statistically significant.

It is not clear why the results failed to achieve statistical significance. It may be that the apparent trend during baseline attenuated the change during the intervention phase. However, this should have influenced only the *t* for a change in slope because there appeared to be discontinuity at the point of intervention—i.e., a change in level. Further, across subsequent phases, the data patterns appeared more obvious and statistical significance might be expected. The failure of these data to achieve statistical significance may have been due to the short durations of individual phases, making any conclusions from the time series analysis equivocal. This point will be discussed below.

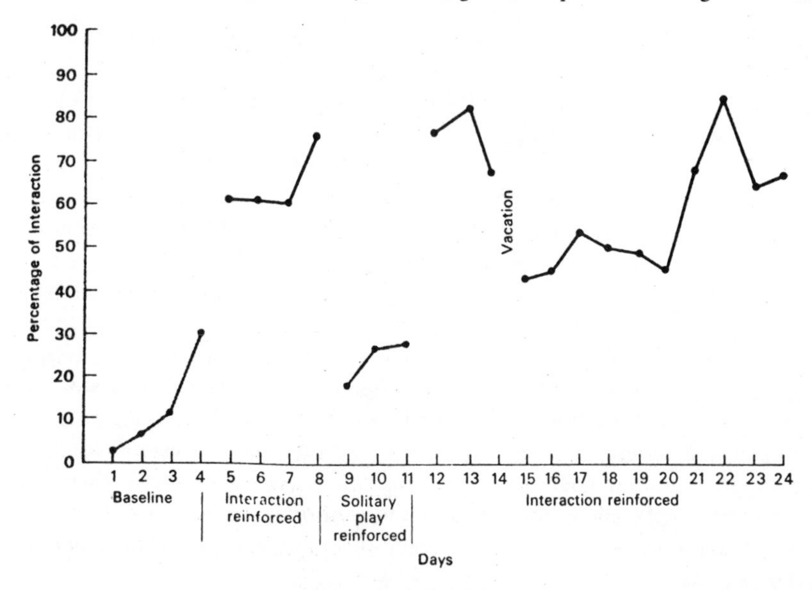

Fig. 8-5. Percentage of time a withdrawn boy spent in social interaction before treatment began, during periods when social behavior toward peers was positively reinforced, and during periods when teachers gave attention for solitary play. (From Harris, F. R., Wolf, M. M., and Baer, D. M. Effects of adult social reinforcement on child behavior. *Young Children*, 1964, **20**, 8-17. Copyright 1964, National Association for the Education of Young Children, 1834 Connecticut Avenue, N. W., Washington, D. C. 20009. Reproduced by permission.)

Considerations in using time series analysis

Among the available statistical analyses, time series analysis is to be recommended because of the manner in which serial dependency is handled. With variations of conventional analyses, the autocorrelation of time series data is either ignored, assumed to be present but disregarded, or recognized as a minimal problem and handled in a relatively cumbersome fashion. In contrast, time series analysis depends upon the serial dependency in the data, adjusts the specific time series in light of the specific dependency, and utilizes it for computing the statistical analyses. In addition, time series analysis assesses the process of change in data by separately evaluating level and slope.

There are a number of additional appealing features of time series analysis. Initially, the analysis is basically consistent with frequently employed statistics in that it can employ a maximum likelihood solution and closely resembles analysis of variance and multiple regression (Jones, Vaught, and Weinrott, 1976). An additional feature is that time series analysis can be readily extended to multiple baseline designs. Of course, multiple baseline designs constitute multiple time series, which can examine the effect of introducing an intervention across different individuals, behaviors, or situations at different points in time. (See Glass, Willson, and Gottman, 1974; Gottman, McFall, and Barnett, 1969.)

As noted by Jones, Vaught, and Weinrott (1976), time series analysis is not intended to replace visual examination of the data, if stable baselines are obtained and the effects of an intervention are immediate and marked. In such cases, statistical analyses are unnecessary. Yet there are a number of situations that may be quite suitable for time series analysis. Specifically, when there is a trend during baseline, time series analysis can evaluate subsequent changes, taking into account the initial trend. Indeed, time series analysis may reveal that an apparent trend in the data during baseline in fact is not a significant trend. Of course, individuals employing intra-subject designs attempt to achieve steady states during baseline that will facilitate subsequent evaluation. While this ideal has been obtained in a variety of situations, steady states are not a necessary requirement for experimentation and evaluation with time series analysis. In cases where there is an initial trend in the data and intervention is required prior to achieving a steady state, the effect of the intervention can be readily evaluated with time series analysis.

As a related point, Jones, Vaught, and Weinrott (1976) noted that as applied clinical research has extended into situations where experimental control is reduced (e.g., open-field settings), the variability in the data may be greater than in quasi-laboratory settings. Visual inspection may provide tenuous conclusions and other evaluation techniques such as time series analysis may be useful. Similarly, effecting change in many naturalistic settings may not be rapid and dramatic because of the experimental "noise" in the system. Statistical evaluation may be more useful in these instances than in situations in which experimental control is greater. Of course, whether social interventions will necessarily show less dramatic changes in behavior than interventions in closed settings (e.g., classrooms, hospital wards) and whether "noise" will restrict visual inspection as a criterion, remain empirical questions. A number of studies in settings in which control is minimal have shown unambiguous results by visual criteria (Kazdin, 1975a).

Potential limitations

Despite the utility and desirable features of time series analysis, various requirements of the analysis may delimit its widespread adoption for clinical investigations. Initially, time series models depend upon a relatively large number of data points to identify the model that best describes the data (Box and Jenkins, 1970). For experimental phases of short duration, reliable results might be difficult to obtain. The identification of the time series model that best describes the data depends upon determining the structure of the series. The structure, which is determined by autocorrelations, can be elucidated only when a sufficient number of data points are available within each phase so that autocorrelations of different lags can be computed. Indeed, as Glass, Willson, and Gottman (1974) have noted, the large sample size in a time series (i.e., the number of data points) is necessary primarily to account for the processes in the series (rather than to achieve statistical power–the usual reason for large Ns in statistical tests).

Another reason to obtain a large number of observations also pertains to autocorrelation. The degrees of freedom for a given number of dependent observations are slightly lower than would be the case with independent observations. Each observation does not present independent information, as described earlier.

Precisely what constitutes a "large number" of observations is not entirely specifiable because it depends in part upon the nature of the dependency, the variability in the data, and other parameters that characterize a given series. However, it should be noted that when researchers ordinarily discuss time series analyses usually they are considering a much longer series than is employed in applied clinical investigations. For example, various authors have noted that in practice at least 50 (Glass, Willson, and Gottman, 1974) and preferably 100 (Box and Jenkins, 1970) observations are required for estimating autocorrelations. Fewer observations can be used and provide for an increasingly conservative test of significance (Jones, Vaught, and Weinrott, 1976). As a rule of thumb, some investigators recommend at least ten data points in each phase (Jones, Vaught, and Reid, 1975). Applied clinical interventions often employ relatively short phases, lasting only a few days, to demonstrate the effect of an intervention on behavior. It is unclear precisely what the limitations are for the use of time series analysis with short phases. However, it is quite possible that statistical significance might not be obtainable from a given series because of too few data points.

Aside from meeting the requirements for time series analysis, a few characteristics of the procedures may temporarily limit extensive use of the analysis. Initially, there are a variety of time series analysis models. Different models can be applied to a given set of data. Since information about time series analysis is not widely disseminated, the conditions dictating the use of specific models may seem unclear. Fortunately, an important characteristic of time series analysis is that degree to which a given model fits the specific data can be tested directly as part of the data analysis (cf., Schnelle and Frank, 1974). Until different time series models are more frequently employed in applied research and the computer programs are widely disseminated, there may be a tendency to view the analyses as statistically esoteric. The availability of computer programs and excellent sources, which relate time series analysis directly to experimental designs discussed in previous chapters (Glass, Willson, and Gottman, 1974), should stimulate further use of the procedures.

8.7. RANDOMIZATION TESTS

Edgington (1967, 1969, 1972) has detailed the logic and use of randomization tests for the statistical evaluation of $N=1$ experiments. The tests are appropriate for time series data even where the observations of a given subject over time are serially dependent. The problem of dependency is ameliorated by the random assignment of different treatments (e.g., A and B treatments) to different measurement occasions.

In a single-subject experiment, randomization tests require that treatment is randomly assigned to occasions upon which response measurements are obtained. Procedurally, at least two "treatments" are required, one of which may be a baseline (A) and the other of which may be an intervention (B). The total number of treatment occasions (e.g., sessions or days) and the specific number of occasions of the total that each treatment will be administered must be specified. The appropriate number of A and B occurrences are randomly assigned to specific occasions. Treatment is administered as planned and observations on each treatment occasion constitute the data.

The null hypothesis is that the client's response is due to his performance on a particular occasion independently of the treatment that is administered on that occasion. The random assignment of treatments to occasions in effect randomly assigns responses of the subject to the treatments. The null hypothesis, given random assignment of treatments to occasions, assumes that the measurements of behavior that are obtained are the same as would have been obtained under any random assignment of treatments to occasions. Thus, the null hypothesis attributes differences between treatments to the "chance" assignment of one treatment rather than the other to particular occasions. To test the null hypothesis, the sampling distribution of differences between the treatments under every equally likely assignment of the same response measures to occasions of treatment A and B is computed. From this distribution, one can determine the probability of obtaining a difference between treatments as large as the one that was actually obtained.[13]

Example. Consider as an example an investigation designed to assess the effect of teacher praise on the attentive behavior of a disruptive student. To

[13] Although the randomization test for a difference between means is discussed throughout this section, there are a number of randomization tests including tests for comparisons of correlation, interactions, and others (see Edgington, 1969).

employ the randomization test for statistical evaluation, the investigator must plan in advance the number of days of the study and the number of days each of the two conditions will be administered. For the example and simplicity of computation, assume that the duration of the study is only 8 days. On each of these days, teacher and student behavior are observed. Also, the investigator wishes to compare the effects of "ordinary" classroom teaching (baseline or A "treatment") and contingent praise (intervention or B "treatment"). The investigator may decide to have an equal number of A and B days (although this is not essential). The condition (A or B) each day is determined randomly in advance of the study with, in our example, the restriction that A and B occur four times each. Consider the major dependent variable as the daily percentage of intervals that the student is attentive.

The predicted difference is that praise (treatment B) will be superior to baseline (treatment A). The one-tailed (directional) null hypothesis is that treatment B is more effective than treatment A.

Under the null hypothesis, any difference between means for the two treatments is due solely to the "chance" difference in performance on the occasions to which treatments were randomly assigned. To determine whether the differences are sufficient to reject the null hypothesis, the means are computed separately for responses under each treatment and the difference between these means is derived. Hypothetical data for the example appear in Table 8-2 (upper portion). The mean difference between A and B conditions is 43·75, as shown in the lower portion of the table. Whether this difference is statistically significant is determined by estimating the probability of obtaining response measurements

TABLE 8-2. Percentage of Intervals of Attentive Behavior Across Days and Treatments (Hypothetical Data)

Days							
A	B	A	A	B	A	B	B
20	50	15	10	60	25	65	70

Comparing Treatment Means

A	B
20	50
15	60
10	65
25	70
$\Sigma_A = 70$	$\Sigma_B = 245$
$\bar{X}_A = 17{\cdot}50$	$\bar{X}_B = 61{\cdot}25$

$$\bar{X}_B > \bar{X}_A = 43{\cdot}75$$

of this discrepant in the predicted direction when treatments have been assigned randomly. The random assignment of treatments to occasions makes equally probable several combinations of the *obtained data*. In fact, 70 combinations (8!/4!4!) are possible.

The question is what proportion of the 70 possible combinations of the obtained data for the eight treatment occasions would provide as large a difference between means as 43·75–i.e., the obtained data.[14] To determine this, a critical region of the sampling distribution must be computed based on the level of confidence chosen. For the example, the ·05 level of confidence is selected. The critical region of data combinations is ·05 x 70 or 3·5. (When a critical region is not an integer, it is recommended to select the larger whole number, which in the example would be 4 [Conover, 1971].) With a critical region of 4, the 4 combinations of the *obtained* data that are least likely under the null hypothesis must be found. Of course, the least likely combination of data is one in which the A and B mean difference in the predicted direction is the greatest possible, given the obtained data. For the example, the critical regions are the four combinations of the obtained data allocated under A and B treatments, which maximize the difference between the two means. Only those four data permutations that maximize the difference between A and B constitute the critical region. The permutations are obtained by recombining the obtained data in such a way that the differences between A and B treatments are maximally discrepant in the predicted direction. Once the combination of data in which the largest mean difference between A and B treatments has been obtained, the data are permuted to find the next least likely combination. The critical region consists of the *n* series of data combinations in the predicted direction that are the least likely to have occurred by chance (where n = the number of combinations that constitute the critical region).

Table 8-3 presents different permutations of the obtained data that reflect the four least likely combinations assuming the one-tailed null hypothesis to be true. Note that for each data combination the difference between means of treatments is computed. The question is whether the difference between means obtained in the original data is equal to or greater than one of the means obtained in the critical region. As is obvious, the obtained mean equals the most extreme

[14] The example selected here is devised for computational simplicity. It is unlikely that an investigator would be interested in only eight occasions for evaluating two different phases (baseline and intervention phases). In addition, it is also unlikely that the non-overlapping distributions of the magnitude included in the example would be subjected to statistical test. To view these points as criticism misses the point of the example. The computational procedures are illustrated here, not the conditions under which statistical tests should be used.

TABLE 8-3. Critical Region for the Obtained Data From the Hypothetical Example*

A				Total for A Occasions	$\bar{X}_A$	B				Total for B Occasions	$\bar{X}_B$	$\bar{X}_B > \bar{X}_A$
20	10	15	25	(70)	17·50	50	60	65	70	(245)	61·25	43·75
20	10	15	50	(95)	23·75	25	60	65	70	(220)	55·00	31·25
50	10	15	25	(100)	25·00	20	60	65	70	(215)	53·75	28·75
60	10	15	20	(105)	26·25	25	50	65	70	(210)	52·50	26·25

*All other combinations of the obtained data (allocated to A and B treatments) are not in the critical region using ·05 as a level of significance for a one-tail test.

value in the critical region that indicates a statistically significant effect ($p = 1/70$ or ·014). In fact, because the data points under treatments A and B did not overlap, there could be no other combination of these data that yields such an extreme mean difference between groups. When the data represent the least probable combination of data (given a one-tailed null hypothesis), the probability is 1 over the total number of data combinations possible.

In the above example, a one-tailed test was performed. For a two-tailed test, the critical region is at both ends (tails) of the distribution. The number of data combinations which constitute the critical region is unchanged for a given level of confidence. However, the number of combinations is divided among the two tails. Because of the division of the critical region in two tails, the probability level of an obtained mean difference is doubled. Thus, if the above example employed a two-tailed test rather than a one-tailed test, the probability level of the obtained difference would be ·028 rather than ·014 (see Edgington, 1969).

Considerations in using randomization tests

Statistical considerations. An advantage of randomization tests is that they do not rely on some of the assumptions of conventional tests. For example, randomization tests are distribution free and make no assumptions about populations (e.g., random sampling from a population, normality of the population distribution).

An interesting feature of randomization tests is the manner in which serial dependency is handled. Dependency is not a problem for these tests and indeed may be irrelevant (Edington, 1974). The model does not ignore that any given data point (response) may be affected by previous responses in the time series. However, the test is based upon the null hypothesis that there would be identical responses across measurement occasions if treatments were presented in a different order or at different times. Every order of presenting treatments should lead to an identical pattern of data (if the null hypothesis is true). Serial dependency in the data will be a constant for the particular sequence of observations. The issue is whether random assignments of treatment designations A and B to the obtained data would be likely to provide such a large difference between means as the obtained difference. Serial dependency does not affect the estimation of the sampling distribution of the statistic from which this inference is drawn. The statistic depends upon the hypothesis that each particular "true" response of the subject would be the same if treatment were presented in any other order. Although a response may well be affected by previous responses, as indicated by autocorrelation, this does not affect the estimation of probability. While serial dependency may occur for a given sequence of responses, it is fixed for that combination of data and therefore does not affect estimations of probability of obtaining the particular set of data given all possible combinations.

Restrictions and practical considerations. A number of restrictions need to be considered that might delimit the utility of randomization tests. Not all of these considerations are unique to randomization tests but they are especially salient with these tests. First, the use of randomization tests for single subject research requires that treatment effects are reversible. Of course, there are a number of occasions in which treatment effects might not be reversible or, if they are, it is undesirable to attempt reversals. In these situations, the random assignment of treatment and non-treatment occasions is not feasible.

The problem of irreversible effects is usually discussed in A-B-A-B designs when behavior does not return to baseline. However, there is a related problem which is not always discussed—namely, the effects of sequentially presenting treatments (or phases). Given that different treatment sessions must be given to the same subject, it is possible that one treatment may influence performance under other treatments. When the response to one treatment is affected by the preceding treatment, statistical inferences about temporally isolated treatments cannot be made. Of course, this problem is not unique to single case experiments; it also exists whenever several subjects are employed across diverse treatments. Balancing the order of treatment presentations within a subject (or across subjects) does not necessarily resolve carryover effects of a given treatment (Lindquist, 1953; Winer, 1971).

A second potential limitation is that when different treatments are being evaluated, such as baseline (A), $treatment_1$ (B), $treatment_2$ (C), it may not be feasible to allow conditions to vary each session or day. For example, considering A and B treatments (baseline, $treatment_1$), where B is token economy, it is not easy to implement a token economy for one day while removing it on the next day. Yet, alternation of conditions is likely to result from a random assignment of A and B treatments to days. In applied clinical research, experimental conditions often are sufficiently complex that they cannot be implemented for a short period only. This would seem to delimit the use of randomization tests. However, there is a readily available solution (Edgington, 1974).

A and B conditions, when implemented, can be conducted for a fixed block of time (e.g., 3 days or a week). Whenever A is implemented, it occurs for 3 consecutive days, and whenever B is assigned it is continued for 3 days. A and B are still assigned to occasions in a random order. However, when one treatment is assigned, it is continued for the entire (fixed) block of time. The result of this procedure is that the minimum length of a given phase is the length of the fixed period (e.g., 3 days). Moreover, if treatment (A or B) is randomly assigned to two consecutive periods, the length of the "phase" is increased to 6 days. Thus, by using fixed periods involving blocks of time, treatments can still be assigned randomly to occasions. The mean for a given block of time for a treatment serves as the observation. Thus, treatments A and B can each be assigned to occasions (where A or B is a fixed block of 3 days). If A and B each were assigned to occasions five different times, there would be 15 days of each condition. For the randomization test, the means for 3 day blocks would be used. So, in fact, the sampling distribution of the statistic would be based on the number of occasions (10) upon which A and B were assigned to occasions, and the number of assigned occasions for each treatment (i.e., $10!/5!5!$).

The net effect of this variation of the randomization test is that the investigator can use blocks of time for given treatments as long as the occasions to which treatments are assigned remain random and the number of occasions in which treatment is assigned serves as the source for computing the sampling distribution (i.e., the mean score for a given treatment block). This solution resolves the practical problem of reversing treatment conditions frequently in the design. However, it does introduce another problem.

Because a block of days of treatment (e.g., A) counts as only one occasion in which treatment is assigned, several blocks will be required to achieve a relatively large number of occasions. A small number of occasions may restrict the possibility of obtaining statistically significant effects when treatments differ in their effects. For example, if the total number of occasions of treatments is five (two occasions of A and three occasions of B), the total combinations of the obtained data are $5!/2!3!$ or 10. Even if the obtained mean difference between A and B treatments is the greatest possible, the probability of achieving this difference is only 1/10 or $p = \cdot 10$, not significant at conventional levels of confidence. To increase the power of the test, a larger number of treatment occasions is desirable. In short, to achieve a larger number of occasions in which A and B treatments are introduced, the investigation would need to be longer if blocks of time are used than would otherwise be the case.

Another factor relevant for considering use of randomization tests pertains to the computation of the critical region for ascertaining whether a particular set of data is statistically significant at a given level of significance. The randomization tests are based upon the sampling distribution of the statistic employed (e.g., difference between means). At a given level of significance, the investigator must compute the number of different ways in which the obtained data could result from random assignment of treatment conditions. In practice, the technique is useful when there is a small number of occasions on which treatment was applied. When the number of occasions for assigning A and B treatments exceeds 10 or 15, even obtaining the possible arrangements of the data on a computer become monumental (Conover, 1971; Edgington, 1969). In applied clinical research, applications of A and B treatments usually occur for several sessions so computations estimating the complete sampling distribution from the obtained data would be cumbersome (cf. Edgington and Strain, 1973). Edgington (1969, pp. 152-155) has suggested the use of "approximate randomization tests" which do not require computation of the entire sampling distribution. However, convenient approximations are readily available.

Convenient approximations to randomization distributions

Perhaps the greatest problem in using randomization tests is carrying out the computations when the number of occasions on which A and B treatments are administered becomes large. Fortunately, there are convenient (and familiar) approximations to the test. The approximations depend upon the same conditions of the randomization tests–namely, random assignment of treatments to occasions. The approximations include the familiar t test and, in a case of more than two treatments, ANOVA.

It is important to note that the use of a t does not refer to quite the same t described in previous sections. Although the formula is computationally identical, there are some important differences conceptually. In the previous discussion of the conventional t, serial dependency in the time series made the test inappropriate. In the present use of t as an approximation, dependency is not a problem. Even if the observations themselves are not independent but the treatments are applied in the random order across all occasions, the t (or F) distribution upon which the published table is based provides a close randomization distribution (Box and Tiao, 1965; Edgington, 1969; Moses, 1952).

Thus, data given in the example provided earlier (Table 8-2) could be readily tested with a t test for independent groups with the degrees of freedom based upon the number of A plus B occasions (i.e., $df = n_1 + n_2 - 2$) (Edgington, 1972). The data in the above example yield the value of $t = 8{\cdot}17$ ($df = 6$) with a probability value ($p < {\cdot}001$), which is less than the probability attained with the exact analysis from the randomization test ($p = {\cdot}014$). Analysis of variance would be an appropriate approximation with three or more treatments (including baseline as a treatment) that are randomly assigned to occasions. As with the t, the degrees of freedom are computed in the usual fashion.

An alternative to the use of the t test to approximate the randomization distribution is the Mann–Whitney U test. To employ this test, the A and B data points are ranked from 1 to the number of treatment occasions without reference to the treatment conditions from which each value is derived. The null hypothesis of no difference between treatments may be rejected if the ranks associated with one treatment tend to be larger than the values of the other sample. The distribution from which this determination is made is available in various tables (Conover, 1971) and need not be computed for each set of data (unless A plus B occasions are relatively large, such as over 20). The Mann–Whitney U is a convenient test that may be used in place of the t given that treatments are assigned randomly to occasions. This test is commonly employed and has been described in several other sources (e.g., Conover, 1971; Kirk, 1968; Siegel, 1956).

Of course, the ability of familiar and computationally simple tests to approximate randomization distributions makes the use of randomization statistics more feasible than the exact analysis. The major problems delimiting use of randomization tests are pragmatic–i.e., randomly assigning treatments to occasions and alternating treatment conditions of short durations.

8.8. THE R_n STATISTIC

Revusky (1967) proposed a statistical test (R_n) to evaluate data obtained in multiple baseline designs. The test was proposed as suitable for situations where treatment effects may be irreversible and the usual A-B-A-B design would be unsuitable. The statistic would be appropriate for multiple baseline designs when data are collected across different individuals, behaviors, or situations, as outlined earlier.

Consider the case where multiple baseline data are gathered across different individuals. Baseline data are gathered for given behavior separately across different clients. Treatment is introduced to each individual at different points in time. In the ordinary use of this design, there is no rule governing the order in which individuals receive treatment. In the use of the statistic R_n proposed by Revusky, it is important that individuals are subjected to the intervention one at a time in *random* order. The performance of an individual subject who receives treatment is compared with "control" subjects for whom treatment has not yet been introduced.

The statistical comparison is achieved by ranking scores of each individual at the point when treatment is introduced for any one of the subjects. Each individual is considered as a subexperiment. When B is introduced for a subject, the performance of all clients (including those for whom treatment is withheld) is ranked. The sum of the ranks across all subexperiments (i.e., each time the treatment is introduced) constitutes the statistic R_n.

Because the client for whom the treatment is assigned at a given point in time is randomly determined, it can be assumed that any combination of ranks at the point of intervention is equally probably if the intervention has no effect. If the behavior of each client changes systematically at the point of intervention, the

TABLE 8-4. Minutes of Work Performance of Five Psychiatric Patients in a Multiple Baseline Design (Hypothetical Data).

	Days									
Patients	1	2	3	4	5	6	7	8	9	10
1	45	30	35	50	40	30*a*	70*b*			
2	60	75	80	60	50	70*a*	50*a*	65*a*	80*b*	
3	20	20	25	10	30	80*b*				
4	55	60	40	45	50	40*a*	75*a*	90*b*		
5	30	25	20	30	20	30*a*	30*a*	40*a*	35*a*	50*b*
					Ranks =	1	2	1	1	1 $\Sigma R = 6$

a = control or baseline days, *b* = experimental or intervention point for a patient. Days 1 through 5 serve as baseline (*a*) days for all subjects and are unmarked.

ranks will reflect this for each subexperiment. Using the R_n statistic, the minimum requirement to detect a difference at the ·05 level is four subjects. (If a multiple baseline design were conducted for one subject across behaviors or situations, a minimum of four behaviors or occasions would be required.) (For a discussion of the probability function upon which R_n is based, the reader is referred to Revusky [1967].)

Example. Application of the ranking procedure can be seen in a hypothetical example employing a multiple baseline design. Consider the purpose of the experimental intervention to increase the amount of time that five psychiatric patients perform work at jobs in the hospital. To fulfill the requirements of the multiple baseline design, data are gathered for the target response separately across each individual. Treatment is introduced to different patients at different points in time.

Table 8-4 provides hypothetical data on the number of minutes worked across 10 days. As is evident from the table, baseline was in effect for 5 days for all patients. On the sixth day, one patient was *randomly* selected to receive the intervention (e.g., contingent token reinforcement for completing work). As indicated in the table, on Day 6, Patient 3 was assigned the experimental intervention (B), whereas all the patients continued under baseline or control (A) conditions. On successive days, a different patient was exposed to the intervention. The patient who received the intervention was always randomly determined, an essential feature for the use of R_n.

The ranked test is applied to each subexperiment (i.e., point at which treatment was introduced). On each occasion that treatment was introduced, which included Days 6 through 10 in the example, the patients were ranked. The lowest rank is given to the patient who has the highest score (if the predicted direction of change is an increase in behavior). In the example, the patient with the

TABLE 8-5. Values for Significance of R_n

(a)
Maximum values of R_n significant at the indicated one-tail probability levels when the experimental scores tend to be smaller than the control scores.

No. of Subjects	Significance Level 0.05	0.025	0.02	0.01	0.005
4	4				
5	6	5	5	5	
6	8	7	7	7	6
7	11	10	10	9	8
8	14	13	13	12	11
9	18	17	16	15	14
10	22	21	20	19	18
11	27	25	24	23	22
12	32	30	29	27	26

(b)
Minimum values of R_n significant at the indicated one-tail probability levels when the experimental scores tend to be larger than the control scores.

No. of Subjects	Significance Level 0.05	0.025	0.02	0.01	0.005
4	10				
5	14	15	15	15	
6	19	20	20	20	21
7	24	25	25	26	27
8	30	31	31	32	33
9	36	37	38	39	40
10	43	44	45	46	47
11	50	52	53	54	55
12	58	60	61	63	64

Each table provides significance for a one-tailed test. (See text for explanation.)

(Table 5a, p. 323, and Table 5b, p. 325, from: Revusky, S. H. Some statistical treatments compatible with individual organism methodology. *Journal of the Experimental Analysis of Behavior*, 1967, **10**, 319-330. Copyright 1967 by Society for the Experimental Analysis of Behavior, Inc., reproduced by permission.)

greatest duration of work on the occasion treatment was introduced would be assigned the rank of 1. When treatment is introduced for the first subject, all subjects are ranked. When treatment is introduced on subsequent occasions, all subjects *except* those who previously received the intervention are ranked.

Although several subjects are ranked on an occasion in which treatment is introduced, all ranks are *not* used for the R_n statistic. On any given occasion, only the rank for the subject for whom treatment was introduced is employed. The ranks for these subjects at the point at which treatment was introduced are summed across occasions. If treatment is not effective, the ranks should be randomly distributed because each rank (of those ranks possible on a given occasion) has an equal probability of occurring. However, if there is a systematic effect of treatment, the point of intervention should result in low ranks for each subject at the point of intervention (assuming the lowest rank was assigned to the most extreme score in the direction of predicted change).

As can be seen in Table 8-4, hypothetical data for the example show that the subject who received the intervention at a given point in time, with the exception of Subject 1, received the lowest rank (i.e., 1) for performance on that occasion. Summing the ranks across all subjects who were exposed to the intervention yields $R_n = 6$. Tables that can be used to determine the significance of ranks for analyses employing different numbers of subjects have been provided by Revusky and are reproduced in Table 8-5. Each table provides a one-tailed test for R_n. (If a two-tailed test is desired, the probability level in the table selected is doubled.) Which table should be consulted depends upon the predicted direction of behavior change and the order in which ranks are applied–i.e., whether the lowest or highest numerical value is ranked as a 1. If the most extreme score in the predicted direction (high or low) is always given the lowest rank (i.e., 1), then Table 8-5a is all that is needed.[15] To return to the above example, $R_n = 6$ for five subjects (one-tailed test) is equal to the tabled value (Table 8-5a) required for the ·05 level. Thus, the data in the hypothetical example permit rejection of the null hypothesis of no treatment effect.

Considerations in using R_n

Use of transformations. If absolute levels of raw scores on the dependent measure differ dramatically among subjects during baseline, the data may have to be transformed. High or low absolute levels of the response across individuals may obscure directional changes in performance. When treatment is introduced for a given subject, the individual may show a dramatic change. Yet the level of the response may not approach the level of another subject who remains in baseline conditions and shows no systematic change. For example, in Table 8-4 compare the hypothetical performance of Patients 2 and 5. The performance of the target behavior of Patient 2 was higher during baseline than was the performance of Patient 5 when treatment was introduced. Had treatment been introduced to Subject 5 prior to Subject 2, the rank assigned to Subject 5 would not have been as low as it was in the example. This would have been an artifact of the differences in absolute levels of performance of the subjects rather than of the ineffectiveness of the intervention.

To ameliorate this potential problem, the raw scores of all subjects can be transformed. The only restriction in using a transformation is that it does not alter the probability that a subject for whom treatment is introduced can (theoretically) assume any rank. A simple transformation that corrects for differences

[15]Revusky posed that the lowest numerical value at the point of intervention be assigned the rank of 1. To determine the significance of R_n, separate tables are used depending on whether the effect of the intervention was expected to decrease or increase the value of the scores (and whether R_n would be expected to be low or high, respectively). The present procedure of assigning ranks–namely, giving the lowest rank to the score that shows the most extreme numerical value in the predicted direction–allows use of R_n with only one table. The difference in assigning ranks in no way affects the test and was selected as a matter of convenience.

in initial baseline is appropriate (Revusky, 1967); for example, consider the use of:

$$\frac{B_i - \bar{A}_i}{\bar{A}_i}$$

where B_i = performance level for subject i when the experimental intervention is introduced, and $\bar{A}_i$ = the mean performance across all baseline days for subject i. This transformation is equivalent to a change in percentage from baseline when treatment is introduced. The raw scores for each subject are transformed on each occasion when the intervention is introduced for any one subject.

As practical matter, it might be useful to routinely employ this simple transformation because statistical sensitivity of the ranks depends upon the relative standing of the treatment subjects to all the subjects that have not received treatment. Moreover, the transformation probably is essential if the test is used in a multiple baseline design across several behaviors of a given subject. Different behaviors for a given subject may differ dramatically in absolute levels and successful comparisons might not be possible without a data transformation.

Statistical vs. treatment considerations. The multiple baseline design, as employed in Table 8-4, shows that treatment is introduced for one subject at a time across subjects. For statistical purposes, all that is required are the data for baseline and the point at which treatment is introduced. It may be possible to show statistically significant effects of treatment following only one session (e.g., day) of treatment for a subject.

In multiple baseline designs, when treatment is introduced, it is usually continued until the end of the experiment (cf. Hall, Fox, Willard, Goldsmith, Emerson, Owen, Davis, and Porcia, 1971). Yet, for use of R_n, as described above, all the available data might not be employed for statistical evaluation. Of course, as indicated below, it might be useful to utilize more data for each subject during the intervention than a single session during which treatment has been introduced.

The statistic depends upon the evaluating behavior of the target subject at the point of introducing the intervention (i.e., initial session or day). In some cases, it may not be meaningful to complete rankings across all subjects on the first day of intervention. For example, extinction might be used to suppress deviant behaviors of obstreperous children. Baseline data may be gathered across four different children. The child who receives the intervention first would be assigned randomly, to meet the requirements of R_n. Extinction characteristically operates on behavior in a gradual fashion. Thus, one might not find an immediate effect of the intervention. Indeed, the beginning of extinction may be accompanied by a burst of deviant responses (Kazdin, 1975a). Assigning the ranks to all individuals on a given day when the intervention is first introduced may not show an effect. In cases where the intervention is not expected to show an immediate effect, an alternative procedure for assigning ranks is available.

Initially, treatment may be evaluated on the mean performance on a given individual for several days after treatment is introduced. For example, treatment could be introduced for a given individual while being withheld from others for a week. The data used for the ranking procedure might be the mean of the intervention period across several days. Using means across days is likely to provide a more stable estimate of actual performance, as noted in the discussion of intrasubject averaging in a previous chapter.

There are further considerations in using the above procedure for handling the data. First, the duration employed for evaluating treatment changes within subjects should be specified in advance. If the treatment effects are predicted to require a certain period of time to emerge, the precise number of days (or some conservative estimate) should be specified. Second, the duration of introducing a treatment effect should be constant across all subjects in the computation of means for the ranking. These two features ensure that randomness will not be influenced by *post hoc* treatment of the data. Such treatment might capitalize on "chance" fluctuations.

8.9. THE SPLIT MIDDLE METHOD OF TREND ESTIMATION

The split middle technique provides a method of describing the rate of behavior change over time for a single individual or group (White, 1971, 1972, 1974). The technique is designed to reveal a linear trend in the data, to characterize present performance, and to make predictions about future performance. By describing the rate of behavior change, the likelihood of an individual achieving a behavioral goal can be determined. Decisions about the adequacy of a given intervention can be made in light of the subject's projected performance at some point in the future. The method is not intended as a tool for determining statistical significance of changes across phases, as in an A-B-A-B design. However, statistical tests have been used (White, 1972) and will be mentioned below.

Most of the work on the split middle method has employed rate of behavior as the dependent measure in evaluating behavior change. Rate of behavior as a measure has a number of advantages (cf., Bijou, Peterson, Harris, Allen, and Johnston, 1969; Skinner, 1950). For the purpose of detecting trends in the data, rate is a particularly sensitive measure because it has no theoretical upper limit (which might create an "artificial ceiling") (White, 1972).

A special chart is used to plot the rate of behavior graphically for the split middle technique. The chart is in semilog units, which is a format selected

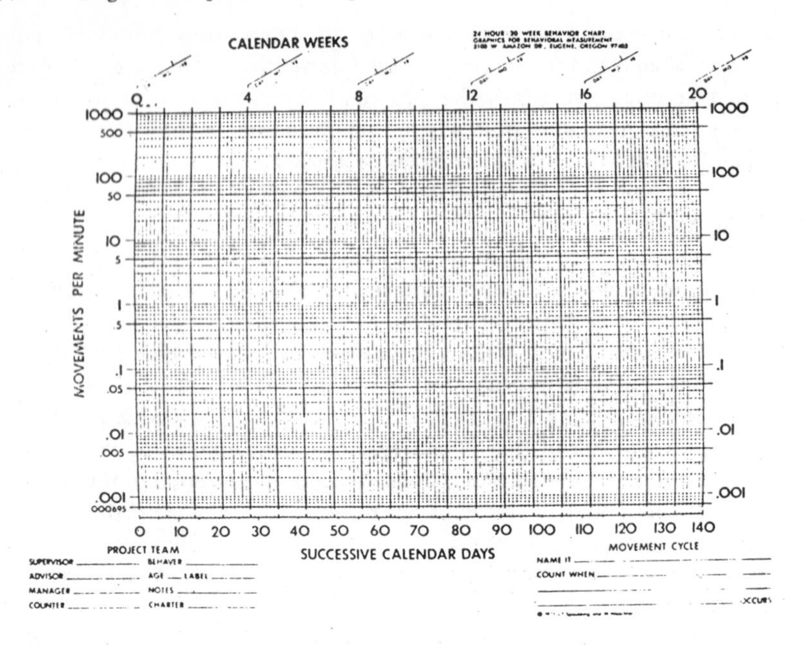

Fig. 8-6. Semilog chart to plot rate of behavior (movements) for split middle trend estimation technique.

because of its utility in achieving predictive validity in estimating linear trends and the ease with which it can be employed by practitioners (White, 1974). (See Figure 8-6 for the chart.)[16] The chart accepts data for behaviors with extremely high or low rates. The rates of behavior (sometimes referred to as movements on the chart) can vary from ·000695 per minute (i.e., one every 24 hours) to 1000 per minute. (The chart also allows 20 weeks of behavior rates to be recorded.) The wide range of rates that can be accepted by the chart facilitates prediction because virtually no data are "lost" due to ceiling and floor effects.

Once the data are plotted, the split middle technique estimates the slope or "line of progress." The line of progress points in the direction of behavior change and indicates the rate of change. This line of progress is also referred to as a "celeration time," a term that derived from the notions of acceleration (if the line of progress is ascending) and deceleration (if the line of progress is descending). The celeration line predicts where the "subject is going" with respect to the target behavior and which goals are likely to be achieved.

[16]This chart has been developed by Behavior Research Company, Kansas City, Kansas 66103. Different versions of the chart are available.

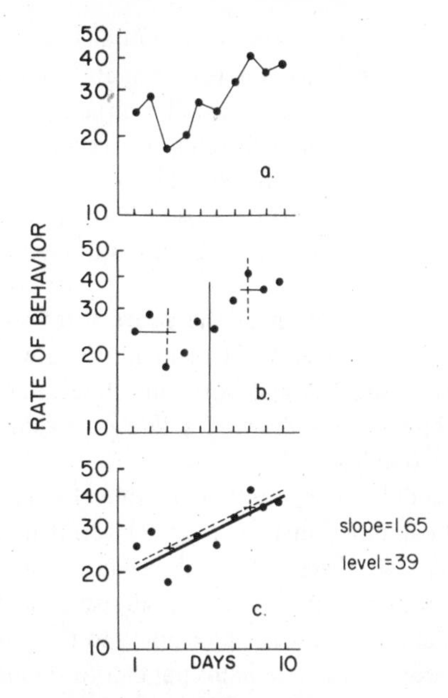

Fig. 8-7. Hypothetical data during one phase of an intra-subject replication design (a), with steps to determine the median data points in each half of the phase (b), and with the original (dotted) and adjusted (solid) celeration line (c).

Calculation of the celeration line

The computations required for calculation of the celeration line are simple and straightforward. Consider the hypothetical data in Fig. 8-7. In Fig. 8-7a, the data are plotted in a magnified portion of the special chart referred to earlier. The portion of the chart that is magnified represents one phase of an A-B-A-B design. For computational purposes, one phase is selected. In practice, a celeration line is drawn separately for each phase.

The initial step in obtaining a celeration line is to divide the phase in half by drawing a vertical line at the median number of session (days). The second step is to divide each of these halves in half again. (When there is an odd number of days, the vertical line is drawn through the data point that is on the median day rather than between two data points.) The dividing lines will always result in an equal number of points on each side of the division. The next step is to determine the median rate of performance for the first and second halves of the phase.

Two potentially confusing points should be resolved. First, although the sessions are divided into quarters, only the first division (halves) is employed at this stage. Second, the median data value within each half of the sessions is selected. These medians are based on the ordinate (dependent variable values) rather than the abscissa (number of days). To obtain the data point that is the median within each half, one merely counts from the bottom (ordinate) up toward the top data point for each half. The data point that constitutes the median value within each half is selected. A horizontal line is drawn through the median at each half of the phase until the line intersects with the vertical line dividing each half.

Figure 8-7b shows a completion of the above three steps, viz., a division of the data into quarters and selection of median values within each half. Within each half of the data, a vertical and horizontal line intersect. The next step is finding the slope, which entails drawing a line connecting the points of intersection between the two halves.

The final step is to determine whether the line that results "splits" all of the data–i.e., is the "split middle" line or slope. The split middle slope is that line that is situated so that 50 percent of the data fall on or above the line and 50 percent fall on or below the line. The line is adjusted to divide the data in this fashion. In practice the line is moved up or down to the point at which all of the data are divided. The adjusted line remains parallel to the original line.

Figure 8-7c shows the original line (dotted) and the line after it has been adjusted to achieve the "split middle" slope (solid). Note that the original line did not divide the data so that an equal number of points fell above and below the line. The adjustment achieves this "middle" slope by altering the level of the line (and not the slope). (In some cases, the original line may not have to be adjusted.)

Expressing the slope and level and evaluating change

The celeration line expresses the rate of behavior change, which can also be expressed numerically. White (1974) has used as the basis of calculating the rate of change per week (7 days), although any time period that might be more meaningful for a given situation can be employed. To calculate the rate of change, a point on the celeration line (day_x) that passes through a given value on the ordinate is determined. The data value on the ordinate for the celeration line 7 days later (i.e., day_{x+7}) is obtained. To compute the rate of change, the numerically larger value (either day_x or day_{x+7}) is divided by the smaller value. If the celeration line is accelerating, the resulting number is labeled with a multiplication sign (x); if the line is decelerating, the number is labeled with a division sign (÷). The procedure can be applied to the data in Fig. 8-7c. At Day 1, the celeration line is at 20. Seven days later, the line is at approximately 33. Applying the above computations, the ratio for the rate of change is 1·65. Because the celeration is accelerating, this indicates that the average rate of responding for a given week is 1·65 times greater than it was for the prior week. The ratio merely expresses the slope of the line.

The level of the slope can be expressed by noting the level of the celeration line on the last day of the phase. In the above example, the level is approximately 39. When separate phases are evaluated (e.g., baseline and intervention), the levels of the celeration lines refer to the last day of the first phase and the first day of the second phase, as will be discussed below.

For each phase in the experimental design, separate celeration lines are drawn. The slope of each line is expressed numerically. The change across phases is evaluated by comparing the levels and slopes. Consider hypothetical data for A and B phases, each with their separate celeration line, in Fig. 8-8. To estimate the change in level, a comparison is made between the last data point in baseline (approximately 22) and the first data point during the intervention (approximately 28). The larger value is divided by the smaller value, yielding a ratio of 1·27. A × or ÷ sign is used to denote an increase or decrease in behavior as a function of the intervention, respectively. The ratio merely expresses how much higher (or lower) the intersection of the different celeration lines are.

Similarly, for a change in slope, the larger slope is divided by the smaller slope, yielding a value in the example of 1·52.[17] (A × or ÷ sign is used to indicate whether the intervention slope was greater or smaller than the baseline slope, respectively.) The change in level and slope summarizes the differences in performance across phases.

Statistical analysis

It should be reiterated that the split middle procedure has been advocated as a technique to describe the process of change in an individual's behavior rather than as a tool to assess statistical significance. However, statistical significance of change across phases can be evaluated once the celeration lines are determined.

To determine whether there is a statistically significant change in behavior across phases, a simple statistical test has been proposed (White, 1972). Again, consider change across A and B phases in an A-B-A-B design. The null hypothesis upon which the test is made is that there is no change in performance across A and B phases. If this hypothesis is true, then the celeration line of the baseline phase should be a valid estimate of the celeration line of the intervention phase. Assuming the intervention had no effect, the split middle slope of baseline

[17] If the slopes are in opposite directions across phases (i.e., one accelerating and one decelerating), the values are multiplied.

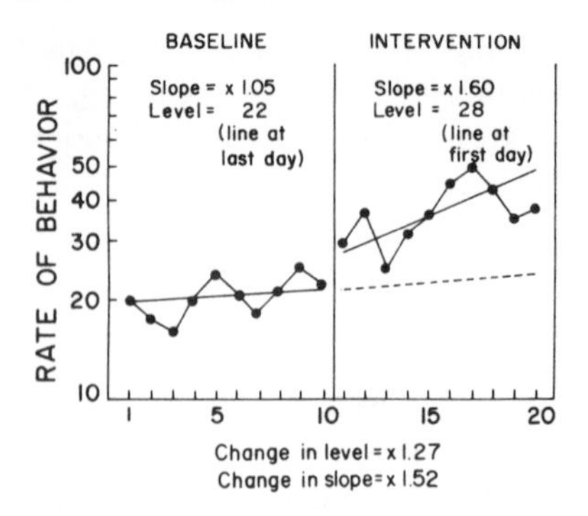

Fig. 8-8. Hypothetical data across Baseline (A) and Intervention (B) phases with separate celeration lines for each phase.

should be the split middle slope of the intervention phase, as well. Thus, 50 percent of the data in the intervention or B phase should fall on or above and 50 percent of the data should fall on or below the slope of baseline when that slope is *projected* into the intervention phase. To complete the statistical test, the slope of baseline is extended through the B phase. Figure 8-8 shows the extension of the celeration line of baseline for the hypothetical set of data. The probability of a data point during the B phase falling above the projected slope of baseline is 50 percent (i.e., $p = \cdot 5$) given the null hypothesis. A binomial test is recommended to determine whether the number of data points (x number of data points) that are above the projected slope is of a sufficiently low probability to reject the null hypothesis.[18] Employing the above procedure to the hypothetical data in Fig. 8-8, there are ten out of ten data points during phase B that fall above the projected slope of baseline. Applying the binomial test to determine the probability of obtaining ten data points above the slope $[p = \binom{10}{10}\frac{1}{2}^{10}]$ yields a value of $p < \cdot 001$. Thus, the null hypothesis that the data during the intervention phase can be represented by the slope during the previous phase is rejected. The results indicated that the data during phase B are significantly different from the data during the A phase. Of course, the results do not reveal

[18]The binomial applied to the split middle slope test would be the probability of attaining x data points above the projected slope:

$$f(x) = \binom{n}{x} p^x q^{n-x} \quad \text{(or simply } \binom{n}{x} p^n\text{)},$$

where n = the number of total data points in phase B; x = the number of data points above (or below) the projected slope; $p = q = \cdot 5$ by definition of the split middle slope; p and q equal the probability of data points appearing above or below the slope, given the null hypothesis.

whether changes resulted primarily from an alteration in the slope or level of performance but only whether there is an overall difference across phases.[19]

Considerations in using the split middle technique

Prediction. The main purpose of the split middle technique is predictive. The method has been emphasized as a clinical tool to determine whether behavior is changing at a sufficient rate relative to some predetermined criterion. If the data suggest that the goal will not be achieved, the program can be altered.

Since a major purpose of the technique is to predict behavior rather than to determine statistical significance of change, it is appropriate to examine the extent to which this purpose is adequately achieved. White (1974) presented data based upon "several thousand" analyses of classroom performance. The analyses determined the accuracy of predicting behavior using the split middle procedure at different points in the future. As might be expected, the extent to which the predictions approximated the actual data depended upon the number of data points upon which the prediction was based and upon the amount of time into the future that was predicted. For example, on the basis of 7 days of data, performance 1 week into the future would be successfully predicted (with a narrow margin of error) 64 percent of the time; for performance 3 weeks into the future, predictions were successful 50 percent of the time. With 11 days of data, predictions 1 week into the future were successful 89 percent of the time; for performance 3 weeks into the future, predictions were successful 81 percent of the time.

Description. The technique of trend estimation focuses on describing performance over time. The technique is likely to appeal to individuals who do not wish to employ inferential statistics but merely wish to describe the rate of change for the single case. In many studies, there has been a great deal of reliance on merely describing mean changes across phases and speculating about changes in slope. The split middle technique provides a way to describe the data in a more analytic fashion than is usually the case.

[19] White (1972) discusses a number of tests to assess the statistical significance of separate aspects of the change that may have occurred during the intervention, such as a change in slope only and change in step (level) at the point of the intervention. These tests also rely on the binomial. As Edgington (1974) has noted, the binomial may not be valid when applied to the split-middle procedure. Given a random arrangement of numerical values distributed across the two phases, the projection of a median line from the first phase is not likely to divide the data of the second phase. If the data points of the first phase show an accelerating or decelerating slope, it is likely that the data points in the second phase will fall below or above, respectively, the projected slope from the first phase. Thus, given a random sequence of numbers, the data points in the second phase are not likely to fall evenly about the projected slope based upon the data points of the first phase if there is a slope in the first phase.

Ease of computation. Certainly an advantage of the split middle method is that it requires little computational effort. Indeed, White (1972) indicated that a celeration line for 20 data points can be calculated in about 10 seconds and is simple enough to be calculated by a second-grade student. Thus, the method should provide a readily available tool for describing slope patterns in the data and predicting outcomes of interventions.

8.10. EVALUATION OF STATISTICAL TESTS: GENERAL CONSIDERATIONS

The above statistical tests are suitable for diverse intra-subject replication designs. A few analyses have particularly wide applicability. For example, time series analysis and split middle techniques can be readily applied to virtually all designs that have distinct phases (e.g., A-B-A-B and multiple baseline designs). The randomization test described earlier is especially suited to the A-B-A-B design and its variations. The R_n statistic is suited to multiple baseline designs. Of course, the applicability of various statistical tests to diverse designs presupposes resolution of specific practical limitations that may restrict utilization of a given test.

Despite the adaptability of various tests to applied situations, there are general considerations and specific sources of caution that apply to many of the statistics and need to be mentioned. Four issues will be discussed, including: the relationship of experimental design and statistical evaluation, the status of statistical tests for intra-subject replication designs, the issue of clinical significance, and practical restrictions for employing statistical tests.

Experimental design and statistical evaluation

The relationship between experimental design and the use of statistical tests should be noted briefly. As is well known, statistics can be applied in situations where the design is incapable of revealing a causal relationship between the intervention and behavior. Drawing a conclusion between the effect of an intervention and behavior assumes an adequate design independent of the techniques used to evaluate the data. This point was not sufficiently clear in presenting statistical tests alone in the present chapter. Many of the statistical techniques discussed could be readily applied to situations with weak experimental designs. The availability of statistical tests should not gainsay the importance of design. As outlined earlier, significant changes in an A-B design may not be sufficient for unambiguously concluding that the intervention (B) was responsible for change. A plethora of rival hypotheses must be excluded (Campbell and Stanley, 1963).

There is always a concern that the use of statistical tests can give the appearance of increased elegance of an "experimental" design from which no conclusions can be reached. In a sense, statistical significance can serve as a distraction (cf. Michael, 1974).

Status of statistical tests for intra-subject replication designs

The status of various statistical tests for $N = 1$ data is not clear for several reasons. Some of the statistical tests themselves have not been widely applied in psychology so that problems associated with their application are not well specified. For example, time series analysis, an efficient means for evaluating intervention effects, has not been applied frequently to experimental designs in applied clinical research. Moreover, it is unclear what effects not meeting various conditions (e.g., large number of data points) will have on the use of time series analysis and the results obtained. Simulation work, which shows the effect of applying different time series models to a time series of short durations, would be exceedingly important for applied operant interventions. In short, the problems of statistical tests for intra-subject replication designs in psychological research have not been sufficiently described. Indeed, the statistical tests themselves in many cases have only been carefully elaborated recently (e.g., Glass, Willson, and Gottman, 1974; White, 1974).

Clinical significance

The issue of clinical versus statistical significance was detailed in Chapter 2 and was mentioned at the beginning of the present chapter. The repetition is justified to emphasize that statistical significance is not a substitute for clinical significance. It might seem appropriate to rely on statistical tests in those instances in which behavior change has not been obvious and some method is required to determine whether there is any effect. However, if treatment effects are not obvious, *a fortiori*, their clinical value probably has not been demonstrated. This condition is not rectified by assessing statistical significance. Many investigators maintain that clinical significance is the only criterion that should be invoked. This is not a minority view in applied operant research. Yet there are situations, outlined at the beginning of the chapter, in which statistical analysis may be valuable.

Sometimes statistics are used to carry the burden of proof to the reader that treatment was effective. While this use of statistics has characterized traditional between-group design strategies, it is unlikely to be accepted for intra-subject replication designs. In many cases, statistical criteria are regarded merely as weak tools to employ when clinical change (the primary criterion) has not been achieved.

Practical restrictions for employing statistical tests

There are practical limitations in obtaining specific conditions that would make various statistical tests appropriate. For example, the R_n statistic requires that treatments are randomly assigned to individuals as part of a multiple baseline design. The random assignment of treatment justifies the statistical test. Yet, in clinical situations random assignment is not possible because of practical restraints. For example, convenience, severity of the target behavior, and limitations of staff all may dictate the order in which individuals receive the intervention. Thus, the R_n statistic often may not be available for applied use. Similarly, randomization tests require that treatments (e.g., baseline, intervention) are randomly assigned to occasions (e.g., days). Yet, treatments often cannot be altered on a daily basis as a function of practical exigencies of the setting. Similarly, alteration of experimental conditions may interfere with clinical goals that require protracted and uninterrupted training.

Some individuals have suggested that statistical techniques for time series data may be most appropriate to laboratory rather than applied situations because of the practical conditions that must be met.[20] For example, time series designs should employ protracted phases for the appropriate use of the time series analysis. Yet, lengthy phases usually are not employed in investigations in applied settings. Indeed, some times phases are explicitly implemented for short periods of time (e.g., "reversal" phases) so as not to interfere with long-term clinical goals.

While the conditions that are required for specific statistical tests might be readily met in laboratory settings, it is precisely those conditions that may obviate the need for statistics. Laboratory investigations in operant work have successfully avoided reliance upon statistical tests because of the control that can be achieved over behavior.

It should be noted, however, that statistical tests can provide assistance when control over the situtation is not ideal. For example, when ideal conditions for data evaluation (e.g., stable baseline rates) cannot be obtained, statistical evaluation or description of the data (e.g., time series analysis, split middle technique) can facilitate interpretation of the outcome. In cases where the ideal conditions for data evaluation with non-statistical criteria cannot be met, statistics make obvious contributions.

[20] Stephanie B. Stolz has suggested this point in a personal communication.

8.11. CONCLUSION

The present chapter has outlined specific statistical tests for intra-subject replication designs and limitations associated with their use. The availability of a variety of statistics provides an investigator with diverse options in evaluating experimental interventions for the single case. However, a few considerations pertaining to the use of statistics expressed throughout this chapter and noted in previous chapters must be reiterated at the conclusion of the discussion of statistics. Initially, the appropriateness of utilizing statistical criteria for the evaluation of applied behavioral interventions remains a major source of controversy. In traditional between-group experimentation, the use of statistics as a legitimate means of evaluation is infrequently questioned. Most of the controversies pertaining to statistics for between-group designs center around deciding when specific tests are inappropriate and the effects of using various tests when the assumptions upon which they are based are violated. In contrast, the major dispute in intra-subject research is whether statistical evaluation *per se* should comprise a part of the evaluation of intervention effects. Certainly, use of the information presented in the present chapter presupposes a resolution of this basic issue.

Another issue important to reiterate is that in many cases the use of statistical tests in the single case experiment may dictate the manner in which a clinical intervention needs to be implemented (e.g., random assignment of treatments or subjects to occasions). Exigencies of clinical settings may delimit the applicability of diverse procedures upon which various statistical tests are based. The range of statistics available to the investigator-clinician may be delimited by practical and ethical constraints.

Notwithstanding the above considerations, statistical tests are likely to be used more frequently in the future. Clinical interventions in applied work are becoming increasingly complex, both in therapeutic focus and experimental design. Settings, populations, and target behaviors focused upon are increasingly diverse and allow somewhat less experimental control than has characterized investigations conducted in closed settings (e.g., hospital ward, classroom). In addition, intra-subject and between-group designs increasingly are combined in applied research (Kazdin, 1973, 1976). Consequently, statistical analyses are likely to proliferate. The present chapter was designed to convey salient statistical options and problems attendant upon their use.

References

Anderson, R. L. Distribution of the serial correlation coefficient. *Annals of Mathematical Statistics*, 1942, **13**, 1-13.

Anderson, T. W. *The statistical analysis of time series.* New York: Wiley, 1971.

Atiqullah, M. On the robustness of analysis of variance. *Bulletin of the Institute of Statistical Research and Training*, 1967, **1**, 77-81.

Bijou, S. W., Peterson, R. F., Harris, F. R., Allen, K. E., and Johnston, M. S. Methodology for experimental studies of young children in natural settings. *Psychological Record*, 1969, **19**, 177-210.

Boneau, C. A. The effects of violations of assumptions underlying the *t* test. *Psychological Bulletin*, 1960, **57**, 49-64.

Box, G. E. P., and Jenkins, G. M. *Time series analysis: Forecasting and control.* San Francisco: Holden-Day, 1970.

Box, G. E. P., and Tiao, G. C. A change in level of non-stationary time series. *Biometrika*, 1965, **52**, 181-192.

Browning, R. M. A same-subject design for simultaneous comparison of three reinforcement contingencies. *Behaviour Research and Therapy*, 1967, **5**, 237-243.

Browning, R. M., and Stover, D. O. *Behavior modification in child treatment: An experimental and clinical approach.* Chicago: Aldine-Atherton, 1971.

Campbell, D. T. From description to experimentation: Interpreting trends as quasi-experiments. In C. W. Harris (Ed.), *Problems in measuring change.* Pp. 212-253. Madison: University of Wisconsin Press, 1963.

Campbell, D. T., and Stanley, J. C. Experimental and quasi-experimental designs for research and teaching. In N. L. Gage (Ed.), *Handbook of research on teaching.* Pp. 171-246. Chigago: Rand-McNally, 1963.

Chassan, J. B. *Research design in clinical psychology and psychiatry.* New York: Appleton-Century-Crofts, 1967.

Cochran, W. G. Some consequences when the assumptions for the analysis of variance are not satisfied. *Biometrics*, 1947, **3**, 22-38.

Conover, W. J. *Practical nonparametric statistics.* New York: Wiley, 1971.

Edgar, E., and Billingsley, F. Believability when $N = 1$. *Psychological Record*, 1974, **24**, 147-160.

Edgington, E. S. Statistical inference from $N = 1$ experiments. *The Journal of Psychology*, 1967, **65**, 195-199.

Edgington, E. S. *Statistical inference: The distribution-free approach.* New York: McGraw-Hill, 1969.

Edgington, E. S. The design of one-subject experiments for testing hypotheses. *Western Psychologist*, 1972, **3**, 33-38.

Edgington, E. S. Personal communication, August 1974.

Edgington, E. S., and Strain, A. R. Randomization tests: Computer time requirements. *Journal of Psychology*, 1973, **85**, 89-95.

Ezekiel, M. and Fox, K. A. *Methods of correlation and regression analysis: Linear and curvilinear.* New York: Wiley, 1959.

Gentile, J. R., Roden, A. H., and Klein, R. D. An analysis of variance model for the intrasubject replication design. *Journal of Applied Behavior Analysis*, 1972, **5**, 193-198.

Glass, G. V. Estimating the effects of intervention into a non-stationary time-series. *American Educational Research Journal*, 1972, **9**, 463-477.

Glass, G. V., and Maguire, T. O. Analysis of time-series quasi-experiments. (Final report, Project No. 6-8329) Boulder: University of Colorado, Laboratory of Educational Research, 1968.

Glass, G. V., Peckham, P. D., and Sanders, J. R. Consequences of failure to meet assumptions underlying the fixed-effects analyses of variance and covariance. *Review of Educational Research*, 1972, **42**, 237-288.

Glass, G. V., Willson, V. L., and Gottman, J. M. *Design and analysis of time-series experiments.* Boulder: Colorado Associated University Press, 1974.

Gottman, J. M. N-of-one and N-of-two research in psychotherapy. *Psychological Bulletin*, 1973, **80**, 93-105.

Gottman, J. M., and Leiblum, S. R. *How to do psychotherapy and how to evaluate it.* New York: Holt, Rinehart and Winston, 1974.

Gottman, J. M., McFall, R. M., and Barnett, J. T. Design and analysis of research using time series. *Psychological Bulletin*, 1969, **72**, 299-306.

Hall, R. V., Fox, R., Willard, D., Goldsmith, L., Emerson, M., Owen, M., Davis, F., and Porcia, E. The teacher as observer and experimenter in the modification of disputing and talk-out behaviors. *Journal of Applied Behavior Analysis*, 1971, **4**, 141-149.

Harris, F. R., Wolf, M. M., and Baer, D. M. Effects of adult social reinforcement on child behavior. *Young Children*, 1964, **20**, 8-17.

Hartmann, D. P. Forcing square pegs into round holes: Some comments on "An analysis-of-variance model for the intrasubject replication design." *Journal of Applied Behavior Analysis*, 1974, **7**, 635-638.

Holtzman, W. H. Statistical models for the study of change in the single case. In C. W. Harris (Ed.), *Problems in measuring change.* Pp. 199-211. Madison: University of Wisconsin Press, 1963.

Jones, R. R., Vaught, R. S., and Reid, J. B. Time series analysis as a substitute for single subject analysis of variance designs. In G. R. Patterson, I. M. Marks, J. D. Matarazzo, R. A. Myers, G. E. Schwartz, and H. H. Strupp (Eds.), *Behavior change, 1974.* Pp. 164-169. Chicago: Aldine, 1975.

Jones, R. R., Vaught, R. S., and Weinrott, M. Time series analysis in operant research. *Journal of Applied Behavior Analysis*, 1976, in press.

Kazdin, A. E. Methodological and assessment considerations in evaluating reinforcement programs in applied settings. *Journal of Applied Behavior Analysis*, 1973, **6**, 517-531.

Kazdin, A. E. *Behavior modification in applied settings.* Homewood, Ill.: Dorsey Press, 1975a.

Kazdin, A. E. Characteristics and trends in applied behavior analysis. *Journal of Applied Behavior Analysis*, 1975b, **8**, 332.

Kazdin, A. E. Methodology of applied behavior analysis. In T. A. Brigham and A. C. Catania (Eds.), *Applied behavior research: Analysis of social and educational processes.* New York: Irvington/Naiburg–Wiley and Sons, 1976, in press.

Kelly, F. J., McNeil, K., and Newman, I. Suggested inferential statistical models for research in behavior modification. *Journal of Experimental Education*, 1973, **41**, 54-63.

Keselman, H. J., and Leventhal, L. Concerning the statistical procedures enumerated by Gentile et al.: Another perspective. *Journal of Applied Behavior Analysis*, 1974, **7**, 643-645.

Kirk, R. E. *Experimental design: Procedures for the behavioral sciences.* Belmont, Calif.: Brooks/Cole, 1968.

Kratochwill, T., Alden, K., Demuth, D., Dawson, D., Panicucci, C., Arntson, P., McMurray, N., Hempstead, J., and Levin, J. A further consideration in the application of an analysis-of-variance model for the intrasubject replication design. *Journal of Applied Behavior Analysis*, 1974, 7, 629-633.

Lindquist, E. F. *Design and analysis of experiments in psychology and education.* Boston: Houghton-Mifflin, 1953.

Maguire, T. O., and Glass, G. V. A program for the analysis of certain time-series quasi-experiments. *Educational and Psychological Measurement*, 1967, **27**, 743-750.

Michael, J. Statistical inference for individual organism research: Mixed blessing or curse? *Journal of Applied Behavior Analysis*, 1974, 7, 647-653.

Moses, L. E. Nonparametric statistics for psychological research. *Psychological Bulletin*, 1952, **49**, 122-143.

Namboodiri, N. K. Experimental designs in which each subject is used repeatedly. *Psychological Bulletin*, 1972, **77**, 54-64.

Revusky, S. H. Some statistical treatments compatible with individual organism methodology. *Journal of the Experimental Analysis of Behavior*, 1967, **10**, 319-330.

Risley, T. R. Behavior modification: An experimental-therapeutic endeavor. In L. A. Hamerlynck, P. O. Davidson, and L. E. Acker (Eds.), *Behavior modification and ideal health services.* Pp. 103-127. Calgary, Alberta, Canada: University of Calgary Press, 1970.

Scheffé, H. *The analysis of variance.* New York: Wiley, 1959.

Schnelle, J. F., and Frank, L. J. A quasi-experimental retrospective evaluation of a prison policy change. *Journal of Applied Behavior Analysis*, 1974, 7, 483-494.

Shine, L. C., and Bower, S. M. A one-way analysis of variance for single-subject designs. *Educational and Psychological Measurement*, 1971, **31**, 105-113.

Sidman, M. *Tactics of scientific research: Evaluating experimental data in psychology.* New York: Basic Books, 1960.

Siegel, S. *Nonparametric statistics for the behavioral sciences.* New York: McGraw-Hill, 1956.

Skinner, B. F. Are theories of learning necessary? *Psychological Review*, 1950, **57**, 193-216.

Skinner, B. F. A case history in scientific method. *American Psychologist*, 1956, **11**, 221-233.

Sween, J., and Campbell, D. T. The interrupted time series as quasi-experiment: Three tests of significance. A Fortran Program for CDC. Vogelback Computing Center, Northwestern University, Evanston, Illinois, August 1965.

Thoresen, C. E., and Elashoff, J. D. Some comments on "An analysis-of-variance model for the intrasubject replication design." *Journal of Applied Behavior Analysis*, 1974, 7, 639-641.

Tversky, A., and Kahneman, D. Belief in the law of small numbers. *Psychological Bulletin*, 1971, **76**, 105-110.

White, O. R. *A glossary of behavioral terminology.* Champaign, Ill.: Research Press, 1971.

White, O. R. A manual for the calculation and use of the median slope–a technique of progress estimation and prediction in the single case. Regional Resource Center for Handicapped Children, University of Oregon, Eugene, Oregon, 1972.

White, O. R. The "split middle"–a "quickie" method of trend estimation. Experimental Education Unit, Child Development and Mental Retardation Center, University of Washington, 1974.

Winer, B. J. *Statistical principles in experimental design.* (Second Edition) New York: McGraw-Hill, 1971.

CHAPTER 9

Beyond the Individual: Replication Procedures

9.1. INTRODUCTION

Replication is at the heart of any science. In all sciences, replication serves at least two purposes: First, to establish the reliability of previous findings and, second, to determine the generality of these findings under differing conditions. These goals, of course, are intrinsically interrelated. Each time that certain results are replicated under different conditions, this not only establishes generality of findings, but also increases confidence in the reliability of these findings. The emphasis of this chapter, however, is on replication procedures for establishing generality of findings.

In Chapter 2 the difficulties in establishing generality of findings in applied research were reviewed and discussed. The problem in generalizing from a heterogenous group to an individual limits generality of findings from this approach. The problem in generalizing from one individual to other individuals who may differ in many ways limits generality of findings from a single case. One answer to this problem is the replication of single case experiments. Through this procedure, the applied researcher can maintain his focus on the individual, but establish generality of findings for those who differ from the individual in the original experiment. Sidman (1960) has outlined two procedures for replicating single case experiments in basic research: replication and systematic replication. In applied research a third type of replication, which we term clinical replication, is relevant.

The purpose of this chapter is to outline the procedure and goals of replication strategies in applied research. Examples of each type of replication series will be presented and criticized. Guidelines for the proper use of these procedures in future series will be suggested from current examples judged to be successful in establishing generality of findings. Finally, the feasibility of large-scale replication series will be discussed in light of the practical limitations inherent in applied research.

9.2. DIRECT REPLICATION

Direct replications of single case experiments have often appeared in professional journals. As noted above, these series are capable of determining both reliability of findings and generality of findings across clients. In most cases, however, the very important issue of generality of findings has not been discussed. Indeed, it seems that most investigators employing single case methodology, as well as editors of journals who judge the adequacy of such endeavors, have been primarily concerned with reliability of findings as a goal in replication series rather than generality of findings. That is, most investigators have been concerned with demonstrating that certain results can or cannot be replicated in subsequent experiments rather than systematically observing the replications themselves to determine generality of findings. However, since any attempt to establish reliability of a finding by replicating the experiment on additional cases also provides information on generality, many applied researchers have conducted direct replication series yielding valuable information on client generality. Examples of several of these series will be presented below.

Definition of direct replication

For our purposes, we agree basically with Sidman's (1960) definition of direct replication as "... repetition of a given experiment by the same investigator" (p. 73). Sidman divides direct replication into two different procedures: repetition of the experiment on the same subject and repetition on different subjects. While repetition on the same subject increases confidence in the reliability of findings and is used occasionally in applied research (see Chapter 5), generality of findings across clients can be ascertained only by replication on different subjects. More specifically, direct replication in applied research refers to administration of a given treatment by the same investigator or group of investigators in a specific setting (e.g., hospital, clinic, or classroom) on a series of clients homogeneous for a particular behavior disorder (e.g., agoraphobia, compulsive hand-washing). While it is recognized that in applied research, clients will always be more heterogeneous on background variables such as age, sex, or presence of additional maladaptive behaviors than in basic research, the conservative approach is to match clients in a replication series as closely as possible on these additional variables. Interpretation of mixed results, where some clients benefit from the procedure and some do not, can then be attributed to as few differences as possible, thereby providing a clearer direction for further experimentation. This point will be discussed more fully below.

Direct replication as we define it can begin to answer questions about generality of findings across clients but cannot address questions concerning generality of findings across therapists or settings. Furthermore, to the extent that clients are homogeneous on a given behavior disorder (such as agoraphobia), a direct replication series cannot answer questions on the results of a given procedure on related behavior disorders such as claustrophobia, although successful results should certainly lead to further replication on related behavior disorders. A close examination of several direct replication series will serve to illustrate the information available concerning generality of findings across clients.

Example one: Two successful replications

The first example concerns one successful experiment and two successful replications of a therapeutic procedure. This early clinical series examined the effects of social reinforcement (praise) on severe agoraphobic behavior in three patients (Agras, Leitenberg, and Barlow, 1968). The procedure was straightforward. All patients were hospitalized. Severity of agoraphobic behavior was measured by observing the distance the patients were able to walk on a course from the hospital to a downtown area. Landmarks were identified at 25-yard intervals for over 1 mile. The patients were asked two or more times a day to walk as far as they could on the course without feeling "undue tension." Their report of distance walked was surreptitiously checked from time to time by an observer to determine reliability. Increases in distance were socially reinforced with praise and approval during treatment phases and ignored during withdrawal phases. In the first patient, increases in time spent away from the center were praised first, but since this resulted in the patient simply standing outside the front door of the hospital for longer periods, the target behavior was changed to distance. Since baseline procedures were abbreviated, this design is best characterized as a B-A-B design (see Chapter 5). The comparison, then, is between treatment (praise) and no treatment (no praise).

It is important to note for purposes of generality across clients that the patients in this experiment were rather heterogeneous as is typically the case in applied research. Although each patient was severely agoraphobic, all had numerous associated fears, including fear of crowds, illness, and death. The extent and severity of these fears differed. One subject was a 36-year-old male with a 15-year agoraphobic history. He was incapacitated to the extent that he could manage a 5-minute drive to work in a rural area only with great difficulty. A second subject was a 23-year-old female with only a 1-year agoraphobic history. The patient, however, could not leave her home unaccompanied. The third subject, a 36-year-old female, also could not leave her home unaccompanied, but had a 16-year agoraphobic history. In fact, this patient had to be sedated and brought to the hospital in an ambulance. In addition, these three patients

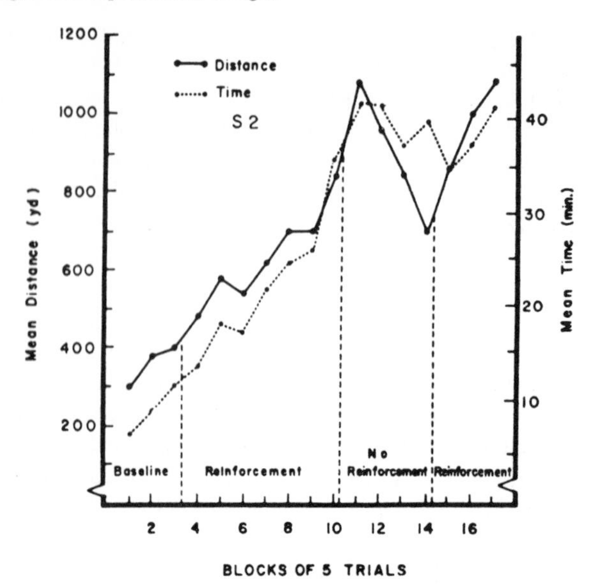

Fig. 9-1. The effects of reinforcement and non-reinforcement upon the performance on an agoraphobic patient (Subject 2). (Fig. 2, p. 425, from: Agras, W. S., Leitenberg, H., and Barlow, D. H. Social reinforcement in the modification of agoraphobia. *Archives of General Psychiatry*, 1968, 19, 423-427. Reproduced by permission.)

presented different background variables such as personality characteristics and cultural variations (one patient was European).

The results from one of the cases (the male) are presented in Fig. 9-1. Praise produced a marked increase in distance walked, and withdrawal of praise resulted in a deterioration in performance. Reintroduction of praise in the final phase produced a further increase in distance walked. These results were replicated on the remaining two patients.

At least three conclusions can be drawn from these data. The first conclusion is that praise is an effective treatment variable in modifying agoraphobic behavior. The second conclusion is that within the limits of these data, the results are reliable and not due to idiosyncracies present in the first experiment, since two replications of the first experiment were successful. The third conclusion, however, is of most interest here. The procedure was clearly effective with three patients of different ages, sex, and duration of agoraphobic behavior as well as different cultural backgrounds. For purposes of generality of findings, this series of experiments would be strengthened by a third replication (a total of four subjects). But the consistency of the results across three quite different patients enables one to draw initially favorable conclusions on the general effectiveness of this procedure across the population of agoraphobic clients through the process of logical generalization (Edgington, 1967).

On the other hand, if one client had failed to improve or improved only slightly such that the result was clinically unimportant, an immediate search would have to be made for procedural or other variables responsible for the lack of generality across clients. Given the flexibility of this experimental design, alterations in procedure (e.g., adding additional reinforcers, changing the criterion for reinforcement) could be made in an attempt to achieve clinically important results. If "mixed" results such as these were observed, further replication would be necessary to determine which procedures were most efficacious for given clients (see Chapter 2, Section 2.2).

In this series, however, these steps were not necessary due to the uniformly successful outcomes, and some preliminary statements about client generality were made. The next step in this series, then, would be an attempt to replicate these results systematically—that is, across different situations and therapists. It is evident that the preliminary series, which was carried out in Burlington, Vermont, does not address questions on effectiveness of techniques in different settings or with different therapists. It is entirely possible that characteristics of the therapist or the particular structure of the "course" that the agoraphobic walked facilitated the favorable results. Thus, these variables must be systematically varied to determine generality of findings across all important clinical domains. In fact, this step has since been taken. Using procedures that were operationally quite similar to those described above, but carrying different labels, Marks (1972) successfully treated a variety of severe agoraphobics in an urban European setting (London) using, of course, different therapists.

Example two: Four successful replications with design alterations during replications

A second example of a direct replication series will be presented since the behavior is clinically important (compulsive rituals), and the issue of client generality within a direct replication series is highlighted since five patients participated in the study (Mills, Agras, Barlow, and Mills, 1973). In this experiment, a new treatment—response prevention—was tested. The basic strategy in this experiment and its replications was an A-B-A design: baseline—response prevention—baseline. During replications, however, the design was expanded somewhat to include controls for instructional and placebo effects. For example, two of the replications were carried out in an A-B-BC-B-A design, where A was baseline, B was a placebo treatment, and C was response prevention.

The addition of new control phases during subsequent replication is not an uncommon strategy in single case design research since each replication is actually a separate experiment that stands alone. When testing a given treatment, however, new variables interacting within the treatment complex that might be responsible for improvement may be identified and "teased out" in later replications. It was noted in Chapter 2 that such flexibility of single case designs allows one to alter experimental procedures *within* a case. Within the context of replication, if a procedure is effective in the first experiment, one has the flexibility to add further, more stringent controls during replication to further ascertain the mechanism of action of a successful treatment. But, to remain a direct replication series within our definition, the major purpose of the series should be to test the effectiveness of a given treatment on a well-defined problem—in this case, compulsive rituals—administered by the same therapeutic team in the same setting. Thus, the treatment, if successful, must remain the same and the comparison is between treatment and no treatment, or treatment and placebo control.

The first four subjects in this experiment were severe compulsive hand-washers. The fifth subject presented with a different ritual. All patients were hospitalized on a research unit. All hand-washers encountered articles or situations throughout the experiment that produced hand-washing. Response prevention consisted of removing the handles from the wash basin wherein all hand-washing occurred. The placebo phase consisted of saline injections and oral placebo medication with instructions suggesting improvement in the rituals, but no response prevention. Once again, the design was either A-B-A, with A representing baseline and B representing response prevention, or A-B-BC-B-A, where A was baseline, B was placebo, and C was response prevention. Both self-report measures (number of urges to wash hands) and an objective measure (occasions when the patient approached the sink, mechanically recorded by a washing pen—see Chapter 4) were administered.

As in the previous series, the patients were relatively heterogeneous. The first subject was a 31-year-old woman with a 2-year history of compulsive hand-washing. Previous to the experiment, she had received over 1 year of both inpatient and outpatient treatment including chemotherapy, individual psychotherapy, desensitization, etc. She performed her ritual ten to twenty times a day, each ritual consisting of eight individual washings and rinsings with alternating hot and cold water. The associated fear was contamination of herself and others through contact with chemicals and dirt. These rituals prevented her from carrying out simple household duties or caring for her child.

The second subject was a 32-year-old woman with a 5-year history of hand-washing. Frequency of hand-washing ranged from 30-60 times per day with an average of thirty-nine during baseline. Unlike the previous subject, these rituals had strong religious overtones concerning salvation, although fear of contamination from dirt was also present. Prior treatments included two series of electric shock treatment, which proved ineffective.

A third subject was a 25-year-old woman who had a 3-year history of the hand-washing compulsion. Situations that produced the hand-washing in this case were associated with illness and death. If an ambulance passed near her home, she engaged in cleansing rituals. Hand-washings averaged thirty per day and the subject was essentially isolated in her home before treatment.

The fourth subject was a 20-year-old male with a history of hand-washing for 1½ years. He had been hospitalized for the previous year and was hand-washing at the rate of twenty to thirty times per day. The fifth subject, whose rituals differ considerably from the first four subjects, will subsequently be described.

Representative results from one case are presented below. Hand-washing remained high during baseline and placebo phases and dropped markedly after

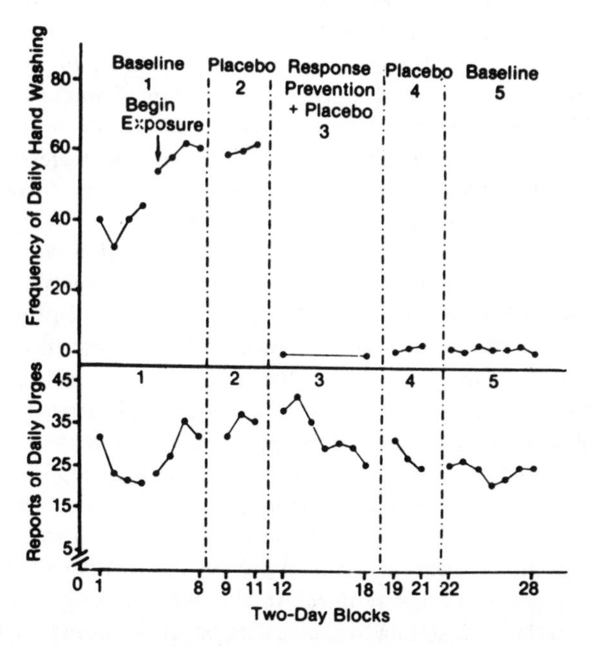

Fig. 9-2. In the upper half of the graph, the frequency of hand-washing across treatment phases is represented. Each point represents the average of 2 days. In the lower portion of the graph, total urges reported by the patient are represented. (Fig. 3, p. 527, from: Mills, H. L., Agras, W. S., Barlow, D. H., and Mills, J. R. Compulsive rituals treated by response prevention: An experimental analysis. *Archives of General Psychiatry*, 1973, 28, 524-529. Copyright 1973, American Medical Association, reproduced by permission.)

response prevention. Subjective reports of urges to wash declined slightly during response prevention and continued into follow-up. This decline continued beyond the data presented in Fig. 9-2 until urges were minimal. These results were essentially replicated in the remaining three hand-washers.

Before discussing issues relative to replication, experimental design considerations in this series deserve comment. The dramatic success of response prevention in this series is obvious, but the continued reduction of hand-washing after response prevention was removed presents some problems in interpretation. Since hand-washing did not recover, it is difficult to attribute its reduction to response prevention using the basic A-B-A withdrawal design. From the perspective of this design, it is possible that some correlated event occurred concurrent with response prevention that was actually responsible for the gains. Fortunately, the aforementioned flexibility in adding new control phases to replication experiments afforded an experimental analysis from a different perspective. In all patients, hand-washing was reasonably stable by history and through both baseline and placebo phases. Hand-washing showed a marked reduction *only* when response prevention was introduced. In these cases, baseline and placebo phases were administered for differing amounts of time. In fact, then, this becomes a multiple baseline design across subjects (see Chapter 6), allowing isolation of response prevention as the active treatment.

Given the apparent irreversibility of rituals after response prevention, future replication will probably rely on the multiple baseline design to systematically extend these findings to other clients in other settings.

Again, this series demonstrates that response prevention works, and replications ensure that this finding is reliable. In addition, the clinical significance of the result is easily observable by inspection, since rituals were entirely eliminated in all four patients. More importantly, however, the fact that this clinical result was consistently present across four patients lends considerable confidence to the notion that this procedure would be effective with other patients, again through the process of logical generalization. It is common sense that confidence in generality of findings across clients increases with each replication, but it is our rule-of-thumb that a point of diminishing returns is reached after one successful experiment and three successful replications for a total of four subjects. At this point, it seems efficient to publish the results so that systematic replication may begin in other settings. An alternative strategy is to administer the procedure in the same setting to clients with behavior disorders demonstrating marked differences from those of the first series. Some behavior disorders such as phobias lend themselves to this method of replication since a given treatment (e.g., systematic desensitization) should theoretically work on many different maladaptive avoidance behaviors subsumed under the label phobia. Within a disorder such as compulsive rituals, this is also feasible since several different types of rituals are encountered in the clinic. The question that can be answered in the original setting then is, will the procedure work on other behavior disorders that are topographically different but presumably maintained by similar psychological processes? In other words, would rituals quite different from handwashing respond to the same procedure? The fifth case in this series was the beginning of a replication along these lines.

The fifth subject was a 15-year-old boy who perfomed a complex set of rituals upon retiring at night and another set of rituals when arising in the morning. The night rituals included checking and rechecking the pillow placement, folding and refolding pajamas, etc. The morning rituals were concerned mostly with dressing. The rituals were extremely time-consuming and disruptive to the family's routine. After a baseline phase in which rituals remained relatively stable, the night rituals were prevented, but the morning rituals were allowed to continue. Here again, response prevention dramatically eliminated night-time rituals. Morning rituals gradually decreased to zero during prevention of night rituals.

The experiment further suggests that response prevention can be effective in the treatment of ritualistic behavior. The implications of this replication, however, are somewhat different than the previous three replications where the behavior in question was topographically similar. Although the treatment was administered by the same therapists in the same setting, this case does *not* represent a direct replication, since the behavior was topographically different. To consider this case as part of a direct replication series, one would have to accept, on an *a priori* basis, the theoretical notion that all compulsive rituals are maintained by similar psychological processes and therefore will respond to the same treatment. Although classification of these behaviors under one name (compulsive rituals) implies this, in fact there is no firm evidence supporting the similarity of these behaviors at this time. As such, it was probably inappropriate to include the fifth case in the present series since the clear implication is that response prevention is applicable to all rituals, but only one case where rituals differed was presented. From the perspective of sound replication procedures, the proper tactic would be to include this case in a second series containing different rituals. This second series would then be the first step in a systematic replication series in that generality of findings across different behaviors would be established in addition to generality of findings across clients.

Example three: Mixed results in three replications

The goal of this experiment was an experimental analysis of a new procedure for increasing heterosexual arousal in homosexuals desiring a change in their sexual orientation (Herman, Barlow, and Agras, 1974b). A chance finding in our laboratories suggested that exposure to an explicitly heterosexual film increased heterosexual arousal in separate measurement sessions (see Chapter 2, Section 2.3). Subsequently, this was tested in an A-B-C-B design, where A was baseline, B was exposure to heterosexual films (the treatment), and C was a control procedure in which the subject was also exposed to erotic films but the content was homosexual. The measures included changes in penile circumference to homosexual and heterosexual slides (recorded in sessions separate from the treatment sessions) as well as reports of behavior outside the laboratory setting. The purpose of the experiment was to analyze the effect on heterosexual arousal of exposure to films with heterosexual content over and above the effects of simply viewing erotic films, a condition obtaining in the control procedure. Thus, the comparison is between treatment and placebo control.

Again, the patients were relatively heterogeneous. The first patient was a 24-year-old male with an 11-year history of homosexuality. During the year preceding treatment, homosexual encounters averaged one to three per day, usually in public restrooms. Also, during this period, the patient had been mugged once, arrested twice, and had attempted suicide. The second patient was a 27-year-old homosexual pedophile with a 10-year history of sexual behavior with young boys. The third patient was an 18-year-old male who had not had homosexual relations for several years, but complained of a high frequency of homosexual urges and fantasies. The fourth patient, a 38-year-old male, reported a 26-year history of homosexual contacts. Homosexual behavior had increased during the previous 4 years, despite the fact that he had recently married. None of the patients reported previous heterosexual experience with the exception of the fourth subject, who had sexual intercourse with his wife approximately twice a week. Intercourse was successful if he employed homosexual fantasies to produce arousal, but he was unable to ejaculate during intercourse. All patients were seen daily, with the exception of the fourth patient, who was seen approximately three times per week.

Representative results from one case, the first patient, are presented below (Fig. 9-3). Heterosexual arousal as measured in separate measurement sessions increased during exposure to the female (heterosexual) film, dropped considerably when the homosexual film was shown, and rose once again when the female film was reintroduced. The results in this case represent clear and clinically important changes in heterosexual arousal, and the experimental analysis isolated the viewing of the heterosexual film as the procedure responsible for increases. Changes in arousal in the laboratory were accompanied by reports of increased heterosexual fantasies and behavior. These results were replicated on Subjects 2 and 3, where similar increases in heterosexual arousal and reports of behavior

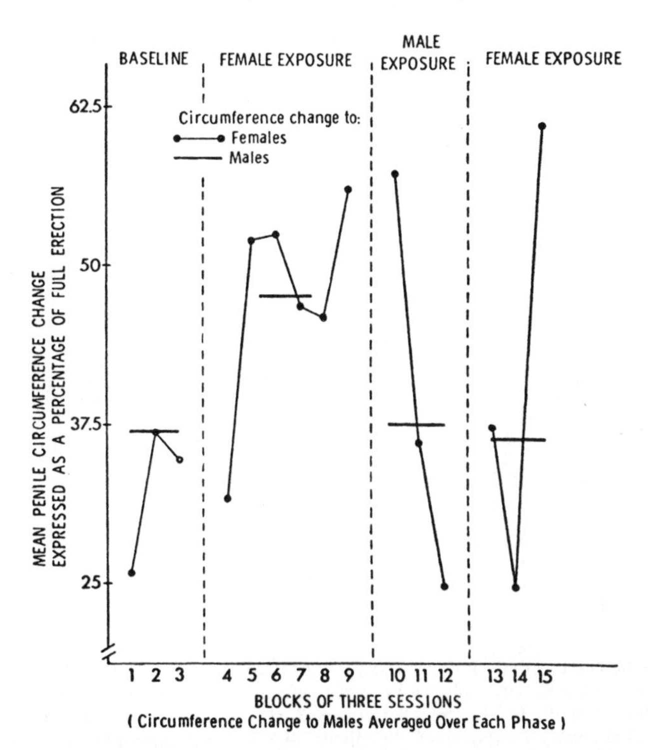

Fig. 9-3. Mean penile circumference change expressed as a percentage of full erection, to nude female (averaged over blocks of three sessions) and nude male (averaged over each phase) slides. (Fig. 1, p. 338, from: Herman, S. H., Barlow, D. H., and Agras, W. S. An experimental analysis of exposure to "explicit" heterosexual stimuli as an effective variable in changing arousal patterns of homosexuals. *Behaviour Research and Therapy*, 1974, **12**, 335-346. Reproduced by permission.)

were noted. But the results from the fourth case differed somewhat, thereby posing difficulties in interpretation in this direct replication series (Fig. 9-4).

In this case, heterosexual arousal increased somewhat during the first treatment phase, but the increase was quite modest. Withdrawing treatment resulted in a slight drop in heterosexual arousal, which increased once again when the heterosexual film was reinstated. This last increase, however, does not become clear until the last point in the phase, which represents only one session. Subsequently, the patient was unable to continue treatment due to prior commitments precluding an extension of this phase, which would have confirmed (or disconfirmed) the increase represented by that one point. Reports of sexual fantasies and behavior were consistent with the modest increases in heterosexual arousal. While some increase in heterosexual fantasies was noted, the patient

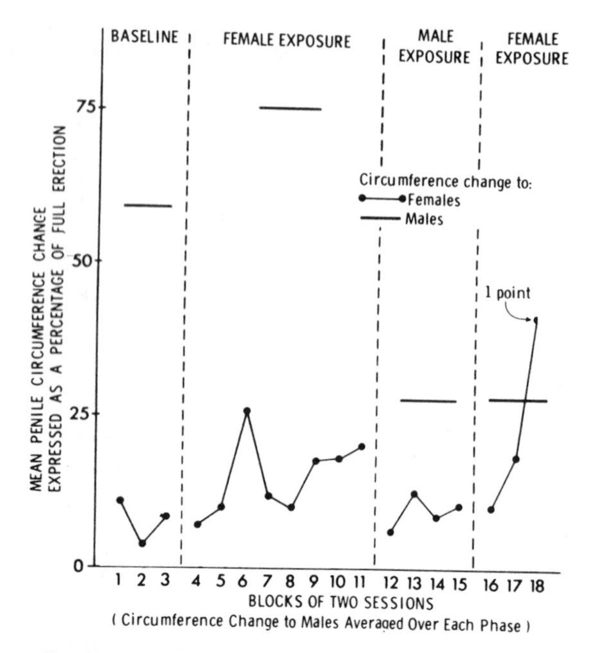

Fig. 9-4. Mean penile circumference change expressed as a percentage of full erection, to nude female (averaged over blocks of two sessions) and nude male (averaged over each phase) slides. (Fig. 4, p. 342 from: Herman, S. H., Barlow, D. H., and Agras, W. S. An experimental analysis of exposure to "explicit" heterosexual stimuli as an effective variable in changing arousal patterns of homosexuals. *Behaviour Research and Therapy*, 1974, **12**, 335-346. Reproduced by permission.)

continued to employ homosexual fantasies occasionally during sexual intercourse with his wife and was still unable to ejaculate.

Again, conclusions in three general areas can be drawn from these data. First, exposure to explicit heterosexual films can be an effective variable for increasing heterosexual arousal as demonstrated by the experimental analysis of the first patient. Second, to the extent that the results were replicated directly on three patients, the data are reliable and are not due to idiosyncracies in the first case. It does *not* follow, however, that generality of findings across clients has been firmly established. Although the results were clear and clinically significant for the first three patients, results from the fourth patient cannot be considered clinically useful due to the weakness of the effect. In this case, a clear distinction arises between the establishment of functional relationships and the establishment of clinically important generality of findings across clients. As in the first three patients, a functional relationship between treatment and heterosexual arousal was demonstrated in the fourth patient. This finding increases our confidence in the reliability of the result. Unlike the first three patients, however, the finding was not clinically useful. The conclusion, then, is that this procedure has only limited generality across clients, and the task remains to pinpoint differences between this patient and the remaining patients to ascertain possible causes for the limitations on client generality.

The authors (Herman, Barlow, and Agras, 1974b) note that the fourth patient differed in at least two ways from the remaining three. One difference falls under the heading of "background" variables and the other is procedural. First, the patient was married and therefore was required to engage in heterosexual intercourse *before* heterosexual arousal or interest was generated. In fact, he reported this to be quite aversive, which may have hampered the development of heterosexual interest during treatment. The remaining patients had experienced no significant heterosexual behavior prior to treatment. Second, this patient was seen less frequently than other patients. At most he was seen three times a week, rather than daily. At times, this dropped to once a week and even once every 3 weeks during periods when other commitments interfered with treatment. It is possible that this factor retarded development of heterosexual interest. To the extent that this was a procedural problem, rather than a variable that the patient brought with him to the experiment, it would have been possible to alter the procedure prior to the beginning of the experiment or even during the experiment (i.e., make daily attendance a requirement for participation in the experiment). If this alteration were undertaken and similar results (the weak effect) had ensued, it might have limited the search for causes of the weak effect to background variables, such as the ongoing aversive heterosexual behavior. Of course, this procedural variable was not thought to be important when the experiment was designed.

The issue of interpreting mixed results and looking for causes of failure illustrates an important principle in replication series. We noted above that subjects in a direct replication series should be as homogeneous as possible. If subjects in a series are not homogeneous, the investigator is gambling (Sidman, 1960). If the procedure is effective across heterogeneous subjects, he has won the gamble. If the results are mixed, he has lost. More specifically, if one subject differs in three or four definable ways from previous subjects, but his data are similar to previous subjects, then the experimenter has won the gamble by demonstrating that a procedure has client generality *despite* these differences. If the results differ in any significant manner, however, as in the example above, the experimenter cannot know which of the three or four variables was responsible for the differences. The task remains, then, to systematically explore the effects of these variables and track down causes of inter-subject variability. In basic research with animals one seldom sees this type of gamble in a direct replication series, since most variables are controlled and subjects are highly homogeneous. In applied research, however, clients always bring a variety of historical experiences, "personality" variables, and other background variables such as age and sex to treatment. To the extent that a given treatment works on three, four, or five clients, the applied researcher has already won a "gamble" even in a direct replication series, since a failure could be attributed to any one of the variables that differentiate one subject from another. In any event, we recommend the conservative approach whenever possible, in that subjects in a direct replication series should be homogeneous for aspects of the target behavior as well as background variables. The issue of "gambling" arises again when one starts a systematic replication series since the researcher must decide on the number of ways he wishes his systematic replication series to differ from the original direct series.

Example four: Mixed results in nine replications

While all subjects demonstrated some improvement in the study described above, the data often are more variable in a direct replication series. Such is the case in the following study where attempts to modify delusional speech in ten paranoid schizophrenics produced mixed results (Wincze, Leitenberg, and Agras, 1972). In this procedure the effects of feedback and token reinforcement on delusional speech were evaluated. Feedback consisted of reading sentences with a high probability of eliciting a particular patient's delusional behavior. If the patient responded delusionally, he would be informed that his response was incorrect and given the correct response. For instance, one patient thought he was Jesus Christ. If he answered affirmatively when asked this question, he would be told that he was not Jesus Christ, who lived 2000 years ago, but rather Mr. M. who was 40 years old. If he answered correctly, he would be so informed. During token reinforcement phases, the patient received tokens redeemable for food and recreational activities contingent on non-delusional speech in the sessions. Sessions consisted of fifteen questions each day. Tokens were also administered to some patients for non-delusional talk on the ward in addition to the contingencies within sessions; but, for our purposes, we will discuss only the effects of feedback and token reinforcement on delusional talk within sessions.

All patients were chronic paranoid schizophrenics who had been hospitalized at least 2 years (ranged from 2 to 35 years). Six males and four females participated, with an age range from 25 to 67. Level of education ranged from eighth grade through college. Thus, these patients were, again, heterogeneous on many background variables.

The experimental design for the first five patients consisted of baseline procedures followed by feedback and then token reinforcement. In some cases,

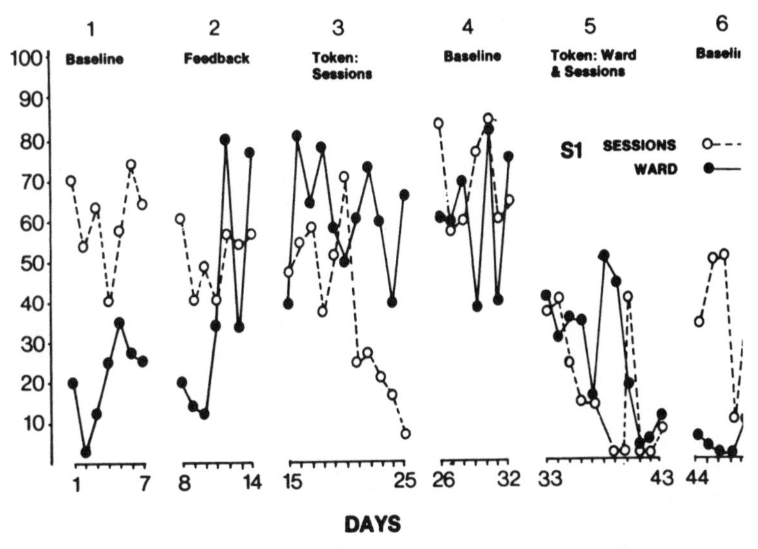

Fig. 9-5. Percentage delusional talk of Subject 1 during therapist sessions and on ward for each experimental day. (Fig. 1, p. 254, from: Wincze, J. P., Leitenberg, H., and Agras, W. S. The effects of token reinforcement and feedback on the delusional verbal behavior of chronic paranoid schizophrenics. *Journal of Applied Behavior Analysis*, 1972, 5, 247-262. Reproduced by permission.)

token reinforcement on the ward in addition to tokens within sessions was introduced toward the end of the experiment. Additional baseline phases were introduced whenever feedback or reinforcement produced marked decreases in delusional talk. For Subjects 6 through 10, the first feedback and token reinforcement in-session phases were withdrawn to examine the effects of token reinforcement when it was presented first in the treatment sequence.

All data were presented individually in the experiment such that any functional relations between treatments and delusional speech were apparent. Individual data from the first patient are presented in Fig. 9-5 to illustrate the manner of presentation. In this particular case, the baseline phase following the first feedback phase was omitted since no improvement was noted during feedback. Results from all patients are summarized in Table 9-1.

In five out of ten cases, feedback alone produced at least a 20 percent decrease in delusional speech within sessions. In two cases, this decrease in delusional speech was clinically impressive both in magnitude and in the consistent trend in behavior throughout the phase (Subjects 2 and 8). In the remaining three patients, the magnitude of the decrease and/or the behavior trend across the feedback phase was relatively weak. For instance, Table 9-1 indicates that the

TABLE 9-1. Mean percentage delusional talk of each S based on last two data points of each phase in therapist sessions and on the ward

Subjects	Phase Sequences							
	Baseline	Feedback	Baseline	Token: Sessions	Baseline	Token: Ward and Sessions	Bonus	Baseline
S1 Sessions	68·1	59·8	–	11·6	61·4	1·6	–	28·2
S1 Ward	26·2	50·4	–	52·9	56·7	7·4	–	11·3
S2 Sessions	83·0	1·6	13·3	–	–	–	–	–
S2 Ward	16·6	5·9	0·0	–	–	–	–	–
S3 Sessions	91·3	73·0	–	3.3	91·3	11·6	5·0	64·7
S3 Ward	27·0	9·9	–	36·3	5·0	21·6	4·6	4·0
S4 Sessions	76·4	66·4	68·1	21·6	61·4	–	29·9	61·4
S4 Ward	27·0	2·6	24·2	4·4	13·3	–	0·0	3·2
S5 Sessions	86·3	51·5	64·7	24·9	59·8	18·3	21·6	38·2
S5 Ward	48·3	79·2	70·6	61·9	51·7	45·1	4·6	29·2
	Baseline	Token: Sessions	Baseline	Feedback	Baseline	Token: Ward and Sessions	Bonus	Baseline
S6 Sessions	79·7	64·7	76·4	68·1		66·4	78·0	83·0
S6 Ward	58·2	79·5	50·7	56·6	–	78·8	69·6	25·7
S7 Sessions	89·6	59·8	69·7	48·1	63·1	48·1	36·5	71·4
S7 Ward	23·0	12·5	19·1	9·1	18·8	14·0	37·4	20·9
S8 Sessions	86·3	18·3	49·8	8·3	0·0	–	–	–
S8 Ward	6·9	3·3	0·0	0·0	0·0	–	–	–
S9 Sessions	79·7	13·3	54·8	5·0	20·0	1·7	–	51·5
S9 Ward	13·4	8·9	44·9	16·3	34·8	3·4	–	14·0
S10 Sessions	83·0	66·4	73·0	64·7	–	66·4	–	–
S10 Ward	16·6	33·1	8·2	11·3	–	58·2	–	–

(Table 2, p. 258, from: Wincze, J. P., Leitenberg, H., and Agras, W. S. The effects of token reinforcement and feedback on the delusional last two data points in the feedback phase for Subject 9 were considerably lower than the last two data points in the preceding baseline phase (a drop of 49·8 percent). But the extreme variability in data across the feedback phase indicates that this was a weak effect. A withdrawal of feedback and a return to baseline procedures was not associated with a clear reversal in delusional speech (at least a 20 percent increase) in any of the five patients who improved, although the finding is particularly important for those two patients who demonstrated improvement of clinical proportions. Thus, it was not demonstrated that feedback was the variable responsible for improvement within treatment sessions.

If the marked improvement of Subjects 2 and 8 were replicated on additional patients, one would be tempted to undertake a further experimental analysis to determine which variables were responsible for the improvement. The lack of replication, however, suggests that this would not be a fruitful line of inquiry.

The results from token reinforcement were quite different. This procedure was administered to nine patients; six (Subjects 1, 3, 4, 5, 8 and 9) improved—an improvement that was confirmed by a return of delusional speech when token reinforcement was removed. Subject 7 also improved, but delusional speech did not reappear when token reinforcement was removed. In all of these patients, the decrease was substantial both in percentage of delusional speech and in trends across the token phase.

Several conclusions can be drawn from these data. In terms of reduction of delusional speech within sessions, the experimental analysis demonstrated that token reinforcement was effective and replication indicated that the finding has some reliability. Generality of findings across clients, however, is limited. Two patients did not improve during administration of token reinforcement. As Sidman (1960) notes, the failure to replicate on all subjects does *not* detract from the successes in the remaining subjects. Token reinforcement is clearly responsible for improvement in those subjects to the extent that the experimental design was sound (internally valid). However, applied researchers cannot stop here, satisfied that the procedure seems to work well enough on most cases, since the practicing clinician would be at a loss to predict which cases would improve with this procedure. In fact, since the authors (Wincze, Leitenberg, and Agras, 1972) note that these two cases actually deteriorated on the ward during this treatment, the search for accurate predictions of success becomes all the more important to the clinician. Thus, a careful search for differences in these cases that might be important should ensue, leading to a more intensive functional investigation and experimental manipulation of those factors that contribute to success or failure.

In view of the additional fact that all subjects in this series demonstrated little generalization of improvement from session to ward behavior, analysis of this treatment is in a very preliminary state and, as Wincze, Leitenberg, and Agras (1972) point out, "... much work needs to be done in order to predict when a given type of behavioral intervention is likely to succeed in a given case" (p. 262).

Finally, it seems important to make a methodological point on the size of this series. While the nine replications in this series yielded a wealth of data, a more efficient approach might have been to stop after four or five replications followed by a functional analysis of failures encountered. In the unlikely event that failures did not occur in the initial replication series, the results would be strong enough to generate systematic replication in other research settings where failures would almost certainly appear, leading to a search for critical differences at this point. If failures did appear in this shorter series, the investigators could immediately begin to determine factors responsible for variant data rather than continuing direct replications that can only have a decreasing yield of information as subjects accumulate.

Guidelines for direct replication

Based on prevailing practice and accumulated knowledge on direct replication, we would suggest the following guidelines in conducting a direct replication series in applied research.

First, therapists and settings should remain constant across replications.

Second, the behavior disorder in question should be topographically similar across clients, such as a specific phobia.

Third, client background variables should be as closely matched as possible, although the ideal goal of identical clients can never be attained in applied research.

Fourth, the procedure employed (treatment) should be uniform across clients, until failures ensue. If failures are encountered during replication, attempts should be made to determine cause of this inter-subject variability through improvised and fast-changing experimental designs (see Chapter 2, Section 2.3). If the search is successful, the necessary alteration in treatment should be tested on additional clients who share the characteristics or behavior of the first client who required the alteration. If the search for sources of variability is not successful, differences in that particular client from other successful clients should be noted for future research.

Fifth, one successful experiment and three successful replications will usually be sufficient to generate systematic replication on topographically different behaviors in the same setting or the same behavior in different settings. This guideline is not as firm as those preceding, since results from a study containing one unusual or significant case may be worth publishing or an investigator may wish to continue direct replication if experimentally successful but clinically "weak" results are obtained. Generally, though, after one experiment and three successful replications it is time to go on to systematic replication.

On the other hand, if direct replication produces "mixed" success and failure, then investigators determine when to stop the series and begin to analyze reasons for failure in what is essentially a new series, since the procedure or treatment presumably will change. If one success is followed by two or three failures, then neither the reliability of the procedure nor the generality of the finding across clients has been established, and it is probably time to find out why. If two or three successes are mixed in with one or two failures, then the reliability of the procedure will be established to some extent, but the investigator must decide when to begin investigating reasons for lack of client generality. In any case, it does not appear to be sound experimental strategy to continue a direct replication series indefinitely, when both successes and failures are occurring.

Finally, broad client generality cannot be established from one experiment and three replications. Although a clinician can observe the extent to which an individual client who responded to treatment in a direct replication series is similar to his client and can proceed accordingly with the treatment, chances are he may have a client with a topographically similar behavior disorder who is different in some clinically important way from those in the series. Fortunately, as systematic replication ensues with other therapists in other settings, many more clients with different background variables are treated, and confidence in generality of findings across clients, which was established in a preliminary manner in the first series, is increased with each new replication.

9.3. CLINICAL REPLICATION

A replication process related to direct replication occurs only in applied research. We will refer to this process as clinical replication. Clinical replication is an advanced replication procedure in which a treatment "package" containing two or more distinct procedures is applied to a succession of clients with multiple behavior or emotional problems that cluster together. This "cluster" of problems is usually labeled (e.g., schizophrenia).

Direct replication was defined as the administration of a given treatment by the same investigator or group of investigators in a specific setting (e.g., hospital, clinic, classroom) on a series of clients homogeneous for a particular behavior disorder such as agoraphobia or compulsive hand-washing. As this definition implies, one treatment procedure is applied to one well-defined problem in succeeding clients.

In applied research many clients present with a number of coexisting problems. If these multiple problems cluster together in any reliable way, the cluster is labeled and the label becomes a diagnostic category. Common examples of diagnostic categories describing multiple behavioral and emotional problems are severe (psychotic) depression, schizophrenia, manic depressive psychosis, and autism.

It is not the purpose of this section to argue the reliability, validity, or merits of different diagnostic categories. But, to the extent that behavioral and emotional problems do cluster together, the goal in applied settings is to treat all problems that are interfering with a client's functioning. In constructing an effective treatment package, however, one must develop and test treatments for one problem at a time, with the eventual goal of combining successful treatments for all coexisting problems. This is the "technique building" strategy suggested by Bergin and Strupp (1972) and described in Chapter 2, Section 2.9. For example, the direct replication series described above tested the effects of a specified treatment on delusional speech, which, of course, is often one component of schizophrenia. If this series were consistently successful, the applied researcher might begin to test treatments for coexisting problems in these patients, such as social isolation or thought disorders, if these were present. When successful procedures are developed for all coexisting problems, the next step would be to establish generality of findings by replicating this treatment "package" on additional patients who present a similar combination of problems. This would be clinical replication.

Definition of clinical replication

We would define clinical replication as the administration of a treatment package containing two or more distinct treatment procedures by the same investigator or group of investigators. These procedures would be administered in a specific setting to a series of clients presenting similar combinations of multiple behavioral and emotional problems, which usually cluster together. Obviously, this type of replication process is advanced in that it is the end result of a systematic "technique building" applied research effort, which should take years.

The usefulness of this effort also depends to some extent on the consistency or reliability of the diagnostic category. If the clustering of the target behaviors is inconsistent, then patients within the series would be so heterogeneous that the same treatment package could not be applied to successive patients. For this reason, and because of the advanced nature of the research effort, clinical replications are presently not common in the literature.

Example: Clinical replication with autistic children

One good example of a clinical replication series is the work of Lovaas and his colleagues with autistic children (e.g., Lovaas, Berberich, Perloff, and Schaeffer, 1966; Lovaas, Schaeffer, and Simmons, 1965; Lovaas and Simmons, 1969). The diagnosis of autism fulfills the requirements of clinical replication in that it subsumes a number of behavioral or emotional problems. Lovaas, Koegel, Simmons, and Long (1973) list eight distinct problems which may contribute to the autistic syndrome, including: (1) apparent sensory deficit, (2) severe affect isolation, (3) self-stimulating behavior, (4) mutism, (5) echolalic speech, (6) deficits in receptive speech, (7) deficits in social and self-help behaviors, and (8) self-injurious behavior. Step by step, they developed and tested treatments for each of these behaviors, such as self-destructive behavior (Lovaas and Simmons, 1969), language acquistion (e.g., Lovaas, Berberich, Perloff, and Schaeffer, 1966), and social and self-help skills (Lovaas, Freitas, Nelson, and Whalen, 1967). These procedures were tested in separate direct replication series on the same initial group of children. The treatment package constructed from these direct replication series was administered to subsequent children, presenting a sufficient number of these behaviors to be labeled autistic.

Lovaas, Koegel, Simmons, and Long (1973) present the results and follow-up data from this clinical replication series for 13 children. Results are presented in terms of response of the group as a whole, as well as individual improvement across the variety of behavioral and emotional problems. While these data are complex, they can be summarized as follows. All children, with no exceptions, demonstrated increases in appropriate behaviors and decreases in inappropriate behaviors. There were marked differences in the amount of improvement. At least one child was returned to a normal school setting, while several children improved very little and required continued institutionalization. In other words, each child improved, but the change was not clinically dramatic for several children.

Since clinical replication is similar to direct replication, it can be analyzed in a similar fashion and conclusions can be made in two general areas. *First*, the treatment package can be effective for behaviors subsumed under the autistic syndrome. This conclusion is based on: (1) the initial experimental analysis of each component of the treatment package in the original direct replication series (e.g., Lovaas and Simmons, 1969) and (2) the withdrawal and reintroduction of this whole package in A-B-A-B fashion in several children (Lovaas, Koegel, Simmons, and Long, 1973). *Second*, replication of this finding across all subjects indicates that the data are reliable and not due to idiosyncrasies in one child. It does not follow, however, that generality across children was established. As in example 3 in the last section (9.2), the results were clear and clinically significant for several children, but the results were weak and clinically unimportant for several children. Thus, the package has only limited generality across clients and the task remains to pinpoint differences between children who improved and those who did not improve. From these differences, possible causes for limitations on client generality should emerge.

In fact, children in this series were markedly heterogeneous. In many respects, this is due to an inherent difficulty in clinical replication noted above—the vagueness and unreliability of many diagnostic categories. As Lovaas, Koegel, Simmons, and Long (1973) point out, "the delineation of 'autism' is one area that will demand considerably more work. It has not been a particularly useful diagnosis. Few people agree on when to apply it" (p. 156). It follows that heterogeneity of clients will most likely be greater than in a direct replication series where the target behavior is well defined and clients can be matched more closely.

Thus, the causes of failure in a series with mixed results are more difficult to ascertain due to the greater number of differences among individuals. Nevertheless, it is necessary to pinpoint these differences and begin the search for intersubject variability. As Lovaas, Koegel, Simmons, and Long (1973) conclude, "finally a major focus of future research should attempt more functional descriptions of autistic children. As we have shown, the children responded in vastly different ways to the treatment we gave them. We paid scant attention to individual differences when we treated the first twenty children. In the future, we will assess such individual differences" (p. 163). In the meantime, child clinicians would do well to examine closely the exemplary series by Lovaas and his associates to determine logical generalization to children under their care.

Guidelines for clinical replication are similar to those for direct replication (see pp. 334-335). Since mixed or weak results are more likely in this series due to the unavoidable heterogeneity of the clients, a clinical replication series will most likely be longer than a direct replication series. In any case, a successful clinical replication series should generate systematic replication as other therapists in other settings attempt to replicate successful results.

9.4. SYSTEMATIC REPLICATION

Sidman (1960) notes that where direct replication helps to establish generality of findings among members of a species, ". . . systematic replication can accomplish this and at the same time extend its generality over a wide range of situations" (p. 111). In applied research, we have noted that direct or clinical replication can begin to establish generality of findings across clients but cannot answer questions concerning applicability of a given procedure in different therapeutic settings or by different therapists. Another limitation of the initial direct replication series is an inability to determine the effectiveness of a procedure proven effective with one type of behavior disorder on a related but topographically different behavior disorder.

Definition of systematic replication

We can define systematic replication in applied research as any attempt to replicate findings from a direct replication series, varying settings, behavior change agents, behavior disorders, or any combination thereof. It would appear that any successful systematic replication series in which one or more of the above-mentioned factors is varied also provides further information on generality of findings across clients since new clients are usually included in these efforts.

Example: Differential attention series

One of the most extensive and advanced systematic replication series to date has been in progress since the early 1960s. The purpose of this series has been to determine the generality of the effectiveness of a single intervention technique, often termed "differential attention." Differential attention consists of attending to a client contingent on the emission of a well-defined desired behavior. Usually such "attention" takes the form of positive interaction with the client consisting of praise, smiling, etc. Absence of the desired behavior results in withdrawal of attention, hence "differential" attention. This series, consisting of over 75 articles, has provided clinicians and other behavior change agents with a great deal of specific information on the effectiveness of this procedure in various settings with different behavior disorders and behavior change agents. Preliminary success in this area has generated a host of books advocating use of this technique in various settings, particularly with children in the home or classroom (e.g., Becker, 1971; Patterson and Gullion, 1968). What is perhaps more important, however, is that articles in this series are beginning to note conditions in which the procedure fails, leading to a clearer specification of the generality of this technique in all relevant domains in the applied area. A brief review of findings from this series in the various important domains of applied research will illustrate the process of systematic replication.

Differential attention: Adult psychotic behaviors

One of the first reports on differential attention appeared in 1959 (Ayllon and Michael, 1959). This report contained several examples of the application of differential attention to psychotic behaviors in a state hospital. The therapists in all cases were psychiatric nurses or aides. The purpose of this early demonstration was to illustrate to personnel in the hospital the possible clinical benefits of differential attention. Thus, differential attention was applied to most cases in an A-B design, with no attempt to experimentally demonstrate its controlling effects. In several cases, however, an experimental analysis was performed. One patient was extremely aggressive and required a great deal of restraint. One behavior incompatible with aggression was sitting or lying on the floor. Four-day baseline procedures revealed a relatively low rate of being on the floor. Social reinforcement by nurses increased the behavior, resulting in decreased aggression. Subsequent withdrawal of social reinforcement produced decreases in the behavior and increases in aggression. Unfortunately, ward personnel could not tolerate this and the patient was restrained once again, aborting a return to social reinforcement. The resultant A-B-A design was sufficient, however, to demonstrate the effects of social reinforcement in this setting for this class of behavior. This early experiment suggested that differential attention could be effective when applied by nurses or aides as therapists. These successes sparked replication by these investigators in additional cases. Other psychotic behavior in adult psychiatric wards modified by differential attention or a combination of differential attention and other procedures included faulty eating behavior (Ayllon and Haughton, 1962) and towel hoarding (Ayllon, 1963). These early studies were the beginning of the systematic replication series, in that topographically different behavior responded to differential attention.

Another problem behavior in adult psychiatric wards considered more "central" to psychiatric psychopathology is psychotic verbal behavior such as delusions or hallucinations. An early example of the application of differential attention to delusions was reported by Rickard, Dignam, and Horner (1960), who attended (smiled, nodded, etc.) to a 60-year-old male during periods of non-delusional speech and withdrew the attention (minimal attention) during delusional speech. Therapists were psychologists. Initially, non-delusional speech increased to almost maximal levels (9 minutes out of a 10-minute session) during periods of attention and decreased during the minimal attention condition. Later, even minimal attention was sufficient to maintain non-delusional speech. A 2-year follow-up (Rickard and Dinoff, 1962) revealed maintenance of these gains and reports of generalization to other hospital settings. Unfortunately, only one patient was included in this experiment, precluding any preliminary conclusion on generality of findings across other patients. Ayllon and Haughton (1964) followed this up with a series of three adult patients in a psychiatric ward who demonstrated bothersome delusional or psychosomatic verbal behavior. In all three cases, differential attention was effective in controlling the behavior as demonstrated by an A-B-C-B design, where A was baseline, B was social attention, and C was withdrawal of attention. Here, as in other reports by Ayllon and his associates, therapists were nurses and/or aides. This early experiment was a good direct replication series in its own right but, more importantly, served to systematically replicate findings from the single case reported by Rickard, Dignam, and Horner (1960). In Ayllon and Haughton's experiment, therapists were nurses or aides, rather than psychologists, and the setting was, of course, a different psychiatric ward. Despite these factors, differential attention again produced control over deviant behavior in adults on a psychiatric ward. This independent systematic replication provides a further degree of confidence in the effectiveness of the technique with psychotic behavior and its generality across therapists and settings.

After these early attempts to control psychotic behavior of adults on psychiatric wards through differential attention, Ayllon and his associates moved on to stronger reinforcers and developed the token economy (Ayllon and Azrin, 1968), abandoning for the most part their work on the exclusive use of differential attention. The impact of this early work was not lost on clinical investigators, however, and the importance of differential attention on adult wards of hospitals was once again demonstrated in a very clever experiment by Gelfand, Gelfand, and Dobson (1967). These investigators observed six psychotic patients in an inpatient psychiatric ward to determine sources of social attention contingent on disruptive or psychotic behavior. At the same time, they noted who was most successful in ignoring behaviors among the groups on the ward (i.e., other patients, nurses' aides, or nurses). Results indicated that other patients reinforced these behaviors least and most effectively ignored them, followed by nurses' aides and nurses. Thus, the personnel most responsible for implementing therapeutic programs, the nurses, were providing the greatest amount of social reinforcement contingent on undesirable behavior. This study does not, of course, demonstrate the controlling effects of differential attention. But, growing out of earlier experimental demonstrations of the effectiveness of this procedure, this study highlighted the potential importance of this factor on inpatient psychiatric units and led to further replication efforts on other wards.

Comment on replication procedures

It is safe to say that the impact of this work on adult wards has been substantial during the last decade, and differential attention to psychotic behavior is now a common therapeutic procedure on many wards. In retrospect, however, there are many methodological faults with this series, leading to large gaps in our knowledge, which could have been avoided had replication been more "systematic." While differential attention was successfully administered on psychiatric wards in several different parts of the country across the range of "therapists" or ward personnel typically employed in these settings and across a variety of "psychotic" behaviors, from motor behavior through inappropriate speech, only a few studies contained experimental analyses. On the other hand, many of the reports would come under the category of case studies (A-B designs with measurement). Certainly, this preliminary series on psychiatric patients would be much improved had each class of behavior (e.g., verbal behavior, withdrawn behavior, inappropriate behavior, aggressive or other motor behaviors) been subjected to a direct replication series with three or four patients and then systematically replicated in other settings with other therapists. This procedure would have, most likely, produced some failures. Reasons for these failures could then have been explored, providing considerably more information to clinicians and ward personnel on the limitations of differential attention. As it stands, Ayllon and Michael (1959) reported a failure but did not describe the patient in any detail or the circumstances surrounding the failure. This type of reporting leads to undue confidence in a procedure among naive clinicians; when failures do occur, disappointment is followed by a tendency to eliminate the procedure entirely from therapeutic programs.

It is only fair to say that the optimism concerning the effectiveness of differential attention was not entirely the result of the early case reports and studies, some of which are mentioned above. During the mid-1960s, the advent of the token economy and the many studies demonstrating the effectiveness of the token economy (e.g., Ayllon and Azrin, 1965) served to heighten interest in positive reinforcement procedures in general. As one of these "reinforcement" procedures, differential attention became popular despite the fact the the replication series on this specific procedure was inadequate to communicate to clinicians the conditions under which it succeeds or fails.

This early series also illustrated a second use of the single case study (A-B). In Chapter 1 we noted that case studies can suggest initially that a new technique is clinically effective, which can lead to more rigorous experimental demonstration and direct replication. In a systematic replication series the single case makes another appearance. Many reports are published that include only one case, but replicate an earlier direct replication series in either an experimental or an A-B form. Usually the reports are from different settings and contain a slight "twist," such as a new form of the behavior disorder or a slight modification of the procedure. While these reports are less desirable from the larger viewpoint of a systematic replication series, the fact is that they are published. When a sufficient number accumulate, these reports can provide considerable information on generality of findings. We will return to this point below.

Differential attention: Other adult behaviors

The early success of differential attention and positive reinforcement procedures in general with psychotic behaviors led to application of this procedure to other types of behavior disorders in other settings. One type of behavior disorder was adult non-psychotic behavior.

Citing earlier work on the effects of positive reinforcement on behavior of psychotic patients and children, Agras, Leitenberg, and Barlow (1968) praised increases in distance walked away from a "safe" setting for three agoraphobics in a study described in Section 9.2. This was one of the first demonstrations of the effectiveness of differential attention with non-psychotic adult behavior disorders, and direct replications within the study began to establish generality of findings across clients.

In a large series, Goldstein (1971) used differential attention procedures to train ten women who were experiencing marital difficulties. These wives attended to desired behaviors emitted by their husbands and ignored undesirable behaviors. Using a time series analysis, statistically significant changes occurred in eight out of ten cases. To the extent that these changes were clinically as well as statistically significant, it is important from the standpoint of systematic replication, since wives rather than professionals acted as behavior change agents and the treatment took place in the home.

Subsequent application of differential attention to other non-psychotic cases were limited to single case reports. Some of these single cases contain a functional analysis of differential attention; others are A-B designs with measurements. For instance, Brookshire (1970) eliminated crying in a 47-year-old male suffering from multiple sclerosis by attending to incompatible verbal behavior. Other single case examples include Brady and Lind's (1961) modification of hysterical blindness through differential attention to a visual task in a hospital setting. A hospital setting was also utilized to test the effectiveness of differential attention on an hysterical conversion reaction, specifically, astasia-abasia, or stumbling and falling while walking (Agras, Leitenberg, Barlow, and Thomson, 1969). Praise combined with ignoring stumbling resulted in improvement in this case. In another setting, these procedures also proved effective on a similar case (Hersen, Gullick, Matherne, and Harbert, 1972). Psychogenic vomiting was treated in a hospital setting by Alford, Blanchard, and Buckley (1972), who ignored vomiting and withdrew social contact immediately after vomiting. Therapists in this case were nurses. The authors cite success of this procedure on vomiting in a child (Wolf, Birnbrauer, Williams, and Lawler, 1965) as a rationale for attempting it with an adult. Other case studies exist for the treatment of non-psychotic adult behaviors by differential attention. Many of the studies describe a slight modification of the procedure or some variation in the behavior disorder.

As in the treatment of psychotic behaviors, differential attention also was combined with other treatment variables such as other forms of positive reinforcement or punishment in many research reports, making it difficult to specify the exclusive effects of differential attention.

One final article of interest was reported by Truax (1966), who reanalyzed tape recordings of Carl Rogers' therapy sessions. He discovered that Rogers responded differently (i.e., positively) to five classes of verbal behavior over a number of therapy sessions, and four of these classes increased in frequency. This is reminiscent of the verbal conditioning studies (e.g., Greenspoon, 1955) and suggests, in a non-experimental A-B fashion, that differential attention is operative in the psychotherapy hour.

Comment on replication procedures

If we consider the application of differential attention to non-psychotic adults as a systematic replication series in itself, the deficits and faults are similar to those encountered in the series with psychotic adults described above. Evidence is accumulating that differential attention *can* be effective in a number of settings (e.g., inpatient, outpatient, or home) when applied by different therapists (e.g., doctors, nurses, or wives) on a number of different behavioral problems. The difficulty here is not with the variety of studies, which is relatively limited now, but rather with the dearth of experimental analyses and direct replication in each new setting. To the extent that only A-B case studies are presented, one cannot be sure that other therapeutic variables are not operating in the setting. When an experimental analysis was performed, often only one case was reported, limiting the further accumulation of knowledge on generality of findings across clients. Of course, if many single cases with experimental analyses appear, confirming the effectiveness of a given procedure, the problem of generality of findings across clients solves itself. Finally, as noted above, reports of single cases, unlike series of cases, cannot include failure or limitation of the procedure, which is all-important to the practicing clinician.

Differential attention: Children's behavior disorders

In fact, differential attention procedures applied to adults, whether psychotic or non-psychotic, comprise only a small part of the work reported in this area. The greatest number of experimental inquiries on the effectiveness of differential attention have been conducted with children, and this series represents what is probably the most comprehensive systematic replication series to date. One of the earliest studies on the application of differential attention to behavior problems of a child was reported by Williams (1959), who instructed parents to withdraw attention from nightly temper tantrums. When an aunt unwittingly attended to tantrum behavior, tantrums increased and were extinguished once again by withdrawing attention. Table 9-2 presents summaries of replication efforts in this series since that time. In the table, it is important to note the variety of clients, problem behaviors, therapists, and settings described in this study, since generality of findings in all relevant domains is entirely dependent on the diversity of setting, clients, etc. employed in such studies.

Most early replication efforts, through 1965, presented an experimental analysis of results from a single case (see Table 9-2). A good example of the early studies is presented by Allen, Hart, Buell, Harris, and Wolf (1964), who reported that differential attention was responsible for increased social interaction with peers in a socially isolated preschool girl. The setting for the demonstration was a classroom, and the behavior change agent, of course, was the teacher. While most of the early studies contained only one case, the experimental demonstration of the effectiveness of differential attention in different settings with different therapists began to provide information on generality of findings across all-important domains. These replications increased confidence in this procedure as a generally effective clinical tool. In addition to isolate behavior, the successful treatment of such problems as regressed crawling (Harris, Johnston, Kelley, and Wolf, 1964), crying (Hart, Allen, Buell, Harris, and Wolf, 1964), and various behavior problems associated with the autistic syndrome (e.g., Davison, 1965) also suggested that this procedure was applicable to a wide variety of behavior problems in children while at the same time providing additional information on generality of findings across therapists and settings.

Although studies of successful application of differential attention to a single case demonstrated that this procedure is applicable in a wide range of situations, a more important development in the series was the appearance of direct replication effects containing three or more cases within the systematic replication series. While reports of single cases are uniformly successful, or they would not be published, exceptions to these reports of success can and do appear in series of cases, and these exceptions or failures begin to define the limits of the applicability of differential attention.

For this reason, it is particularly impressive that many series of three or more cases reported consistent success across many different clients, with such behavior disorders as inappropriate social behavior in disturbed hospitalized children (e.g., Laws, Brown, Epstein, and Hocking, 1971), disruptive behavior in the elementary classroom (e.g., Cormier, 1969; Hall, Fox, Willard, Goldsmith, Emerson, Owen, Davis, and Porcia, 1971; Hall, Lund, and Jackson, 1968) or high school classroom (e.g., Schutte and Hopkins, 1970), chronic thumbsucking

TABLE 9-2. Summary of studies on differential attention with children

Authors	Client(s)	N	Behavior	Setting	Therapist	Experimental Analysis
Williams (1959)	18-mo.-old female	1	tantrums	home	parents	no
Zimmerman and Zimmerman (1962)	11-yr.-old males	2	unproductive class-room behavior	residential treatment center	teachers	no
Allen, Hart, Buell, Harris, and Wolf (1964)	4-yr.-old female	1	isolate behavior	lab. preschool	teacher	yes
Harris, Johnston, Kelley, and Wolf (1964)	3-yr.-old female	1	crawling behavior	university nursery school	teacher	yes
Hart, Allen, Buell, Harris, and Wolf (1964)	4-yr.-old males	2	crying	university lab preschool	teachers	yes
Davison (1965)	10-yr.-old males	2	autistic behavior	private day-care center	undergraduates	no
Wahler, Winkel, Peterson, and Morrison (1965)	4- to 6-yr.-old males	3	oppositional behavior	lab. playroom	mother	yes
Allen and Harris (1966)	5-yr.-old female	1	scratching behavior	lab. preschool and home	mother	no
Hawkins, Peterson, Schweid, and Bijou (1966)	4-yr.-old male	1	tantrums and oppo-sitional behavior	home	mother	yes
Johnston, Kelley, Harris, and Wolf (1966)	3-yr.-old male	1	unusually low rate of physical activity	preschool	teacher	yes
Allen, Henke, Harris, Baer, and Reynolds (1967)	4½-yr.-old male	1	short attention span	lab preschool	teachers	yes
Etzel and Gerwitz (1967)	6- and 20-wk.-old inf.	2	crying	lab.	professional*	yes
Hall and Broden (1967)	5- and 6-yr.-old males and 9-yr.-old female with CNS dysfunction	3	behavior considered by staff to be inter-fering with their devel-opmental progress	experimental edu-cational unit	parents and teachers	yes

Authors	Client(s)	N	Behavior	Setting	Therapist	Experimental Analysis
Sloane, Johnston, and Bijou (1967)	4-yr.-old male	1	extreme aggression, temper tantrums, and excessive fantasy play	remedial nursery school	teachers	yes
Carlson, Arnold, Becker, and Madsen	8-yr.-old female	1	tantrums	classroom	teacher	no
Ellis (1968)	4- and 5-yr.-old males	5	aggressive behavior	lab. school	teacher and helper	yes
Hall, Lund, and Jackson (1968)	elementary school pupils	6	disruptive and dawdling study behavior	poverty area classroom	teachers	yes
Hall, Panyan, Rabon, and Broden (1968)	3 classroom (1st, 6th 7th grades)	24	study behavior	classroom	teachers	yes
Hart, Reynolds, Baer, Brawley, and Harris (1968)	5-yr.-old female	1	uncooperative play	preschool	teacher	yes
Madsen, Becker, and Thomas (1968)	elementary school pupils	3	classroom disruption	classroom	teachers	yes
Reynolds and Risley (1968)	4-yr.-old females	1	low frequency of talking	preschool	teacher	yes
Thomas, Becker and Armstrong (1968)	6- to 11-yr.-old males and females	10	disruptive behavior	classroom	teacher	yes
Thomas, Nielson, Kuypers, and Becker (1968)	6-yr.-old male	1	disruptive behavior	classroom	teacher	yes
Wahler and Pollio (1968)	8-yr.-old male	1	excessive dependency and lack of aggressive behavior	university clinic	parents and therapist	yes
Ward and Baker (1968)	1st grade children	4	disruptive behavior	classroom	teacher	yes
Zeilberger, Sampen, and Sloan (1968)	4½-yr.-old male	1	disobedience and aggressive behavior	home	mother	yes
Brawley, Harris, Allen, Fleming, and Peterson (1969)	7-yr.-old male	1	autistic behavior	hospital day-care unit	professional*	yes

Authors	Client(s)	N	Behavior	Setting	Therapist	Experimental Analysis
Cormier (1969)	6th and 8th grade classes	18	disruptive behavior and lack of motivation	classroom	teachers	yes
McCallister, Stachowiak, Baer, and Conderman (1969)	high school English class	25	inappropriate talking and turning around	classroom	teacher	yes
Wahler (1969a)	elementary school age males	2	oppositional behavior	home	parents	yes
Wahler (1969b)	5- and 8-yr.-old males	2	oppositional and disruptive behavior	home and classroom	parent and teacher	yes
Broden, Bruce, Mitchell, Carter, and Hall (1970a)	2nd grade males	2	disruptive behavior	poverty area classroom	teacher	yes
Broden, Hall, Dunlap, and Clark (1970b)	7th and 8th grade males and females	13	disruptive classroom behavior	special education class	teacher	yes
Conger (1970)	9-yr.-old male	1	encopresis	home	mother	yes
Goodlet, Goodlet, and Dredge (1970)	5- and 7-yr.-old males	2	disruptive behavior	university lab classroom	teacher	yes
Schutte and Hopkins (1970)	4- to 6-yr.-old females	5	instruction following	classroom	teacher	yes
Smeets (1970)	18-yr.-old male	1	rumination and regurgitation	hospital room	teacher	yes
Wahler, Sperling, Thomas, Teeter, and Luper (1970)	4- and 9-yr.-old males	2	"beginning" stuttering and mildly deviant behavior	hearing and speech center	parents	yes
Wright, Clayton, and Edger (1970)	severely retarded children	15	negative behaviors	state residential institution	ward technicians	no
Buys (1971)	9 problem and 9 control elementary school pupils	18	deviant classroom behavior	classroom	teacher	yes
Corte, Wolf, and Locke (1971)	profoundly retarded adolescents	4	self-injurious behavior	hospital training lab	professional*	no

Authors	Client(s)	N	Behavior	Setting	Therapist	Experimental Analysis
Hall, Fox, Willard, Goldsmith, Emerson, Owen, Davis, and Porcia (1971)	individual pupils and classroom groups from 1st grade–Junior High School	Ex. # & N 1. 1 2. 1 3. 1 4. 1 5. 30 6. 27	disruptive and talking out behavior	classroom–white middle class and black poverty	teacher	yes
Laws, Brown, Epstein, Jordan, and Hocking (1971)	severely disturbed 8- and 9-yr.-old males	3	behavior that interfered with speech and language	state hospital	speech therapist	yes
Skiba, Pettigrew, and Alden (1971)	8-yr.-old females	3	thumbsucking	classroom	teacher	yes
Thomas and Adams (1971)	well-behaved and remedial primary school pupils	16	task-related behavior and lowering sound levels	classroom	teacher	yes
Veenstra (1971)	5- to 14-yr.-old siblings	4	disruptive behavior	home	mother	yes
Vukelich and Hake (1971)	18-yr.-old severely retarded female	1	choking and grabbing	state hospital	ward staff	yes
Yawkey (1971)	7-yr.-old female 7-yr.-old male	2	poor attending behavior	classroom	teacher	yes
Barnes, Wootton, and Wood (1972)	3- and 4-yr.-old males females	24	immature play	mental health center	public health nurse	yes
Hall, Axelrod, Tyler, Grief, Jones, and Robertson (1972)	4- and 8-yr.-old males and 5- and 10-yr.-old females	4	whining and failure to wear orthodontic device	home	parents	yes
Herbert and Baer (1972)	5-yr.-old male and female	2	inappropriate behavior in home	home	mother	yes
Kirby and Shields (1972)	13-yr.-old male	1	non-attending and poor arithmetic	classroom	teacher	yes
Sajwaj, Twardooz, and Burke (1972)	7-yr.-old retarded male	1	excessive conversation with the teacher	remedial preschool	teacher	yes
Twardosz and Sajwaj (1972)	4-yr.-old hyperactive retarded male	1	sitting	remedial preschool	teacher	yes

*Professional usually refers to Ph.D. Psychologist or Psychiatrist

(Skiba, Pettigrew, and Alden, 1971), disruptive behavior in the home (Veenstra, 1971; Wahler, Winkel, Peterson, and Morrison, 1965), and disruptive behavior in brain-injured children (Hall and Broden, 1967). These improvements occurred in many different settings such as elementary and high school classes, hospitals and homes, as well as kindergartens and various preschools. Therapists included professionals, teachers, aides, parents, and nurses, among others (see Table 9-2).

The consistency of their success was impressive, but as these series of cases accumulated, the inevitable but extremely valuable reports of failures began to appear. Almost from the beginning, investigators noted that differential attention was not effective with self-injurious behavior in children. For instance, Tate and Baroff (1966) noted that in the length of time necessary for differential attention to work, severe injury would result. In place of differential attention, a strong aversive stimulus, electric shock, proved effective in suppressing this behavior. Later, Corte, Wolf, and Locke (1971) found that differential attention was totally ineffective on mild self-injurious behavior in retarded children but, again, electric shock proved effective. Since there are no reports of success in the literature using differential attention for self-injurious behavior, it is unlikely that these cases would have been published at all if differential attention had not proven effective on other behavior disorders. Thus, this is an example of a systematic replication series setting the stage for reports of limitations of a procedure. More subtle limitations of the procedure are reported in series of cases wherein the technique worked in some cases, but not in others. In an early series, Wahler, Winkel, Peterson, and Morrison (1965) trained mothers of young, oppositional children with differential attention procedures. The setting was an experimental preschool. In two out of three cases the mothers were quite successful in modifying oppositional behavior in their children, and an experimental analysis isolated differential attention as the important ingredient. In a third child, however, this procedure was not effective and an additional punishment (timeout) procedure was necessary. The authors do not offer any explanation for this discrepancy, and there are no obvious differences in the cases based on descriptions in the article that could account for the failure. The authors did not seem concerned with the discrepancy, probably because it was an early effort in the replication series and the goal was to control the oppositional behavior, which was accomplished when timeout was added. This study was important, however, since it contained the first hint that differential attention might not be effective with some cases of oppositional behavior.

In a later series, after differential attention was well established as an effective procedure, further failures to replicate did elicit concern from the investigator (Wahler, 1968, 1969a). Wahler trained parents of children with severe oppositional behavior in differential attention procedures. Results indicated that this procedure was ineffective across five children, but the addition of timeout again produced the desired changes. Replication in two more cases of oppositional behavior confirmed that differential attention was only effective when combined with a timeout procedure. In the best tradition of science, Wahler (1969a) did not gloss over the failure of differential attention, although his treatment "package" was ultimately successful. Contemplating reasons for the failure, Wahler hypothesized that in cases of severe oppositional behavior, parental reinforcement value may be extremely low; that is, attention from parents is not as reinforcing. After treatment, using the combination of timeout and attention, oppositional behavior was under control, even though timeout was no longer used. Employing a test of parental reinforcement values, Wahler demonstrated that the treatment "package" increased the reinforcing value of parental attention, allowing the gain to be maintained. This was the first clear suggestion that therapist variables are important in the application of differential attention, and that with oppositional children particularly, differential attention alone may be ineffective due to low reinforcing value of parental attention.

While differential attention occasionally has been found ineffective in other settings, such as the classroom (O'Leary, Becker, Evans, and Saudargas, 1969), some recent data raise doubts about the effectiveness of differential attention in settings where it had been demonstrated successfully. Herbert, Pinkston, Hayden, Sajwaj, Pinkston, Cordua, and Jackson (1973) trained mothers in the use of differential attention in two separate geographical locations (Kansas and Mississippi). Although preschools were the settings in both locations, the design and function of the preschools were quite dissimilar. Clients were children with a variety of disruptive and deviant behaviors, including hyperactivity, oppositional behavior, and other inappropriate social behaviors. These young children presented different background variables from familial retardation through childhood autism and Down's syndrome, and came from differing socioeconomic backgrounds. The one similarity among the six cases (two from Mississippi, four from Kansas) was that differential attention from parents was not only ineffective but detrimental in many cases, in that deviant behavior increased and dangerous and surprising side effects appeared. Deleterious effects of this procedure were confirmed in extensions of A-B-A designs, where behavior worsened under differential attention and improved when the procedure was withdrawn. These results were, of course, surprising to the authors, and discovery of similar results in two settings through personal communication prompted the combining of the data into a single publication. What is perhaps more disturbing is the inability of the investigators to pinpoint factors that may be responsible for their failures. Despite a seemingly thorough search among client, therapist, and setting variables, no clear differences among clients, settings, or therapists between the failures and the success reported in previous experiments emerged. As the authors note, ". . . the results were not peculiar to a particular setting, certain parent-child activities, observation code, or recording system, experimenter or parent training procedure. Subject characteristics also were not predictory of the results obtained" (Herbert, Pinkston, Sajwaj, Pinkston, Cordua, and Jackson, 1973, p. 26). Since the clients and procedure are clearly described, these findings are open to examination by other investigators, who may discover explanations for the failures that were overlooked by the authors. At the very least, these data will generate further research into the conditions that limit or facilitate the effects of differential attention. For the present, however, we would agree with the authors' conclusion that the failure in the differential attention series ". . . suggests that the assumption of broad generality is certainly questionable. Until the limiting conditions have been identified for the successful application of such 'elementary' behavior modification techniques, those engaged in widespread social intervention programs should be advised to proceed cautiously. At least, quantitative data on the effects of those procedures are absolutely essential" (p. 28).

Comment on replication

In our view, these data are a sign of the maturity of a systematic replication series. Only when a procedure is proven successful through many replications, do negative results assume this importance. But these failures do not detract from the previous successful replications. The effectiveness of differential attention has been established repeatedly in previous experiments. These data do, however, indicate that there is much that we do not know about differential attention.

In conclusion, this advanced systematic replication series on differential attention has generated a great deal of confidence among clinicians. The evidence indicates that it can be effective with adults and children with a variety of behavioral problems in most any setting. The clinically oriented books widely advocating its use (e.g., Patterson and Gullion, 1968) have made this procedure available to numerous professionals concerned with behavior change, as well as to the consuming public. However, the process of establishing generality of findings across all relevant domains is a slow one indeed, and it will most likely be years before we know all we should about this treatment or other treatments currently undergoing systematic replication. Until the time that the process of systematic replication reveals the precise limitations of a procedure, clinicians and other behavior change agents should proceed with caution, but also with hope and confidence that this powerful process will ultimately establish the conditions under which a given treatment is effective or ineffective.

Guidelines for systematic replication

The formulation of guidelines for conducting systematic replication is more difficult than for direct replication due to the variety of experimental efforts that comprise a systematic replication series. However, in the interest of providing some structure to future systematic replication, we will attempt to provide an outline of the general procedures necessary for sound systematic replication in applied research. These procedures or guidelines fall into four categories.

1. Earlier we defined systematic replication in applied research as any attempt to replicate findings from a direct replication series varying settings, behavior change agents, behavior disorders, or some combination thereof. Ideally, then, the systematic replication should begin with sound direct replication where the reliability of a procedure is established and the beginnings of client generality is ascertained. If results in the initial experiment and three or more replications are uniformly successful, then the important work of testing the effectiveness of the procedure in other settings with other therapists, etc. can begin. If a series begins with a report of a single case (as it so often does), then the first order of business is to initiate a direct replication series on this procedure so that the search for exceptions can begin.

2. Investigators evaluating systematic replication should clearly note the differences among their clients, therapists, or settings with those in the original experiment. In a conservative systematic replication, only one or possibly two variables indicate that the investigator is "gambling" somewhat (Sidman, 1960). That is, if the experiment is successful, the series will take a large step forward in establishing generality of findings. If the experiment fails, the investigator cannot know which of the differing variables or combination of variables was responsible for the change and must go back and retrace his steps. Whether a scientist takes the gamble depends on the setting and his own inclinations; there is no guideline one could suggest here without also limiting the creativity of the scientific process. But it *is* important to be fully aware of previous efforts in the series and to list the number of ways in which the current experiment differs from past efforts so that other investigators and clinicians can hypothesize along with the experimenter on which differences were important in the event of failure. In fact, most good scientists already do this (e.g., Herbert, Pinkston, Hayden, Sajwaj, Pinkston, Cordua, and Jackson 1973).

3. Systematic replication is essentially a search for exceptions. If no exceptions are found as replications proceed, then wide generality of findings is

established. However, the purpose of systematic replication is to define the conditions under which a technique will succeed or fail, and this means a search for exception or failure. Thus, any experimental tactics that hinder the finding and reporting of exceptions are of less value than an experimental design that will highlight failure. Of those experimental procedures typically found in a systematic replication series (e.g., see Table 9-2), two fall into this category: the experimental analysis containing only one case and the group study.

As noted above, the report of a single case, particularly when accompanied by and experimental analysis, can be a valuable addition to a series in that it describes another setting, behavior disorder, etc. where the procedure was successful. Reports of single cases also may lead to direct and systematic replication as the differential attention series. Unfortunately, however, failures in a single case are seldomly published in journals. Among the numerous successful reports contained in the differential attention series, none reported a failure, although it is our guess that differential attention has failed on many occasions and these failures simply have not been reported.

The group study suffers from the same limitation since failures are lost in the group average. Again, group studies can play an important role in systematic replication in that demonstration that a technique is successful with a given group, as opposed to individuals in the group, may serve an important function (see Section 2.9). In the differential attention series, several investigators thought it important to demonstrate the the procedure could be effective in a classroom as a whole (e.g., Ward and Baker, 1968). These data contributed to generality of findings across several domains. The fact remains, however, that failures will not be detected (unless the whole experiment fails, in which case it would not be published), thus leading us no closer to the goal of defining the conditions in which a successful technique fails.

4. Finally, the question arises: When is a systematic replication series over? For direct replication series, it was possible to make some tentative recommendations on a number of subjects given experimental findings. With systematic replication, no such recommendations are possible. In applied research, we would have to agree with Sidman's (1960) conclusion concerning basic research that a series is never over, since scientists will always attempt to find exceptions to a given principle, as well they should. It may be safe to say that a series is over when no exception to a proven therapeutic principle can be found, but, as Sidman points out, this is entirely dependent on the complexity of the problem and the inductive reasoning of clinical researchers who will have to judge in the light of new and emerging knowledge which conditions could provide exceptions to old principles.

Fortunately, clinicians do not have to wait for the end of a series to apply interim findings to their clients. In these series, knowledge is cumulative. A clinician may apply a procedure from an advanced series, such as differential attention, with more confidence than procedures from less advanced and more complex series, such as those on sexual deviation (Barlow, 1974) or anorexia nervosa (Agras, Barlow, Chapin, Abel, and Leitenberg, 1974). However, it is still possible through inspection of these data to utilize those new procedures with a degree of confidence dependent on the degree to which the experimental clients, therapists, and settings are similar to those facing the clinician. At the very least, this is a good beginning to the often discouraging and sometimes painful process of clinical trial and error.

9.5. REPLICATION OF SINGLE CASE EXPERIMENTS: SOME ADVANTAGES

In view of the reluctance of clinical researchers to carry out the large-scale replication studies required in traditional experimental design (Bergin and Strupp, 1972), one might be puzzled with the seeming enthusiasm with which investigators undertake replication efforts using single case designs as evidenced by the differential attention series and other less advanced series. A quick examination of Table 9-2 demonstrates that there is most likely little or no savings in time or money when compared to the large-scale collaborative factorial designs initially proposed by Bergin and Strupp (1972). No fewer clients are involved and, in all likelihood, more applied researchers and settings are involved. Why, then, does this replication tactic succeed when Bergin and Strupp concluded that the alternative could not be implemented? In our view, there are four very important but rather simple reasons.

First, the effort is *decentralized.* Rather than funding the type of large collaborative factorial study necessary to determine generality of findings at a cost of millions of dollars, the replication efforts are carried out in many settings such that funding, when available, is dispersed. This, of course, is more practical for government or other funding sources who are not reluctant to award $10,000 to each of 100 investigators, but would be quite reluctant to award $1,000,000 to one group of investigators. Often, of course, these "small" studies involving three or four subjects are unfunded. Also, rather than administering a large collaborative study from a central location where all scientists or therapists are to carry out prescribed duties, each scientist administers his own replication effort based on his ideas and views of previous findings. What is lost here is some efficiency, since there is no guarantee that the next obvious step in the replication series will be carried out at the logical time. What is gained is the freedom and creativity of individual scientists to attack the problem in there own ways.

Second, systematic replication will continue because the professional contingencies are favorable to its success. The professional contingencies in this case are publications and the accompanying professional recognition. Initial efforts in a series experimentally demonstrating success of a technique on a single case are publishable. Direct replications are publishable. Systematic replications are publishable each time the procedure is successful in a different setting or with a different behavior disorder, etc. Finally, after a procedure has been proven effective, failures or exceptions to the success are publishable. It is a well-established principle in psychology that intermittent reinforcement, preferably on a short variable interval schedule, is more effective in maintaining behavior (in this case the replication series) than the schedule arrangement for a large group study where years may pass before publishable data are available.

Third, the experimental analysis of the single case is close to the clinic. As noted in Chapter 1, this approach tends to merge the role of scientist and practitioner. Many an important series has started only after the clinician confronted an interesting case. Subsequently, measures were developed and an experimental analysis of the treatment performed (Mills, Agras, Barlow, and Mills, 1973). As a result, the data increase one's understanding of the problem, but the client also receives and benefits from treatment. If one plans to treat the patient, it is an easy enough matter to develop measures and perform the necessary experimental analyses. This ability to work with ease within the clinical setting, more than any other reason, may ensure the future of meaningful systematic replication efforts.

Finally, as noted above, the results of the series are cumulative and each new replicative effort has some immediate "pay-off" for the practicing clinician. Since this is the ultimate goal of the applied researcher, it is far more satisfactory than participating in a multiyear collaborative study where knowledge or benefit to the clinician is a distant goal.

Nevertheless, the advancement of a systematic replication series is a long and arduous road full of pitfalls and "dead-ends." In the face of the immediate demands on clinicians and behavior change agents to provide services to society, it is tempting to "grab the glimmer of hope" provided by treatments that prove successful in preliminary reports on case studies. That these hopes have been repeatedly dashed as therapeutic techniques and schools of therapy have come and gone supplies the most convincing evidence that the slow but inexorable process of the scientific method is the only way to meaningful advancement in our knowledge. Although we are a long way from the sophistication of the physical sciences, the single case experimental design with adequate replication may provide us with the methodology necessary to overcome the complex problems of human behavior disorders.

References

Agras, W. S., Barlow, D. H., Chapin, H. N., Abel, G. G., and Leitenberg, H. Behavior modification of anorexia nervosa. *Archives of General Psychiatry*, 1974, **30**, 279-286.

Agras, W. S., Leitenberg, H., and Barlow, D. H. Social reinforcement in the modification of agoraphobia. *Archives of General Psychiatry*, 1968, **19**, 423-427.

Agras, W. S., Leitenberg, H., Barlow, D. H., and Thomson, L. E. Instructions and reinforcement in the modification of neurotic behavior. *American Journal of Psychiatry*, 1969, **125**, 1435-1439.

Alford, G. S., Blanchard, E. B., and Buckley, M. Treatment of hysterical vomiting by modification of social contingencies: A case study. *Journal of Behavior Therapy and Experimental Psychiatry*, 1972, **3**, 209-212.

Allen, K. E., and Harris, F. R. Elimination of a child's excessive scratching by training the mother in reinforcement procedures. *Behaviour Research and Therapy*, 1966, **4**, 79-84.

Allen, K. E., Hart, B., Buell, J. S., Harris, F. R., and Wolf, M. M. Effects of social reinforcement on isolate behavior of a nursery school child. *Child Development*, 1964, **35**, 511-518.

Allen, K. E., Henke, L. B., Harris, F. R., Baer, D. M., and Reynolds, N. J. Control of hyperactivity by social reinforcement of attending behavior. *Journal of Educational Psychology*, 1967, **58**, 231-237.

Ayllon, T. Interview treatment of psychotic behavior by stimulus satiation and food reinforcement. *Behaviour Research and Therapy*, 1963, **1**, 53-61.

Ayllon, T., and Azrin, N. H. The measure and reinforcement of behavior of psychotics. *Journal of the Experimental Analysis of Behavior*, 1965, **8**, 357-382.

Ayllon, T., and Azrin, N. H. *The token economy: A motivational system for therapy and rehabilitation.* New York: Appleton-Century-Crofts, 1968.

Ayllon, T., and Haughton, E. Control of the behavior of schizophrenic patients by food. *Journal of the Experimental Analysis of Behavior*, 1962, **5**, 343-352.

Ayllon, T., and Haughton, E. Modification of symptomatic verbal behavior of mental patients. *Behaviour Research and Therapy*, 1964, **2**, 87-91.

Ayllon, T., and Michael, J. The psychiatric nurse as a behavioral engineer. *Journal of the Experimental Analysis of Behavior*, 1959, **2**, 323-334.

Barlow, D. H. The treatment of sexual deviation: Toward a comprehensive behavioral approach. In K. S. Calhoun, H. E. Adams, and K. M. Mitchell, (Eds.), *Innovative treatment methods in psychopathology.* Pp. 121-148. New York: Wiley, 1974.

Barnes, K. E., Wooton, M., and Wood, S. The public health nurse as an effective therapist-behavior modifier of preschool play behavior. *Community Mental Health Journal*, 1972, **8**, 3-7.

Becker, W. C. *Parents are teachers: A child management program.* Champaign, Ill.: Research Press, 1971.

Bergin, A. E., and Strupp, H. H. New directions in psychotherpay research. *Journal of Abnormal Psychology*, 1970, **76**, 13-26.

Bergin, A. E., and Strupp, H. H. *Changing frontiers in the science of psychotherapy.* New

York: Aldine-Atherton, 1972.
Brady, J. P., and Lind, D. L. Experimental analysis of hysterical blindness. *Archives of General Psychiatry*, 1961, **4**, 331-339.
Brawley, E. R., Harris, F. R., Allen, K. E., Fleming, R. S., and Peterson, R.F. Behavior modification of an autistic child. *Behavioral Science*, 1969, **14**, 87-97.
Broden, M., Bruce, C., Mitchell, M. A., Carter, V., and Hall, R. V. Effects of teacher attention on attending behavior of two boys at adjacent desks. *Journal of Applied Behavior Analysis*, 1970a, **3**, 205-211.
Broden, M., Hall, R. V., Dunlap, A., and Clark, R. Effects of teacher attention and a token reinforcement system in a junior high school special education class. *Exceptional Children*, 1970b, **36**, 341-349.
Brookshire, R. H. Control of "involuntary" crying behavior emitted by a multiple sclerosis patient. *Journal of Community Disorders*, 1970, **1**, 386-390.
Buys, C. J. Effects of teacher reinforcement on classroom behaviors and attitudes. *Dissertation Abstracts International*, 1971, **31**, (9-A), 4884-4885.
Carlson, C. S., Arnold, C. R., Becker, W. C., and Madsen, C. H. The elimination of tantrum behavior of a child in an elementary classroom. *Behaviour Research and Therapy*, 1968, **6**, 117-119.
Conger, J. C. The treatment of encopresis by the management of social consequences. *Behavior Therapy*, 1970, **1**, 386-390.
Cormier, W. H. Effects of teacher random and contingent social reinforcement on the classroom behavior of adolescents. *Dissertation Abstracts International*, 1969, **31**, 1615-1616.
Corte, H. E., Wolf, M. M., and Locke, B. J. A comparison of procedures for eliminating self-injurious behavior of retarded adolescents. *Journal of Applied Behavior Analysis*, 1971, **4**, 201-213.
Davison, G. C. The training of undergraduates as social reinforcers for autistic children. In L. P. Ullmann and L. Krasner (Eds.), *Case studies in behavior modification*. Pp, 146-148. New York: Holt, Rinehart and Winston, 1965.
Edgington, E. S. Statistical inference from N=1 experiments. *Journal of Psychology*, 1967, **65**, 195-199.
Ellis, D. P. The design of a social structure to control aggression. *Dissertation Abstracts International*, 1968, **29** (2-A), 672.
Etzel, B. C., and Gerwitz, J. L. Experimental modification of caretaker-maintained high-rate operant crying in a 6- and 20-week-old infant (*Infans tyrannotearus*): Extinction of crying with reinforcement of eye contact and smiling. *Journal of Experimental Child Psychology*, 1967, **5**, 303-317.
Gelfand, D. M., Gelfand, S., and Dobson, W. R. Unprogrammed reinforcement of patients' behavior in a mental hospital. *Behaviour Research and Therapy*, 1967, **5**, 201-207.
Goldstein, M. K. Behavior rate change in marriages: Training wives to modify husbands' behavior. *Dissertation Abstracts International*, 1971, **32**, 559.
Goodlet, G. R., Goodlet, M. M., and Dredge, M. Modification of disruptive behavior of two young children and follow-up one year later. *Journal of School Psychology*, 1970, **8**, 60-63.
Greenspoon, J. The reinforcing effect of two spoken sounds on the frequency of two responses. *American Journal of Psychology*, 1955, **68**, 409-416.
Hall, R. V., Axelrod, S., Tyler, L., Grief, E., Jones, F. C., and Robertson, R. Modification of behavior problems in the home with a parent as observer and experimenter. *Journal of Applied Behavior Analysis*, 1972, **5**, 53-64.
Hall, R. V., and Broden, M. Behavior changes in brain-injured children through social reinforcement. *Journal of Experimental Child Psychology*, 1967, **5**, 463-479.
Hall, R. V., Fox, R., Willard, D., Goldsmith, L., Emerson, M., Owen, M., Davis, F., and Porcia, E. The teacher as observer and experimenter in the modification of disputing

and talking-out behaviors. *Journal of Applied Behavior Analysis*, 1971, **4**, 141-149.

Hall, R. V., Lund, D., and Jackson, D. Effects of teacher attention on study behavior. *Journal of Applied Behavior Analysis*, 1968, **1**, 1-12.

Hall, R. V., Panyan, M., Rabon, D., and Broden, M. Instructing beginning teachers in reinforcement procedures which improve classroom control. *Journal of Applied Behavior Analysis*, 1968, **1**, 315-322.

Harris, F. R., Johnston, M. K., Kelley, C. S., and Wolf, M. M. Effects of positive social reinforcement on regressed crawling of a nursery school child. *Journal of Educational Psychology*, 1964, **55**, 35-41.

Hart, B. M., Allen, K. E., Buell, J. S., Harris, F. R., and Wolf, M. M. Effects of social reinforcement on operant crying. *Journal of Experimental Child Psychology*, 1964, **1**, 145-153.

Hart, B. M., Reynolds, N. J., Baer, D. M., Brawley, E. R., and Harris, F. R. Effect of contingent social reinforcement on the cooperative play of a preschool child. *Journal of Applied Behavior Analysis*, 1968, **1**, 73-76.

Hawkins, R. P., Peterson, R. F., Schweid, E., and Bijou, S. W. Behavior therapy in the home: Amelioration of problem parent-child relations with the parent in a therapeutic role. *Journal of Experimental Child Psychology*, 1966, **4**, 99-107.

Herbert, E. W., and Baer, D. M. Training parents as behavior modifiers: Self-recording of contingent attention. *Journal of Applied Behavior Analysis*, 1972, **5**, 139-149.

Herbert, E. W., Pinkston, E. M., Hayden, M. L., Sajwaj, T. E., Pinkston, S., Cordua, G., and Jackson, C. Adverse effects of differential parental attention. *Journal of Applied Behavior Analysis*, 1973, **6**, 15-30.

Herman, S. H., Barlow, D. H., and Agras, W. S. An experimental analysis of classical conditioning as a method of increasing heterosexual arousal in homosexuals. *Behavior Therapy*, 1974a, **5**, 33-47.

Herman, S. H., Barlow, D. H., and Agras, W. S. An experimental analysis of exposure to "explicit" heterosexual stimuli as an effective variable in changing arousal patterns of homosexuals. *Behaviour Research and Therapy*, 1974b, **12**, 335-346.

Hersen, M., Gullick, E. L., Matherne, P. M., and Harbert, T. L. Instructions and reinforcement in the modification of a conversion reation. *Psychological Reports*, 1972, **31**, 719-722.

Johnston, M. K., Kelley, C. S., Harris, F. R., and Wolf, M. M. An application of reinforcement principles to development of motor skills of a young child. *Child Development*, 1966, **37**, 379-387.

Laws, D. R., Brown, R. A., Epstein, J., and Hocking, N. Reduction of inappropriate social behavior in disturbed children by an untrained paraprofessional therapist. *Behavior Therapy*, 1971, **2**, 519-533.

Kirby, F. D., and Shields, F. Modification of arithmetic response rate and attending behavior in a seventh-grade student. *Journal of Applied Behavior Analysis*, 1972, **5**, 79-84.

Lovaas, O. I., Berberich, J. P., Perloff, B. I., and Schaeffer, B. Acquisition of imitative speech in schizophrenic children. *Science*, 1966, **151**, 705-707.

Lovaas, O. I., Freitas, L., Nelson, K., and Whalen, C. The establishment of imitation and its use for the development of complex behavior in schizophrenic children. *Behaviour Research and Therapy*, 1967, **5**, 171-181.

Lovaas, O. I., Koegel, R., Simmons, J. Q., and Long, J. D. Some generalization and follow-up measures on autistic children in behavior therapy. *Journal of Applied Behavior Analysis*, 1973, **5**, 131-166.

Lovaas, O. I., Schaeffer, B., and Simmons, J. Q. Experimental studies in childhood schizophrenia: Building social behaviors by use of electric shock. *Journal of Experimental Studies in Personality*, 1965, **1**, 99-109.

Lovaas, O. I., and Simmons, J. Q. Manipulation of self-destruction in three retarded children. *Journal of Applied Behavior Analysis*, 1969, **2**, 143-157.

McCallister, L. W., Stachowiak, J. G., Baer, D. M., and Conderman, L. The application of operant conditioning techniques in a secondary school classroom. *Journal of Applied Behavior Analysis*, 1969, **2**, 277-285.

Madsen, C. H., Becker, W. C., and Thomas, D. R. Rules, praise, and ignoring. Elements of elementary classroom control. *Journal of Applied Behavior Analysis*, 1968, **1**, 139-150.

Marks, I. M. Flooding (implosion) and allied treatments. In W. S. Agras (Ed.), *Behavior modification: Principles and clinical applications*. Pp. 151-213. Boston: Little, Brown, 1972.

Mills, H. L., Agras, W. S., Barlow, D. H., and Mills, J. R. Compulsive rituals treated by response prevention: An experimental analysis. *Archives of General Psychiatry*, 1973, **28**, 524-529.

O'Leary, K. D., Becker, W. C., Evans, M. B., and Saudargas, R. A. A token reinforcement program in a public school: A replication and systematic analysis. *Journal of Applied Behavior Analysis*, 1969, **2**, 3-13.

Patterson, G. R., and Gullion, M. E. *Living with children: New methods for parents and teachers.* Champaign, Ill.: Research Press, 1968.

Reynolds, N. J., and Risley, T. R. The role of social and material reinforcers in increasing talking of a disadvantaged preschool child. *Journal of Applied Behavior Analysis*, 1968, **1**, 253-262.

Rickard, H. C., Dignam, P. J., and Horner, R. F. Verbal manipulation in a psychotherapeutic relationship. *Journal of Clinical Psychology*, 1960, 164-367.

Rickard, H. C., and Dinoff, M. A follow-up note on "verbal manipulation in a psychotherapeutic relationship." *Psychological Reports*, 1962, **11**, 506.

Sajwaj, T., Twardosz, S., and Burke, M. Side effects of extinction procedures in a remedial preschool. *Journal of Applied Behavior Analysis*, 1972, **5**, 163-175.

Schutte, R. C., and Hopkins, B. L. The effects of teacher attention following instructions in a kindergarten class. *Journal of Applied Behavior Analysis*, 1970, **3**, 117-122.

Sidman, M. *Tactics of scientific research: Evaluating experimental data in psychology*. New York: Basic Books, 1960.

Skiba, E. A., Pettigrew, E., and Alden, S. E. A behavioral approach to the control of thumbsucking in the classroom. *Journal of Applied Behavior Analysis*, 1971, **4** 121-125.

Sloane, H. N., Johnston, M. K., and Bijou, S. W. Successive modification of aggressive behavior and aggressive fantasy play by management of contingencies. *Journal of Child Psychology and Psychiatry*, 1967, **8**, 217-226.

Smeets, P. M. Withdrawal of social reinforcers as a means of controlling rumination and regurgitation in a profoundly retarded person. *Training School Bulletin*, 1970, **67**, 158-163.

Tate, B. G., and Baroff, G. S. Aversive control of self-injurious behavior in a psychotic boy. *Behaviour Research and Therapy*, 1966, **4**, 281-287.

Thomas, J. D., and Adams, M. A. Problems in teacher use of selected behaviour modification techniques in the classroom. *New Zealand Journal of Educational Studies*, 1971, **6**, 151-165.

Thomas, D. R., Becker, W. C., and Armstrong, M. Production and elimination of disruptive classroom behavior by systematically varying teacher's behavior. *Journal of Applied Behavior Analysis*, 1968, **1**, 35-45.

Thomas, D. R., Nielsen, T. J., Kuypers, D. S., and Becker, W. C. Social reinforcement and remedial instruction in the elimination of a classroom behavior problem. *Journal of Special Education*, 1968, **2**, 291-305.

Truax, C. B. Reinforcement and non-reinforcement in Rogerian psychotherapy. *Journal of Abnormal Psychology*, 1966, **71**, 1-9.

Twardosz, S., and Sajwaj, T. Multiple effects of a procedure to increase sitting in a hyperactive, retarded boy. *Journal of Applied Behavior Analysis*, 1972, **5**, 73-78.

Veenstra, M. Behavior modification in the home with the mother as the experimenter: The effect of differential reinforcement on sibling negative response rates. *Child Development*, 1971, **42**, 2079-2083.

Vukelich, R., and Hake, D. F. Reduction of dangerously aggressive behavior in a severely retarded resident through a continuation of positive reinforcement procedures. *Journal of Applied Behavior Analysis*, 1971, **4**, 215-225.

Wahler, R. G. Behavior therapy for oppositional children: Love is not enough. Paper read at Eastern Psychological Association, Washington, D. C., April 1968.

Wahler, R. G. Oppositional children: A quest for parental reinforcement control. *Journal of Applied Behavior Analysis*, 1969a, **2**, 159-170.

Wahler, R. G. Setting generality: Some specific and general effects of child behavior therapy. *Journal of Applied Behavior Analysis*, 1969b, **2**, 239-246.

Wahler, R. G., and Pollio, H. R. Behavior and insight: A case study in behavior therapy. *Journal of Experimental Research in Personality*, 1968, **3**, 45-56.

Wahler, R. G., Sperling, K. A., Thomas, M. R., Teeter, N. C., and Luper, H. L. The modification of childhood stuttering: Some response-response relationships. *Journal of Experimental Child Psychology*, 1970, **9**, 411-428.

Wahler, R. G., Winkel, G. H., Peterson, R. F., and Morrison, D. C. Mothers as behavior therapists for their own children. *Behaviour Research and Therapy*, 1965, **3**, 113-124.

Ward, M. H., and Baker, B. L. Reinforcement therapy in the classroom. *Journal of Applied Behavior Analysis*, 1968, **1**, 323-328.

Williams, C. D. Case report, the elimination of tantrum behavior by extinction procedures. *Journal of Abnormal and Social Psychology*, 1959, **59**, 269.

Wincze, J. P., Leitenberg, H., and Agras, W. S. The effects of token reinforcement and feedback on the delusional verbal behavior of chronic paranoid schizophrenics. *Journal of Applied Behavior Analysis*, 1972, **5**, 247-262.

Wolf, M. M., Birnbrauer, J. S., Williams, T., and Lawler, J. A note on apparent extinction of the vomiting behavior of a retarded child. In L. P. Ullmann and L. Krasner (Eds.), *Case studies in behavior modification.* Pp 364-366. New York: Holt, Rinehart and Winston, 1965.

Wright, J., Clayton, J., and Edgar, C. L. Behavior modification with low-level mental retardates. *Psychological Record*, 1970, **20**, 465-471.

Yawkey, T. D. Conditioning independent work behavior in reading with seven-year-old children in a regular early childhood classroom. *Child Study Journal*, 1971, **2**, 23-34.

Zeilberger, J., Sampen, S. E., and Sloane, H. N. Modification of a child's problem behaviors in the home with the mother as therapist. *Journal of Applied Behavior Analysis*, 1968, **1**, 47-53.

Zimmerman, E. H., and Zimmerman, J. The alteration of behavior in a special classroom situation. *Journal of the Experimental Analysis of Behavior*, 1962, **5**, 59-60.

Author Index

Subject Index